D1604820

The Learning Tree
at Saint Mary's College
47 Madeleva Hall - Room 251
Notre Dame, IN 46556

BIG IDEAS MATH®
COURSE 1
A Bridge to Success

Ron Larson

Laurie Boswell

Erie, Pennsylvania
BigIdeasLearning.com

Big Ideas Learning, LLC
1762 Norcross Road
Erie, PA 16510-3838
USA

For product information and customer support, contact Big Ideas Learning at **1-877-552-7766** or visit us at ***BigIdeasLearning.com***.

Cover Image
Songquan Deng/Shutterstock.com

Copyright © 2014 by Big Ideas Learning, LLC. All rights reserved.

No part of this work may be reproduced or transmitted in any form or by any means, electronic or mechanical, including, but not limited to, photocopying and recording, or by any information storage or retrieval system, without prior written permission of Big Ideas Learning, LLC unless such copying is expressly permitted by copyright law. Address inquiries to Permissions, Big Ideas Learning, LLC, 1762 Norcross Road, Erie, PA 16510.

Big Ideas Learning and *Big Ideas Math* are registered trademarks of Larson Texts, Inc.

Printed in the U.S.A.

ISBN 13: 978-1-68033-120-2
ISBN 10: 1-68033-120-5

2 3 4 5 6 7 8 9 10 WEB 17 16 15

AUTHORS

Ron Larson is a professor of mathematics at Penn State Erie, The Behrend College, where he has taught since receiving his Ph.D. in mathematics from the University of Colorado. Dr. Larson is well known as the lead author of a comprehensive program for mathematics that spans middle school, high school, and college courses. His high school and Advanced Placement books are published by Holt McDougal. Ron's numerous professional activities keep him in constant touch with the needs of students, teachers, and supervisors. Ron and Laurie Boswell began writing together in 1992. Since that time, they have authored over two dozen textbooks. In their collaboration, Ron is primarily responsible for the pupil edition and Laurie is primarily responsible for the teaching edition of the text.

Laurie Boswell is the Head of School and a mathematics teacher at the Riverside School in Lyndonville, Vermont. Dr. Boswell received her Ed.D. from the University of Vermont in 2010. She is a recipient of the Presidential Award for Excellence in Mathematics Teaching. Laurie has taught math to students at all levels, elementary through college. In addition, Laurie was a Tandy Technology Scholar, and served on the NCTM Board of Directors from 2002 to 2005. She currently serves on the board of NCSM, and is a popular national speaker. Along with Ron, Laurie has co-authored numerous math programs.

ABOUT THE BOOK

Big Ideas Math: A Bridge to Success was developed using the same research-based strategy of a balanced approach that has become synonymous with the *Big Ideas Math* series. This approach opens doors to abstract thought, reasoning, and inquiry as students persevere to answer the Essential Questions that introduce each section.

From start to finish, this program was designed with the student in mind. Students are subtly introduced to "Habits of Mind" that help them internalize concepts for a greater depth of understanding. These habits serve students well not only in mathematics, but across all curricula throughout their academic careers. *Big Ideas Math: A Bridge to Success* exposes students to highly motivating and relevant problems. Woven throughout the program are the depth and rigor students need to prepare for career-readiness and other college-level courses.

We consider the Big Ideas Math series to be the crowning jewel of 30 years of achievement in writing educational materials.

Ron Larson

Laurie Boswell

TEACHER REVIEWERS

- Lisa Amspacher
  Milton Hershey School
  Hershey, PA

- Mary Ballerina
  Orange County Public Schools
  Orlando, FL

- Lisa Bubello
  School District of Palm
     Beach County
  Lake Worth, FL

- Sam Coffman
  North East School District
  North East, PA

- Kristen Karbon
  Troy School District
  Rochester Hills, MI

- Laurie Mallis
  Westglades Middle School
  Coral Springs, FL

- Dave Morris
  Union City Area
     School District
  Union City, PA

- Bonnie Pendergast
  Tolleson Union High
     School District
  Tolleson, AZ

- Valerie Sullivan
  Lamoille South
     Supervisory Union
  Morrisville, VT

- Becky Walker
  Appleton Area School District
  Appleton, WI

- Zena Wiltshire
  Dade County Public Schools
  Miami, FL

STUDENT REVIEWERS

- Mike Carter
- Matthew Cauley
- Amelia Davis
- Wisdom Dowds
- John Flatley
- Nick Ganger

- Hannah Iadeluca
- Paige Lavine
- Emma Louie
- David Nichols
- Mikala Parnell
- Jordan Pashupathi

- Stephen Piglowski
- Robby Quinn
- Michael Rawlings
- Garrett Sample
- Andrew Samuels
- Addie Sedelmyer
- Tyler Steffy
- Erin Taylor
- Reid Wilson

CONSULTANTS

- **Patsy Davis**
  Educational Consultant
  Knoxville, Tennessee

- **Bob Fulenwider**
  Mathematics Consultant
  Bakersfield, California

- **Linda Hall**
  Mathematics Assessment Consultant
  Norman, Oklahoma

- **Ryan Keating**
  Special Education Advisor
  Gilbert, Arizona

- **Michael McDowell**
  Project-Based Instruction Specialist
  Fairfax, California

- **Sean McKeighan**
  Interdisciplinary Advisor
  Norman, Oklahoma

- **Bonnie Spence**
  Differentiated Instruction Consultant
  Missoula, Montana

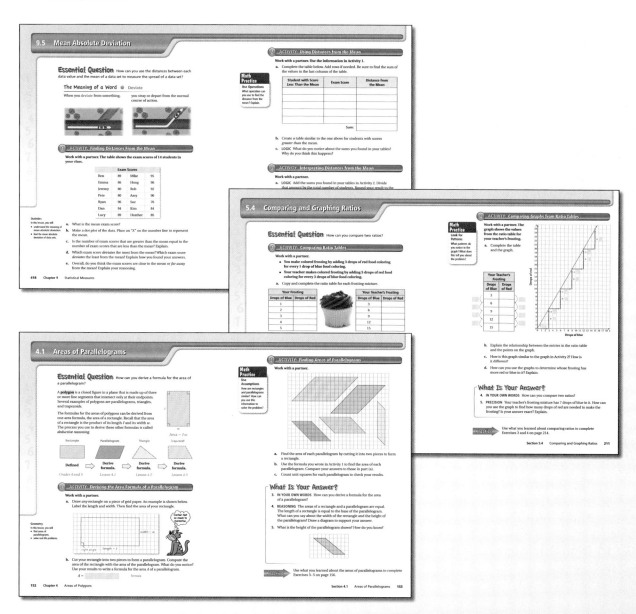

Mathematical Practices

Make sense of problems and persevere in solving them.
- Multiple representations are presented to help students move from concrete to representative and into abstract thinking
- *Essential Questions* help students focus and analyze
- *In Your Own Words* provide opportunities for students to look for meaning and entry points to a problem

Reason abstractly and quantitatively.
- Visual problem solving models help students create a coherent representation of the problem
- Opportunities for students to decontextualize and contextualize problems are presented in every lesson

Construct viable arguments and critique the reasoning of others.
- *Error Analysis*; *Different Words, Same Question*; and *Which One Doesn't Belong* features provide students the opportunity to construct arguments and critique the reasoning of others
- *Inductive Reasoning* activities help students make conjectures and build a logical progression of statements to explore their conjecture

Model with mathematics.
- Real-life situations are translated into diagrams, tables, equations, and graphs to help students analyze relations and to draw conclusions
- Real-life problems are provided to help students learn to apply the mathematics that they are learning to everyday life

Use appropriate tools strategically.
- *Graphic Organizers* support the thought process of what, when, and how to solve problems
- A variety of tool papers, such as graph paper, number lines, and manipulatives, are available as students consider how to approach a problem
- Opportunities to use the web, graphing calculators, and spreadsheets support student learning

Attend to precision.
- *On Your Own* questions encourage students to formulate consistent and appropriate reasoning
- Cooperative learning opportunities support precise communication

Look for and make use of structure.
- *Inductive Reasoning* activities provide students the opportunity to see patterns and structure in mathematics
- Real-world problems help students use the structure of mathematics to break down and solve more difficult problems

Look for and express regularity in repeated reasoning.
- Opportunities are provided to help students make generalizations
- Students are continually encouraged to check for reasonableness in their solutions

Go to *BigIdeasMath.com* for more information on the Mathematical Practices.

Mathematical Content
Chapter Coverage

Ratios and Proportional Relationships
- Understand ratio concepts and use ratio reasoning to solve problems.

The Number System
- Apply and extend previous understandings of multiplication and division to divide fractions by fractions.
- Compute fluently with multi-digit numbers and find common factors and multiples.
- Apply and extend previous understandings of numbers to the system of rational numbers.

Expressions and Equations
- Apply and extend previous understandings of arithmetic to algebraic expressions.
- Reason about and solve one-variable equations and inequalities.
- Represent and analyze quantitative relationships between dependent and independent variables.

Geometry
- Solve real-world and mathematical problems involving area, surface area, and volume.

Statistics and Probability
- Develop understanding of statistical variability.
- Summarize and describe distributions.

1 Numerical Expressions and Factors

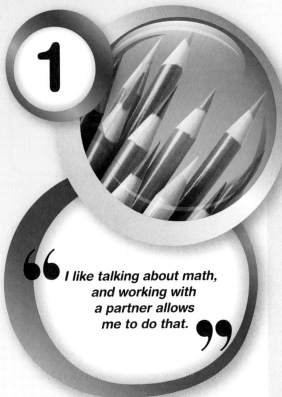

I like talking about math, and working with a partner allows me to do that.

	What You Learned Before	1
Section 1.1	**Whole Number Operations**	
	Activity	2
	Lesson	4
Section 1.2	**Powers and Exponents**	
	Activity	10
	Lesson	12
Section 1.3	**Order of Operations**	
	Activity	16
	Lesson	18
	Study Help/Graphic Organizer	22
	1.1–1.3 Quiz	23
Section 1.4	**Prime Factorization**	
	Activity	24
	Lesson	26
Section 1.5	**Greatest Common Factor**	
	Activity	30
	Lesson	32
Section 1.6	**Least Common Multiple**	
	Activity	36
	Lesson	38
	Extension: Adding and Subtracting Fractions	42
	1.4–1.6 Quiz	44
	Chapter Review	45
	Chapter Test	48
	Cumulative Assessment	49

Fractions and Decimals

	What You Learned Before	53
Section 2.1	**Multiplying Fractions**	
	Activity	54
	Lesson	56
Section 2.2	**Dividing Fractions**	
	Activity	62
	Lesson	64
Section 2.3	**Dividing Mixed Numbers**	
	Activity	70
	Lesson	72
	Study Help/Graphic Organizer	76
	2.1–2.3 Quiz	77
Section 2.4	**Adding and Subtracting Decimals**	
	Activity	78
	Lesson	80
Section 2.5	**Multiplying Decimals**	
	Activity	84
	Lesson	86
Section 2.6	**Dividing Decimals**	
	Activity	92
	Lesson	94
	2.4–2.6 Quiz	100
	Chapter Review	101
	Chapter Test	104
	Cumulative Assessment	105

"With my eBook, I get to decide when I use technology and when I use print."

3 Algebraic Expressions and Properties

" I like that the Essential Question helps me begin thinking about the lesson. "

	What You Learned Before 109
Section 3.1	**Algebraic Expressions**
	Activity 110
	Lesson 112
Section 3.2	**Writing Expressions**
	Activity 118
	Lesson 120
	Study Help/Graphic Organizer 124
	3.1–3.2 Quiz 125
Section 3.3	**Properties of Addition and Multiplication**
	Activity 126
	Lesson 128
Section 3.4	**The Distributive Property**
	Activity 132
	Lesson 134
	Extension: Factoring Expressions ... 140
	3.3–3.4 Quiz 142
	Chapter Review 143
	Chapter Test 146
	Cumulative Assessment 147

Areas of Polygons

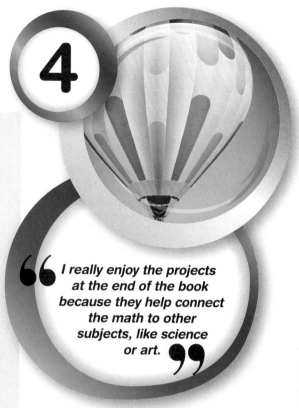

	What You Learned Before	151
Section 4.1	**Areas of Parallelograms**	
	Activity	152
	Lesson	154
Section 4.2	**Areas of Triangles**	
	Activity	158
	Lesson	160
	Study Help/Graphic Organizer	164
	4.1–4.2 Quiz	165
Section 4.3	**Areas of Trapezoids**	
	Activity	166
	Lesson	168
	Extension: Areas of Composite Figures	172
Section 4.4	**Polygons in the Coordinate Plane**	
	Activity	174
	Lesson	176
	4.3–4.4 Quiz	180
	Chapter Review	181
	Chapter Test	184
	Cumulative Assessment	185

"I really enjoy the projects at the end of the book because they help connect the math to other subjects, like science or art."

5 Ratios and Rates

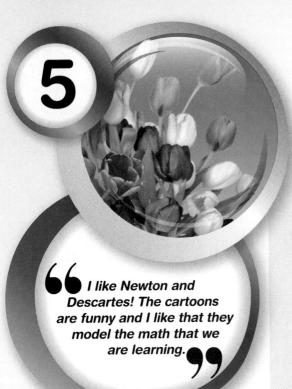

I like Newton and Descartes! The cartoons are funny and I like that they model the math that we are learning.

	What You Learned Before	189
Section 5.1	**Ratios**	
	Activity	190
	Lesson	192
Section 5.2	**Ratio Tables**	
	Activity	196
	Lesson	198
Section 5.3	**Rates**	
	Activity	204
	Lesson	206
Section 5.4	**Comparing and Graphing Ratios**	
	Activity	210
	Lesson	212
	Study Help/Graphic Organizer	216
	5.1–5.4 Quiz	217
Section 5.5	**Percents**	
	Activity	218
	Lesson	220
Section 5.6	**Solving Percent Problems**	
	Activity	224
	Lesson	226
Section 5.7	**Converting Measures**	
	Activity	232
	Lesson	234
	5.5–5.7 Quiz	238
	Chapter Review	239
	Chapter Test	242
	Cumulative Assessment	243

Integers and the Coordinate Plane

6

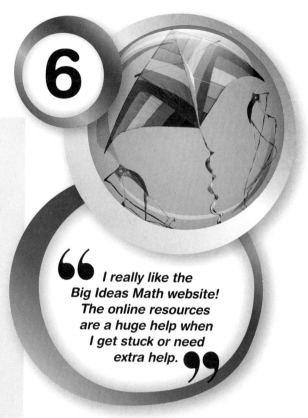

	What You Learned Before	247
Section 6.1	**Integers**	
	Activity	248
	Lesson	250
Section 6.2	**Comparing and Ordering Integers**	
	Activity	254
	Lesson	256
Section 6.3	**Fractions and Decimals on the Number Line**	
	Activity	260
	Lesson	262
	Study Help/Graphic Organizer	266
	6.1–6.3 Quiz	267
Section 6.4	**Absolute Value**	
	Activity	268
	Lesson	270
Section 6.5	**The Coordinate Plane**	
	Activity	274
	Lesson	276
	Extension: Reflecting Points in the Coordinate Plane	282
	6.4–6.5 Quiz	284
	Chapter Review	285
	Chapter Test	288
	Cumulative Assessment	289

"I really like the Big Ideas Math website! The online resources are a huge help when I get stuck or need extra help."

xiii

7 Equations and Inequalities

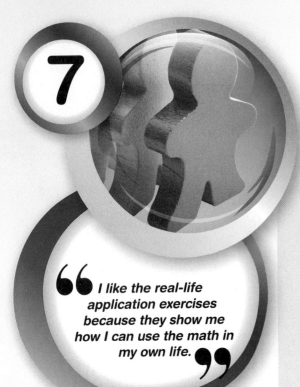

"I like the real-life application exercises because they show me how I can use the math in my own life."

	What You Learned Before	293
Section 7.1	**Writing Equations in One Variable**	
	Activity	294
	Lesson	296
Section 7.2	**Solving Equations Using Addition or Subtraction**	
	Activity	300
	Lesson	302
Section 7.3	**Solving Equations Using Multiplication or Division**	
	Activity	308
	Lesson	310
Section 7.4	**Writing Equations in Two Variables**	
	Activity	314
	Lesson	316
	Study Help/Graphic Organizer	322
	7.1–7.4 Quiz	323
Section 7.5	**Writing and Graphing Inequalities**	
	Activity	324
	Lesson	326
Section 7.6	**Solving Inequalities Using Addition or Subtraction**	
	Activity	332
	Lesson	334
Section 7.7	**Solving Inequalities Using Multiplication or Division**	
	Activity	338
	Lesson	340
	7.5–7.7 Quiz	344
	Chapter Review	345
	Chapter Test	348
	Cumulative Assessment	349

Surface Area and Volume

8

	What You Learned Before	353
Section 8.1	**Three-Dimensional Figures**	
	Activity	354
	Lesson	356
Section 8.2	**Surface Areas of Prisms**	
	Activity	360
	Lesson	362
	Study Help/Graphic Organizer	366
	8.1–8.2 Quiz	367
Section 8.3	**Surface Areas of Pyramids**	
	Activity	368
	Lesson	370
Section 8.4	**Volumes of Rectangular Prisms**	
	Activity	374
	Lesson	376
	8.3–8.4 Quiz	380
	Chapter Review	381
	Chapter Test	384
	Cumulative Assessment	385

"*I like playing the games in the Game Closet! They are a fun way to practice concepts we are learning in class.*"

9 Statistical Measures

"With the BigIdeasMath.com website I don't have to worry if I forget my book or my workbook at school."

	What You Learned Before	389
Section 9.1	**Introduction to Statistics**	
	Activity	390
	Lesson	392
Section 9.2	**Mean**	
	Activity	396
	Lesson	398
Section 9.3	**Measures of Center**	
	Activity	402
	Lesson	404
	Study Help/Graphic Organizer	410
	9.1–9.3 Quiz	411
Section 9.4	**Measures of Variation**	
	Activity	412
	Lesson	414
Section 9.5	**Mean Absolute Deviation**	
	Activity	418
	Lesson	420
	9.4–9.5 Quiz	424
	Chapter Review	425
	Chapter Test	428
	Cumulative Assessment	429

Data Displays

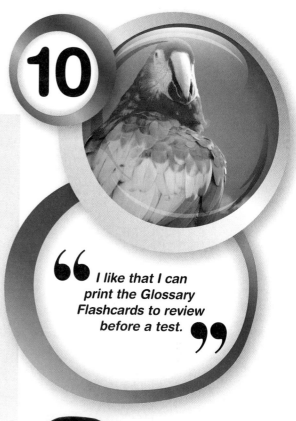

	What You Learned Before	433
Section 10.1	**Stem-and-Leaf Plots**	
	Activity	434
	Lesson	436
Section 10.2	**Histograms**	
	Activity	440
	Lesson	442
	Study Help/Graphic Organizer	448
	10.1–10.2 Quiz	449
Section 10.3	**Shapes of Distributions**	
	Activity	450
	Lesson	452
	Extension: Choosing Appropriate Measures	456
Section 10.4	**Box-and-Whisker Plots**	
	Activity	458
	Lesson	460
	10.3–10.4 Quiz	466
	Chapter Review	467
	Chapter Test	470
	Cumulative Assessment	471

"*I like that I can print the Glossary Flashcards to review before a test.*"

Appendix A: My Big Ideas Projects

Section A.1	**Literature Project**	A2
Section A.2	**History Project**	A4
Section A.3	**Art Project**	A6
Section A.4	**Science Project**	A8

Selected Answers	A10
Key Vocabulary Index	A45
Student Index	A46
Mathematics Reference Sheet	B1

How to Use Your Math Book

- Read the **Essential Question** in the activity.

  Discuss the **Math Practice** question with your partner.

  Work with a partner to decide **What Is Your Answer?**

  Now you are ready to do the **Practice** problems.

- Find the **Key Vocabulary** words, **highlighted in yellow**.

  Read their definitions. Study the concepts in each **Key Idea**.

  If you forget a definition, you can look it up online in the

  Multi-Language Glossary at BigIdeasMath.com.

- After you study each **EXAMPLE**, do the exercises in the **On Your Own**.

  Now You're Ready to do the exercises that correspond to the example.

  As you study, look for a **Study Tip** or a **Common Error**.

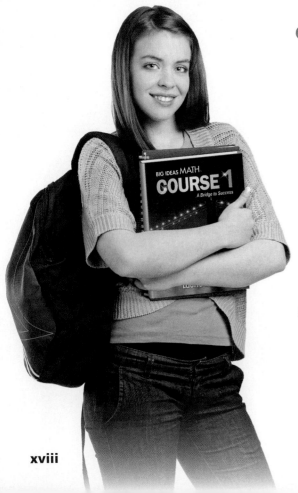

- The exercises are divided into 3 parts.

  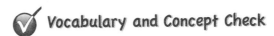 Vocabulary and Concept Check

   Practice and Problem Solving

   Fair Game Review

  If an exercise has a ① next to it, look back at Example 1 for help with that exercise.

  More help is available at  **Check It Out Lesson Tutorials BigIdeasMath.com**.

- To help study for your test, use the following.

   Quiz    Study Help

  Chapter Review   Chapter Test

xviii

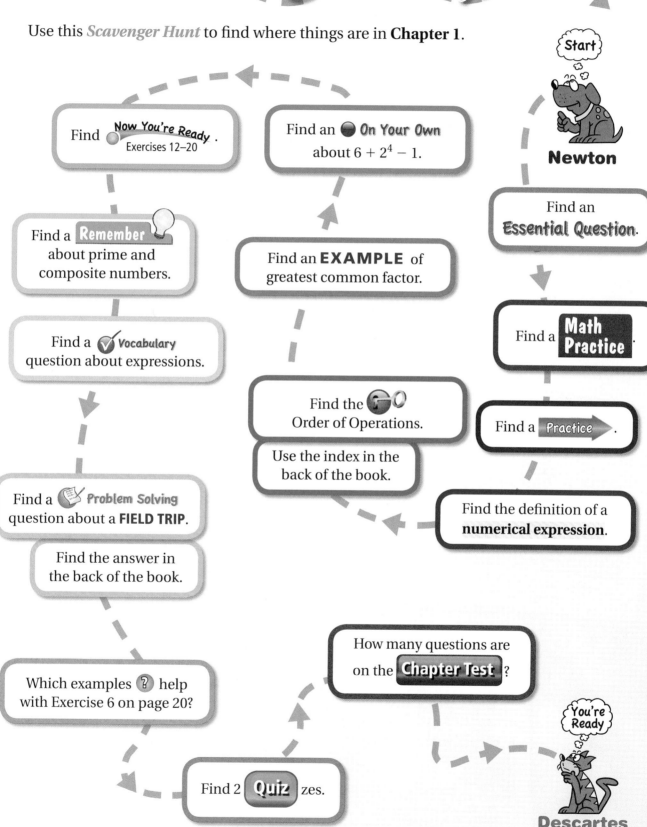

1 Numerical Expressions and Factors

- **1.1** Whole Number Operations
- **1.2** Powers and Exponents
- **1.3** Order of Operations
- **1.4** Prime Factorization
- **1.5** Greatest Common Factor
- **1.6** Least Common Multiple

"Dear Sir: You say that x^3 is called x-cubed."

"And you say that x^2 is called x-squared."

"So, why isn't x^1 called x-lined?"

"My sign on adding fractions with unlike denominators is keeping the hyenas away."

"See, it's working."

What You Learned Before

"Because 6 is composite, you can have 2 piles of 3 blocks, 3 piles of 2 blocks, or..."

- **Identifying Prime and Composite Numbers**

    Example 1 **Determine whether 26 is prime or composite.**
    Because the factors of 26 are 1, 2, 13, and 26, it is composite.

    Example 2 **Determine whether 37 is prime or composite.**
    Because the only factors of 37 are 1 and 37, it is prime.

    Try It Yourself
    Determine whether the number is prime or composite.

    1. 5
    2. 14
    3. 17
    4. 23
    5. 28
    6. 33
    7. 43
    8. 57
    9. 64

- **Adding and Subtracting Mixed Numbers with Like Denominators**

    Example 3 **Find $2\frac{3}{5} + 4\frac{1}{5}$.**

    $2\frac{3}{5} + 4\frac{1}{5} = \frac{2 \cdot 5 + 3}{5} + \frac{4 \cdot 5 + 1}{5}$ Rewrite the mixed numbers as improper fractions.

    $= \frac{13}{5} + \frac{21}{5}$ Simplify.

    $= \frac{13 + 21}{5}$ Add the numerators.

    $= \frac{34}{5}$, or $6\frac{4}{5}$ Simplify.

    Try It Yourself
    Add or subtract.

    10. $4\frac{1}{9} + 2\frac{7}{9}$
    11. $6\frac{1}{11} + 3\frac{6}{11}$
    12. $3\frac{7}{8} + 4\frac{3}{8}$
    13. $5\frac{8}{13} - 1\frac{2}{13}$
    14. $7\frac{1}{4} - 3\frac{3}{4}$
    15. $4\frac{1}{6} - 2\frac{5}{6}$

1.1 Whole Number Operations

Essential Question How do you know which operation to choose when solving a real-life problem?

1 ACTIVITY: Choosing an Operation

Work with a partner. The double bar graph shows the history of a citywide cleanup day.

- Copy each question below.
- Underline a key word or phrase that helps you know which operation to use to answer the question. State the operation. Why do you think the key word or phrase indicates the operation you chose?
- Write an expression you can use to answer the question.
- Find the value of your expression.

Whole Numbers

In this lesson, you will
- determine which operation to perform.
- divide multi-digit numbers.

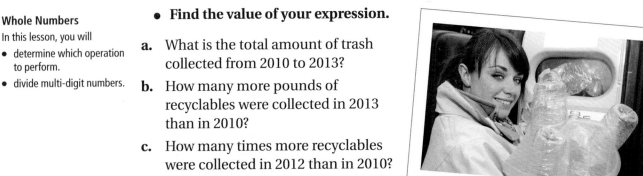

a. What is the total amount of trash collected from 2010 to 2013?

b. How many more pounds of recyclables were collected in 2013 than in 2010?

c. How many times more recyclables were collected in 2012 than in 2010?

d. The amount of trash collected in 2014 is estimated to be twice the amount collected in 2011. What is that amount?

2 ACTIVITY: Checking Answers

Math Practice

Communicate Precisely
What key words should you use so that your partner understands your explanation?

Work with a partner.

a. Explain how you can use estimation to check the reasonableness of the value of your expression in Activity 1(a).

b. Explain how you can use addition to check the value of your expression in Activity 1(b).

c. Explain how you can use estimation to check the reasonableness of the value of your expression in Activity 1(c).

d. Use mental math to check the value of your expression in Activity 1(d). Describe your strategy.

3 ACTIVITY: Using Estimation

Work with a partner. Use the map. Explain how you found each answer.

a. Which two lakes have a combined area of about 33,000 square miles?

b. Which lake covers an area about three times greater than the area of Lake Erie?

c. Which lake covers an area that is about 16,000 square miles greater than the area of Lake Ontario?

d. Estimate the total area covered by the Great Lakes.

Lake Superior 31,698 mi²
Lake Huron 23,011 mi²
Lake Michigan 22,316 mi²
Lake Ontario 7320 mi²
Lake Erie 9922 mi²

What Is Your Answer?

4. **IN YOUR OWN WORDS** How do you know which operation to choose when solving a real-life problem?

5. In a *magic square*, the sum of the numbers in each row, column, and diagonal is the same and each number from 1 to 9 is used only once. Complete the magic square. Explain how you found the missing numbers.

?	9	2
?	5	?
8	?	?

Practice

Use what you learned about choosing operations to complete Exercises 8–11 on page 7.

1.1 Lesson

Check It Out
Lesson Tutorials
BigIdeasMath.com

Recall the four basic operations: addition, subtraction, multiplication, and division.

Operation	Words	Algebra
Addition	the *sum* of	$a + b$
Subtraction	the *difference* of	$a - b$
Multiplication	the *product* of	$a \times b \quad a \cdot b$
Division	the *quotient* of	$a \div b \quad \dfrac{a}{b} \quad b\overline{)a}$

EXAMPLE 1 Adding and Subtracting Whole Numbers

The bar graph shows the attendance at a three-day art festival.

a. What is the total attendance for the art festival?

You want to find the total attendance for the three days. In this case, the phrase *total attendance* indicates you need to find the sum of the daily attendances.

Line up the numbers by their place values, then add.

$$\begin{array}{r} \overset{1\,1\,1}{}2570 \\ 3145 \\ +\ 3876 \\ \hline 9591 \end{array}$$

∴ The total attendance is 9591 people.

b. What is the increase in attendance from Day 1 to Day 2?

You want to find how many more people attended on Day 2 than on Day 1. In this case, the phrase *how many more* indicates you need to find the difference of the attendances on Day 2 and Day 1.

Line up the numbers by their place values, then subtract.

$$\begin{array}{r} \overset{10}{2\,\cancel{0}\,14} \\ \cancel{3}\cancel{1}45 \\ -\ 2570 \\ \hline 575 \end{array}$$

∴ The increase in attendance from Day 1 to Day 2 is 575 people.

Art Festival Attendance

Day 1: 2570
Day 2: 3145
Day 3: 3876

EXAMPLE 2 Multiplying Whole Numbers

A school lunch contains 12 chicken nuggets. Ninety-five students buy the lunch. What is the total number of chicken nuggets served?

You want to find the total number of chicken nuggets in 95 groups of 12 chicken nuggets. The phrase *95 groups of 12* indicates you need to find the product of 95 and 12.

$$\begin{array}{r} 12 \\ \times\ 95 \\ \hline 60 \\ 108 \\ \hline 1140 \end{array}$$

Multiply 12 by the ones digit, 5.
Multiply 12 by the tens digit, 9.
Add.

Study Tip

In Example 2, you can use estimation to check the reasonableness of your answer.
$12 \times 95 \approx 12 \times 100$
    $= 1200$
Because $1200 \approx 1140$, the answer is reasonable.

∴ There were 1140 chicken nuggets served.

On Your Own

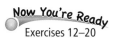
Exercises 12–20

Find the value of the expression. Use estimation to check your answer.

1. $1745 + 682$
2. $912 - 799$
3. 42×118

EXAMPLE 3 **Dividing Whole Numbers: No Remainder**

You make 24 equal payments for a go-kart. You pay a total of $840. How much is each payment?

You want to find the number of groups of 24 in $840. The phrase *groups of 24 in $840* indicates you need to find the quotient of 840 and 24.

Use long division to find the quotient. Decide where to write the first digit of the quotient.

$$\begin{array}{r} ? \\ 24 \overline{)840} \end{array}$$ Do not use the hundreds place because 24 is greater than 8.

$$\begin{array}{r} ? \\ 24 \overline{)840} \end{array}$$ Use the tens place because 24 is less than 84.

So, divide the tens and write the first digit of the quotient in the tens place.

$$\begin{array}{r} 3 \\ 24 \overline{)840} \\ -72 \\ \hline 12 \end{array}$$

Divide 84 by 24: There are three groups of 24 in 84.
Multiply 3 and 24.
Subtract 72 from 84.

Next, bring down the 0 and divide the ones.

dividend / divisor = quotient

So, quotient × divisor = dividend.

$$\begin{array}{r} 35 \\ 24 \overline{)840} \\ -72\downarrow \\ \hline 120 \\ -120 \\ \hline 0 \end{array}$$

Divide 120 by 24: There are five groups of 24 in 120.

Multiply 5 and 24.
Subtract 120 from 120.

The quotient of 840 and 24 is 35.

∴ So, each payment is $35.

Check Find the product of the quotient and the divisor.

$$\begin{array}{r} 35 \\ \times\ 24 \\ \hline 140 \\ 70 \\ \hline 840 \end{array}$$ quotient
divisor

dividend ✓

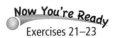
Exercises 21–23

Find the value of the expression. Use estimation to check your answer.

4. $234 \div 9$
5. $\dfrac{986}{58}$
6. $840 \div 105$

7. Find the quotient of 9920 and 320.

When you use long division to divide whole numbers and you obtain a remainder, you can write the quotient as a mixed number using the rule

$$\text{dividend} \div \text{divisor} = \text{quotient} + \dfrac{\text{remainder}}{\text{divisor}}.$$

EXAMPLE 4   Real-Life Application

A 301-foot-high swing at an amusement park can take 64 people on each ride. A total of 8983 people ride the swing today. All the rides are full except for the last ride. How many rides are given? How many people are on the last ride?

To find the number of rides given, you need to find the number of groups of 64 people in 8983 people. The phrase *groups of 64 people in 8983 people* indicates you need to find the quotient of 8983 and 64.

Divide the place-value positions from left to right.

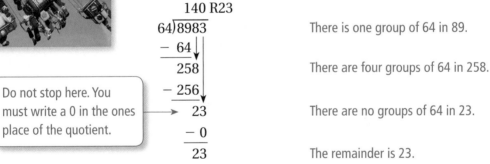

Do not stop here. You must write a 0 in the ones place of the quotient.

There is one group of 64 in 89.

There are four groups of 64 in 258.

There are no groups of 64 in 23.

The remainder is 23.

The quotient is $140\dfrac{23}{64}$. This indicates 140 groups of 64, with 23 remaining.

∴ So, 141 rides are given, with 23 people on the last ride.

On Your Own

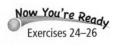
Exercises 24–26

Find the value of the expression. Use estimation to check your answer.

8. $\dfrac{6096}{30}$
9. $45{,}691 \div 28$
10. $3215 \div 430$

11. **WHAT IF?** In Example 4, 9038 people ride the swing. What is the least number of rides possible?

1.1 Exercises

Vocabulary and Concept Check

VOCABULARY Determine which operation the word or phrase represents.

1. sum
2. times
3. the quotient of
4. decreased by
5. total of
6. minus

7. **VOCABULARY** Use the division problem shown to tell whether the number is the divisor, dividend, or quotient.

   a. 884
   b. 26
   c. 34

$$34\overline{)884}^{\,26}$$

Practice and Problem Solving

The bar graph shows the attendance at a food festival. Write an expression you can use to answer the question. Then find the value of your expression.

8. What is the total attendance at the food festival from 2010 to 2013?

9. How many more people attended the food festival in 2012 than in 2011?

10. How many times more people attended the food festival in 2013 than in 2010?

11. The festival projects that the total attendance for 2014 will be twice the attendance in 2012. What is the projected attendance for 2014?

Find the value of the expression. Use estimation to check your answer.

① 12. $2219 + 872$

13. 5351
    $+1730$

14. $3968 + 1879$

15. $7694 - 5232$

16. $9165 - 4729$

17. 2416
    -1983

② 18. 84
    $\times 37$

19. 124×56

20. 419×236

③ 21. $837 \div 27$

22. $\dfrac{588}{84}$

23. $7440 \div 124$

④ 24. $6409 \div 61$

25. $8241 \div 173$

26. $\dfrac{33{,}505}{160}$

Section 1.1 Whole Number Operations 7

ERROR ANALYSIS Describe and correct the error in finding the value of the expression.

27.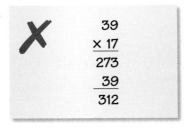
```
      39
    × 17
    273
     39
    312
```

28.
```
       19
   12)1308
      -12
      108
     -108
        0
```

Determine the operation you would use to solve the problem. Do not answer the question.

29. Gymnastic lessons cost $30 per week. How much will 18 weeks of gymnastic lessons cost?

30. The scores on your first two tests were 82 and 93. By how many points did your score improve?

31. You are setting up tables for a banquet for 150 guests. Each table seats 12 people. What is the minimum number of tables you will need?

32. A store has 15 boxes of peaches. Each box contains 45 peaches. How many peaches does the store have?

33. Two shirts cost $18 and $25. What is the total cost of the shirts?

34. A gardener works for 14 hours during a week and charges $168. How much does the gardener charge for each hour?

Find the perimeter and area of the rectangle.

35.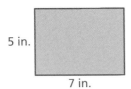
5 in.
7 in.

36.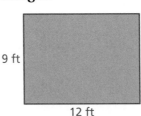
9 ft
12 ft

37.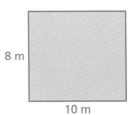
8 m
10 m

38. **BOX OFFICE** The number of tickets sold for the opening weekend of a movie is 879,575. The movie was shown in 755 theaters across the nation. What was the average number of tickets sold at each theater?

39. **LOGIC** You find that the product of 93 and 6 is 558. How can you use addition to check your answer? How can you use division to check your answer?

40. **NUMBER SENSE** Without calculating, decide which is greater: 3999 ÷ 129 or 3834 ÷ 142. Explain.

8 Chapter 1 Numerical Expressions and Factors

41. **REASONING** In a division problem, can the remainder be greater than the divisor? Explain.

42. **WATER COOLER** You change the water jug on the water cooler. How many cups can be completely filled before you need to change the water jug again?

43. **ARCADE** You have $9, one of your friends has $10, and two of your other friends each have $13. You combine your money to buy arcade tokens. You use a coupon to buy 8 tokens for $1. The cost of the remaining tokens is four for $1. You and your friends share the tokens evenly. How many tokens does each person get?

44. **BOOK SALE** You borrow bookcases like the one shown to display 943 books at a book sale. You plan to put 22 books on each shelf. No books will be on top of the bookcases.

   a. How many bookcases must you borrow to display all the books?

   b. You fill the shelves of each bookcase in order, starting with the top shelf. How many books are on the third shelf of the last bookcase?

45. **MODELING** The siding of a house is 2250 square feet. The siding needs two coats of paint. The table shows information about the paint.

Can Size	Cost	Coverage
1 quart	$18	80 square feet
1 gallon	$29	320 square feet

   a. What is the minimum cost of the paint needed to complete the job?

   b. How much paint is left over?

46. Use the digits 3, 4, 6, and 9 to complete the division problem. Use each digit once.

$$\boxed{}\,\boxed{},000 \div \boxed{}00 = \boxed{}0$$

Fair Game Review *What you learned in previous grades & lessons*

Plot the ordered pair in a coordinate plane. *(Skills Review Handbook)*

47. (1, 3)   48. (0, 4)   49. (6, 0)   50. (4, 2)

51. **MULTIPLE CHOICE** Which of the following numbers is *not* prime? *(Skills Review Handbook)*

   Ⓐ 1   Ⓑ 2   Ⓒ 3   Ⓓ 5

1.2 Powers and Exponents

Essential Question How can you use repeated factors in real-life situations?

*As I was going to St. Ives
I met a man with seven wives
Each wife had seven sacks
Each sack had seven cats
Each cat had seven kits
Kits, cats, sacks, wives
How many were going to St. Ives?* Nursery Rhyme, 1730

1 ACTIVITY: Analyzing a Math Poem

Work with a partner. Here is a "St. Ives" poem written by two students. Answer the question in the poem.

As I was walking into town
I met a ringmaster with five clowns
Each clown had five magicians
Each magician had five bunnies
Each bunny had five fleas
Fleas, bunnies, magicians, clowns
How many were going into town?

Numerical Expressions

In this lesson, you will
- write expressions as powers.
- find values of powers.

Number of clowns: 5   =
Number of magicians: 5 × 5   =
Number of bunnies: 5 × 5 × 5   =
Number of fleas: 5 × 5 × 5 × 5 =

So, the number of fleas, bunnies, magicians, and clowns is   . Explain how you found your answer.

10   Chapter 1   Numerical Expressions and Factors

2 ACTIVITY: Writing Repeated Factors

Math Practice

Repeat Calculations
What patterns do you notice with each problem? How does this help you write exponents?

Work with a partner. Copy and complete the table.

Repeated Factors	Using an Exponent	Value
a. 4×4		
b. 6×6		
c. $10 \times 10 \times 10$		
d. $100 \times 100 \times 100$		
e. $3 \times 3 \times 3 \times 3$		
f. $4 \times 4 \times 4 \times 4 \times 4$		
g. $2 \times 2 \times 2 \times 2 \times 2 \times 2$		

h. In your own words, describe what the two numbers in the expression 3^5 mean.

3 ACTIVITY: Writing and Analyzing a Math Poem

Work with a partner.

a. Write your own "St. Ives" poem.
b. Draw pictures for your poem.
c. Answer the question in your poem.
d. Show how you can use exponents to write your answer.

What Is Your Answer?

4. **IN YOUR OWN WORDS** How can you use repeated factors in real-life situations? Give an example.

5. **STRUCTURE** Use exponents to complete the table. Describe the pattern.

10	100	1000	10,000	100,000	1,000,000
10^1	10^2				

Practice

Use what you learned about exponents to complete Exercises 4–6 on page 14.

Section 1.2 Powers and Exponents 11

1.2 Lesson

Key Vocabulary
power, p. 12
base, p. 12
exponent, p. 12
perfect square, p. 13

A **power** is a product of repeated factors. The **base** of a power is the repeated factor. The **exponent** of a power indicates the number of times the base is used as a factor.

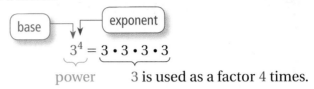

Power	Words
3^2	Three *squared*, or three to the second
3^3	Three *cubed*, or three to the third
3^4	Three to the fourth

EXAMPLE 1 Writing Expressions as Powers

Math Practice

Choose Tools
Why are calculators more efficient when finding the values of expressions involving exponents?

Write each product as a power.

a. $4 \cdot 4 \cdot 4 \cdot 4 \cdot 4$

Because 4 is used as a factor 5 times, its exponent is 5.

∴ So, $4 \cdot 4 \cdot 4 \cdot 4 \cdot 4 = 4^5$.

b. $12 \times 12 \times 12$

Because 12 is used as a factor 3 times, its exponent is 3.

∴ So, $12 \times 12 \times 12 = 12^3$.

On Your Own

Now You're Ready
Exercises 4–12

Write the product as a power.

1. $6 \cdot 6 \cdot 6 \cdot 6 \cdot 6 \cdot 6$ **2.** $15 \times 15 \times 15 \times 15$

EXAMPLE 2 Finding Values of Powers

Find the value of each power.

a. 7^2 **b.** 5^3

$7^2 = 7 \cdot 7$ Write as repeated multiplication. $5^3 = 5 \cdot 5 \cdot 5$

$= 49$ Simplify. $= 125$

12   Chapter 1   Numerical Expressions and Factors   Multi-Language Glossary at BigIdeasMath.com

The square of a whole number is a **perfect square**.

EXAMPLE 3 Identifying Perfect Squares

Determine whether each number is a perfect square.

a. 64

Because $8^2 = 64$, 64 is a perfect square.

b. 20

No whole number squared equals 20. So, 20 is not a perfect square.

On Your Own

Now You're Ready
Exercises 14–21 and 25–32

Find the value of the power.

3. 6^3 **4.** 9^2 **5.** 3^4 **6.** 18^2

Determine whether the number is a perfect square.

7. 25 **8.** 2 **9.** 99 **10.** 100

Remember
The *area* of a figure is the amount of surface it covers. Area is measured in square units.

The area of a square is equal to its side length squared.

Area $= 3^2 = 9$ square units

EXAMPLE 4 Real-Life Application

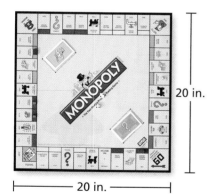

20 in.

20 in.

A MONOPOLY® game board is a square with a side length of 20 inches. What is the area of the game board?

Use a verbal model to solve the problem.

$$\text{area of game board} = (\text{side length})^2$$
$$= 20^2 \quad \text{Substitute 20 for side length.}$$
$$= 400 \quad \text{Multiply.}$$

∴ The area of the game board is 400 square inches.

On Your Own

11. What is the area of the square traffic sign in square inches? in square feet?

24 in.

STOP AHEAD

24 in.

1.3 Order of Operations

Essential Question What is the effect of inserting parentheses into a numerical expression?

1 ACTIVITY: Comparing Different Orders

Work with a partner. Find the value of the expression by using different orders of operations. Are your answers the same? *(Circle yes or no.)*

a. Add, then multiply.   Multiply, then add.   Same?
   $3 + 4 \times 2 =$ $3 + 4 \times 2 =$ Yes   No

b. Add, then subtract.   Subtract, then add.   Same?
   $5 + 3 - 1 =$ $5 + 3 - 1 =$ Yes   No

c. Divide, then multiply. Multiply, then divide. Same?
   $12 \div 3 \cdot 2 =$ $12 \div 3 \cdot 2 =$ Yes   No

d. Divide, then add.   Add, then divide.   Same?
   $16 \div 4 + 4 =$ $16 \div 4 + 4 =$ Yes   No

e. Multiply, then subtract. Subtract, then multiply. Same?
   $8 \times 4 - 2 =$ $8 \times 4 - 2 =$ Yes   No

f. Multiply, then divide. Divide, then multiply. Same?
   $8 \cdot 4 \div 2 =$ $8 \cdot 4 \div 2 =$ Yes   No

g. Subtract, then add.   Add, then subtract.   Same?
   $13 - 4 + 6 =$ $13 - 4 + 6 =$ Yes   No

h. Multiply, then add.   Add, then multiply.   Same?
   $1 \times 2 + 3 =$ $1 \times 2 + 3 =$ Yes   No

Numerical Expressions
In this lesson, you will
- evaluate numerical expressions with whole-number exponents.

16   Chapter 1   Numerical Expressions and Factors

2 ACTIVITY: Using Parentheses

Work with a partner. Use all the symbols and numbers to write an expression that has the given value.

	Symbols and Numbers	Value	Expression
a.	( ), +, ÷, 3, 4, 5	3	
b.	( ), −, ×, 2, 5, 8	11	
c.	( ), ×, ÷, 4, 4, 16	16	
d.	( ), −, ÷, 3, 8, 11	1	
e.	( ), +, ×, 2, 5, 10	70	

3 ACTIVITY: Reviewing Fractions and Decimals

Math Practice

Use Operations
How do you know which operation to perform first?

Work with a partner. Evaluate the expression.

a. $\dfrac{3}{4} - \left(\dfrac{1}{4} + \dfrac{1}{2}\right)$ =

b. $\left(\dfrac{5}{6} - \dfrac{1}{6}\right) - \dfrac{1}{12}$ =

c. $7.4 - (3.5 - 3.1)$ =

d. $10.4 - (8.6 + 0.9)$ =

e. $(\$7.23 + \$2.32) - \$5.40$ =

f. $\$124.60 - (\$72.41 + \$5.67)$ =

What Is Your Answer?

4. In an expression with two or more operations, why is it necessary to agree on an order of operations? Give examples to support your explanation.

5. **IN YOUR OWN WORDS** What is the effect of inserting parentheses into a numerical expression?

Practice — Use what you learned about the order of operations to complete Exercises 3–5 on page 20.

Section 1.3 Order of Operations 17

1.3 Lesson

Key Vocabulary
numerical expression, *p. 18*
evaluate, *p. 18*
order of operations, *p. 18*

A **numerical expression** is an expression that contains only numbers and operations. To **evaluate**, or find the value of, a numerical expression, use a set of rules called the **order of operations**.

 **Key Idea**

Order of Operations
1. Perform operations in **P**arentheses.
2. Evaluate numbers with **E**xponents.
3. **M**ultiply or **D**ivide from left to right.
4. **A**dd or **S**ubtract from left to right.

EXAMPLE 1 Using Order of Operations

a. Evaluate $12 - 2 \times 4$.

$12 - 2 \times 4 = 12 - 8$ Multiply 2 and 4.
$ = 4$ Subtract 8 from 12.

b. Evaluate $7 + 60 \div (3 \times 5)$.

$7 + 60 \div (3 \times 5) = 7 + 60 \div 15$ Perform operation in parentheses.
$ = 7 + 4$ Divide 60 by 15.
$ = 11$ Add 7 and 4.

EXAMPLE 2 Using Order of Operations with Exponents

Evaluate $30 \div (7 + 2^3) \times 6$.

Evaluate the power in parentheses first.

$30 \div (7 + 2^3) \times 6 = 30 \div (7 + 8) \times 6$ Evaluate 2^3.
$ = 30 \div 15 \times 6$ Perform operation in parentheses.
$ = 2 \times 6$ Divide 30 by 15.
$ = 12$ Multiply 2 and 6.

Study Tip
Remember to multiply and divide from left to right. In Example 2, you should divide before multiplying because the division symbol comes first when reading from left to right.

On Your Own

Now You're Ready
Exercises 6–14

Evaluate the expression.

1. $7 \cdot 5 + 3$
2. $(28 - 20) \div 4$
3. $6 \times 15 - 10 \div 2$
4. $6 + 2^4 - 1$
5. $4 \cdot 3^2 + 18 - 9$
6. $16 + (5^2 - 7) \div 3$

The symbols × and · are used to indicate multiplication. You can also use parentheses to indicate multiplication. For example, 3(2 + 7) is the same as 3 × (2 + 7).

EXAMPLE 3 **Using Order of Operations**

a. Evaluate 9 + 7(5 − 2).

9 + 7(5 − 2) = 9 + 7(3)	Perform operation in parentheses.
= 9 + 21	Multiply 7 and 3.
= 30	Add 9 and 21.

b. Evaluate $15 - 4(6 + 1) \div 2^2$.

$15 - 4(6 + 1) \div 2^2 = 15 - 4(7) \div 2^2$	Perform operation in parentheses.
$= 15 - 4(7) \div 4$	Evaluate 2^2.
$= 15 - 28 \div 4$	Multiply 4 and 7.
$= 15 - 7$	Divide 28 by 4.
$= 8$	Subtract 7 from 15.

EXAMPLE 4 **Real-Life Application**

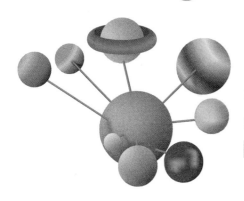

You buy foam spheres, paint bottles, and wooden rods to construct a model of our solar system. What is your total cost?

Item	Quantity	Cost per Item
Spheres	9	$2
Paint	6	$3
Rods	8	$1

Use a verbal model to solve the problem.

cost of 9 spheres + cost of 6 paint bottles + cost of 8 rods

9 · 2 + 6 · 3 + 8 · 1

9 · 2 + 6 · 3 + 8 · 1 = 18 + 18 + 8	Multiply.
= 44	Add.

∴ Your total cost is $44.

On Your Own

Exercises 18–23

Evaluate the expression.

7. $50 + 6(12 \div 4) - 8^2$ 8. $5^2 - 5(10 - 5)$ 9. $\dfrac{8(3 + 4)}{7}$

10. **WHAT IF?** In Example 4, you add the dwarf planet Pluto to your model. Use a verbal model to find your total cost assuming you do not need more paint. Explain.

Section 1.3   Order of Operations

1.3 Exercises

Vocabulary and Concept Check

1. **WRITING** Why does $12 - 8 \div 2 = 8$, but $(12 - 8) \div 2 = 2$?

2. **REASONING** Describe the steps in evaluating the expression $8 \div (6 - 4) + 3^2$.

Practice and Problem Solving

Find the value of the expression.

3. $(4 \times 15) - 3$
4. $10 - (7 + 1)$
5. $18 \div (6 + 3)$

Evaluate the expression.

6. $5 + 18 \div 6$
7. $(11 - 3) \div 2 + 1$
8. $45 \div 9 \times 12$
9. $6^2 - 3 \cdot 4$
10. $42 \div (15 - 2^3)$
11. $4^2 \cdot 2 + 8 \cdot 7$
12. $3^2 + 12 \div (6 - 3) \times 8$
13. $(10 + 4) \div (26 - 19)$
14. $(5^2 - 4) \cdot 2 - 18$

ERROR ANALYSIS Describe and correct the error in evaluating the expression.

15.
✗ $9 + 2 \times 3 = 11 \times 3$
$= 33$

16.
✗ $19 - 6 + 12 = 19 - 18$
$= 1$

17. **POETRY** You need to read 20 poems in 5 days for an English project. Each poem is 2 pages long. Evaluate the expression $20 \times 2 \div 5$ to find how many pages you need to read each day.

Evaluate the expression.

18. $9^2 - 8(6 + 2)$
19. $(3 - 1)^3 + 7(6) - 5^2$
20. $8\left(1\dfrac{1}{6} + \dfrac{5}{6}\right) \div 4$
21. $7^2 - 2\left(\dfrac{11}{8} - \dfrac{3}{8}\right)$
22. $8(7.3 + 3.7) - 14 \div 2$
23. $2^4(5.2 - 3.2) \div 4$

24. **MONEY** You have four $10 bills and eighteen $5 bills in your piggy bank. How much money do you have?

25. **THEATER** Before a show, there are 8 people in a theater. Five groups of 4 people enter, and then three groups of 2 people leave. Evaluate the expression $8 + 5(4) - 3(2)$ to find how many people are in the theater.

4($10) + 18($5)

Evaluate the expression.

26. $\dfrac{6(3+5)}{4}$

27. $\dfrac{12^2 - 4(6) + 1}{11^2}$

28. $\dfrac{26 \div 2 + 5}{3^2 - 3}$

29. **FIELD TRIP** Eighty students are going on a field trip to a history museum. The total cost includes
    - 2 bus rentals and
    - $10 per student for lunch.

    What is the total cost per student?

30. **OPEN-ENDED** Use all four operations without parentheses to write an expression that has a value of 100.

31. **SHOPPING** You buy 6 notebooks, 10 folders, 1 pack of pencils, and 1 lunch box for school. After using a $10 gift card, how much do you owe? Explain how you solved the problem.

32. **LITTER CLEANUP** Two groups collect litter along the side of a road. It takes each group 5 minutes to clean up a 200-yard section. How long does it take to clean up 2 *miles*? Explain how you solved the problem.

33. Copy each statement. Insert +, −, ×, or ÷ symbols to make each statement true.

    a. 27 ☐ 3 ☐ 5 ☐ 2 = 19

    b. 9^2 ☐ 11 ☐ 8 ☐ 4 ☐ 1 = 60

    c. 5 ☐ 6 ☐ 15 ☐ 9 = 24

    d. 14 ☐ 2 ☐ 7 ☐ 3 ☐ 9 = 10

Fair Game Review What you learned in previous grades & lessons

Add or subtract. *(Skills Review Handbook)*

34. $5.2 + 0.5$

35. $8 - 1.9$

36. $12.6 - 3$

37. $0.7 + 0.2$

38. **MULTIPLE CHOICE** You are making two recipes. One recipe calls for $2\dfrac{1}{3}$ cups of flour. The other recipe calls for $1\dfrac{1}{4}$ cups of flour. How much flour do you need to make both recipes? *(Skills Review Handbook)*

    Ⓐ $1\dfrac{1}{12}$ cups   Ⓑ $3\dfrac{1}{12}$ cups   Ⓒ $3\dfrac{2}{7}$ cups   Ⓓ $3\dfrac{7}{12}$ cups

1 Study Help

You can use an **information frame** to help you organize and remember concepts. Here is an example of an information frame for powers.

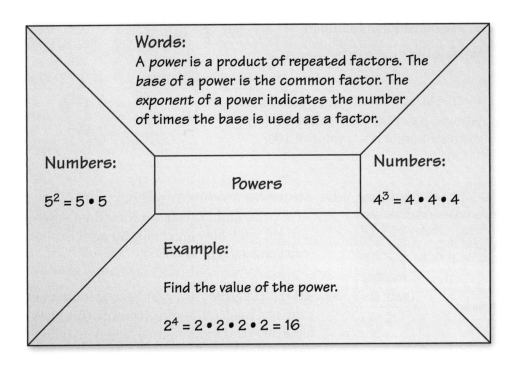

On Your Own

Make information frames to help you study these topics.

1. adding whole numbers
2. subtracting whole numbers
3. multiplying whole numbers
4. dividing whole numbers
5. order of operations

After you complete this chapter, make information frames for the following topics.

6. prime factorization
7. greatest common factor (GCF)
8. least common multiple (LCM)
9. least common denominator (LCD)

"Dear Mom, I am sending you an information frame card for Mother's Day!"

22   Chapter 1   Numerical Expressions and Factors

1.1–1.3 Quiz

Find the value of the expression. Use estimation to check your answer. *(Section 1.1)*

1. $4265 + 3896$
2. $5327 - 2624$
3. 276×49
4. $648 \div 72$

Find the value of the power. *(Section 1.2)*

5. 3^3
6. 11^2

Determine whether the number is a perfect square. *(Section 1.2)*

7. 36
8. 15

Evaluate the expression. *(Section 1.3)*

9. $6 + 21 \div 7$
10. $\dfrac{4(12-3)}{12}$
11. $16 \div 2^3 + 6 - 2$
12. $2 \times 14 \div (3^2 - 2)$

13. **AUDITORIUM** An auditorium has a total of 592 seats. There are 37 rows of seats, and each row has the same number of seats. How many seats are there in a single row? *(Section 1.1)*

14. **SOFTBALL** The bases on a softball field are square. What is the area of each base? *(Section 1.2)*

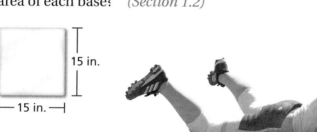

15. **DUATHLON** In an 18-mile duathlon, you run, then bike 12 miles, and then run again. The two runs are the same distance. Find the distance of each run. *(Section 1.3)*

16. **AMUSEMENT PARK** Tickets for an amusement park cost $10 for adults and $6 for children. Find the total cost for 2 adults and 3 children. *(Section 1.3)*

1.4 Prime Factorization

Essential Question Without dividing, how can you tell when a number is divisible by another number?

1 ACTIVITY: Finding Divisibility Rules for 2, 3, 5, and 10

Work with a partner. Copy the set of numbers (1–50) as shown.

1	2	3	4	5	6	7	8	9	10
11	12	13	14	15	16	17	18	19	20
21	22	23	24	25	26	27	28	29	30
31	32	33	34	35	36	37	38	39	40
41	42	43	44	45	46	47	48	49	50

a. Highlight all the numbers that are divisible by 2.
b. Put a box around the numbers that are divisible by 3.
c. Underline the numbers that are divisible by 5.
d. Circle the numbers that are divisible by 10.
e. **STRUCTURE** In parts (a)–(d), what patterns do you notice? Write four rules to determine when a number is divisible by 2, 3, 5, and 10.

Common Factors and Multiples

In this lesson, you will
- use divisibility rules to find prime factorizations of numbers.

2 ACTIVITY: Finding Divisibility Rules for 6 and 9

Work with a partner.

a. List ten numbers that are divisible by 6. Write a rule to determine when a number is divisible by 6. Use a calculator to check your rule with large numbers.

b. List ten numbers that are divisible by 9. Write a rule to determine when a number is divisible by 9. Use a calculator to check your rule with large numbers.

24   Chapter 1   Numerical Expressions and Factors

3 ACTIVITY: Rewriting a Number Using 2s, 3s, and 5s

Work with three other students. Use the following rules and only the prime factors 2, 3, and 5 to write each number below as a product.

- Your group should have four sets of cards: a set with all 2s, a set with all 3s, a set with all 5s, and a set of blank cards. Each person gets one set of cards.
- Begin by choosing two cards to represent the given number as a product of two factors. The person with the blank cards writes any factors that are not 2, 3, or 5.
- Use the cards again to represent any number written on a blank card as a product of two factors. Continue until you have represented each handwritten card as a product of two prime factors.
- You may use only one blank card for each step.

a. **Sample:** 108

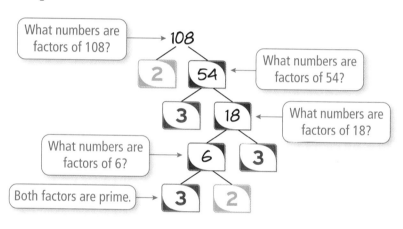

$108 = 2 \cdot 3 \cdot 3 \cdot 3 \cdot 2$

Math Practice

Interpret Results

How do you know your answer makes sense?

b. 80   c. 162   d. 300

e. Compare your results with those of other groups. Are your steps the same for each number? Is your final answer the same for each number?

What Is Your Answer?

4. **IN YOUR OWN WORDS** Without dividing, how can you tell when a number is divisible by another number? Give examples to support your explanation.

5. Explain how you can use your divisibility rules from Activities 1 and 2 to help with Activity 3.

Practice

Use what you learned about divisibility rules to complete Exercises 4–7 on page 28.

Section 1.4 Prime Factorization 25

1.4 Lesson

Because 2 is factor of 10 and 2 • 5 = 10, 5 is also a factor of 10. The pair 2, 5 is called a **factor pair** of 10.

EXAMPLE 1 Finding Factor Pairs

Key Vocabulary
factor pair, p. 26
prime factorization, p. 26
factor tree, p. 26

The brass section of a marching band has 30 members. The band director arranges the brass section in rows. Each row has the same number of members. How many possible arrangements are there?

Use the factor pairs of 30 to find the number of arrangements.

30 = 1 • 30	There could be 1 row of 30 or 30 rows of 1.
30 = 2 • 15	There could be 2 rows of 15 or 15 rows of 2.
30 = 3 • 10	There could be 3 rows of 10 or 10 rows of 3.
30 = 5 • 6	There could be 5 rows of 6 or 6 rows of 5.
30 = 6 • 5	The factors 5 and 6 are already listed.

Study Tip
When making an organized list of factor pairs, stop finding pairs when the factors begin to repeat.

∴ There are 8 possible arrangements: 1 row of 30, 30 rows of 1, 2 rows of 15, 15 rows of 2, 3 rows of 10, 10 rows of 3, 5 rows of 6, or 6 rows of 5.

On Your Own

Now You're Ready
Exercises 8–15

List the factor pairs of the number.

1. 18
2. 24
3. 51

4. **WHAT IF?** The woodwinds section of the marching band has 38 members. Which has more possible arrangements, the brass section or the woodwinds section? Explain.

Key Idea

Prime Factorization

The **prime factorization** of a composite number is the number written as a product of its prime factors.

You can use factor pairs and a **factor tree** to help find the prime factorization of a number. The factor tree is complete when only prime factors appear in the product. A factor tree for 60 is shown.

Remember
A *prime number* is a whole number greater than 1 with exactly two factors, 1 and itself. A *composite number* is a whole number greater than 1 with factors other than 1 and itself.

```
      60
     /  \
    2  • 30
        /  \
       2  • 15
            /  \
           3  • 5
```

$60 = 2 \cdot 2 \cdot 3 \cdot 5$, or $2^2 \cdot 3 \cdot 5$

26 Chapter 1 Numerical Expressions and Factors

EXAMPLE 2 Writing a Prime Factorization

Write the prime factorization of 48.

Choose any factor pair of 48 to begin the factor tree.

Study Tip
Notice that beginning with different factor pairs results in the same prime factorization. Every composite number has only one prime factorization.

Tree 1

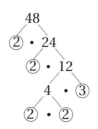

Find a factor pair and draw "branches."

Circle the prime factors as you find them.

Find factors until each branch ends at a prime factor.

Tree 2

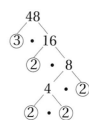

$48 = 2 \cdot 2 \cdot 3 \cdot 2 \cdot 2$

$48 = 3 \cdot 2 \cdot 2 \cdot 2 \cdot 2$

∴ The prime factorization of 48 is $2 \cdot 2 \cdot 2 \cdot 2 \cdot 3$, or $2^4 \cdot 3$.

EXAMPLE 3 Using a Prime Factorization

What is the greatest perfect square that is a factor of 1575?

Because 1575 has many factors, it is not efficient to list all of its factors and check for perfect squares. Use the prime factorization of 1575 to find any perfect squares that are factors.

```
         1575
         /  \
       25 · 63
       /\   /\
      5·5  7· 9
              /\
             3·3
```

$1575 = 3 \cdot 3 \cdot 5 \cdot 5 \cdot 7$

The prime factorization shows that 1575 has three factors other than 1 that are perfect squares.

$3 \cdot 3 = 9 \qquad 5 \cdot 5 = 25 \qquad (3 \cdot 5) \cdot (3 \cdot 5) = 15 \cdot 15 = 225$

∴ So, the greatest perfect square that is a factor of 1575 is 225.

On Your Own

Now You're Ready
Exercises 16–23 and 29–32

Write the prime factorization of the number.

5. 20   **6.** 88   **7.** 90   **8.** 462

9. What is the greatest perfect square that is a factor of 396? Explain.

Section 1.4 Prime Factorization

1.4 Exercises

Vocabulary and Concept Check

1. **VOCABULARY** What is the prime factorization of a number?
2. **VOCABULARY** How can you use a factor tree to help you write the prime factorization of a number?
3. **WHICH ONE DOESN'T BELONG?** Which factor pair does not belong with the other three? Explain your reasoning.

   2, 28   4, 14   6, 9   7, 8

Practice and Problem Solving

Use divisibility rules to determine whether the number is divisible by 2, 3, 5, 6, 9, and 10. Use a calculator to check your answer.

4. 1044
5. 1485
6. 1620
7. 1709

List the factor pairs of the number.

① 8. 15
9. 22
10. 34
11. 39
12. 45
13. 54
14. 59
15. 61

Write the prime factorization of the number.

② 16. 16
17. 25
18. 30
19. 26
20. 84
21. 54
22. 65
23. 77

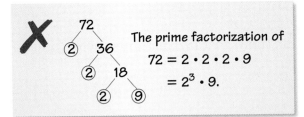

24. **ERROR ANALYSIS** Describe and correct the error in writing the prime factorization.

25. **FACTOR RAINBOW** You can use a factor rainbow to check whether a list of factors is correct. To create a factor rainbow, list the factors of a number in order from least to greatest. Then draw arches that link the factor pairs. For perfect squares, there is no connecting arch in the middle. So, just circle the middle number. A factor rainbow for 12 is shown. Create factor rainbows for 6, 24, 36, and 48.

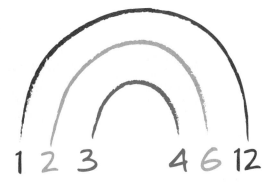

28   Chapter 1   Numerical Expressions and Factors

Find the number represented by the prime factorization.

26. $2^2 \cdot 3^2 \cdot 5$ **27.** $3^2 \cdot 5^2 \cdot 7$ **28.** $2^3 \cdot 11^2 \cdot 13$

Find the greatest perfect square that is a factor of the number.

③ 29. 244 **30.** 650 **31.** 756 **32.** 1290

33. CRITICAL THINKING Is 2 the only even prime number? Explain.

34. BASEBALL The coach of a baseball team separates the players into groups for drills. Each group has the same number of players. Is the total number of players on the baseball team *prime* or *composite*? Explain.

35. SCAVENGER HUNT A teacher divides 36 students into equal groups for a scavenger hunt. Each group should have at least 4 students but no more than 8 students. What are the possible group sizes?

36. PERFECT NUMBERS A *perfect number* is a number that equals the sum of its factors, not including itself. For example, the factors of 28 are 1, 2, 4, 7, 14, and 28. Because 1 + 2 + 4 + 7 + 14 = 28, 28 is a perfect number. What are the perfect numbers between 1 and 28?

37. BAKE SALE One table at a bake sale has 75 cookies. Another table has 60 cupcakes. Which table allows for more rectangular arrangements when all the cookies and cupcakes are displayed? Explain.

38. MODELING The stage manager of a school play creates a rectangular acting area of 42 square yards. String lights will outline the acting area. To the nearest whole number, how many yards of string lights does the manager need to enclose this area?

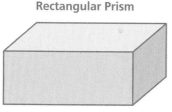

Rectangular Prism
Volume = 40 cubic inches

39. Volume The volume of a rectangular prism can be found using the formula *volume = length × width × height*. Using only whole number dimensions, how many different prisms are possible? Explain.

Fair Game Review What you learned in previous grades & lessons

Find the difference. *(Skills Review Handbook)*

40. 192 − 47 **41.** 451 − 94 **42.** 3210 − 815 **43.** 4752 − 3504

44. MULTIPLE CHOICE You buy 168 pears. There are 28 pears in each bag. How many bags of pears do you buy? *(Skills Review Handbook)*

Ⓐ 5 Ⓑ 6 Ⓒ 7 Ⓓ 28

1.5 Greatest Common Factor

Essential Question How can you find the greatest common factor of two numbers?

A **Venn diagram** uses circles to describe relationships between two or more sets. The Venn diagram shows the names of students enrolled in two activities. Students enrolled in both activities are represented by the overlap of the two circles.

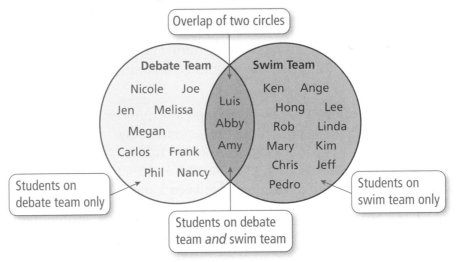

1 ACTIVITY: Identifying Common Factors

Work with a partner. Copy and complete the Venn diagram. Identify the *common factors* of the two numbers.

a. 36 and 48

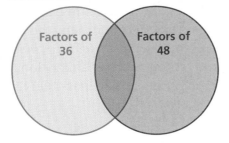

b. 16 and 56

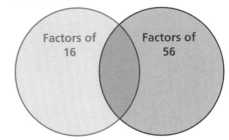

Common Factors

In this lesson, you will
- use diagrams to identify common factors.
- find greatest common factors.

c. 30 and 75

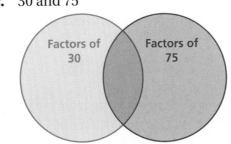

d. 54 and 90

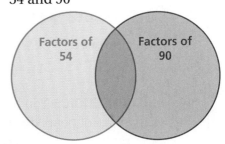

e. Look at the Venn diagrams in parts (a)–(d). Explain how to identify the *greatest common factor* of each pair of numbers. Then circle it in each diagram.

30   Chapter 1   Numerical Expressions and Factors

2 ACTIVITY: Interpreting a Venn Diagram of Prime Factors

Work with a partner. The Venn diagram represents the prime factorization of two numbers. Identify the two numbers. Explain your reasoning.

a.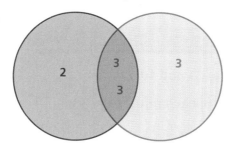

b.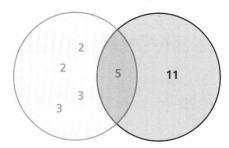

3 ACTIVITY: Identifying Common Prime Factors

Math Practice

Interpret a Solution

What does the diagram of the resulting prime factorization mean?

Work with a partner.

a. Write the prime factorizations of 36 and 48. Use the results to complete the Venn diagram.

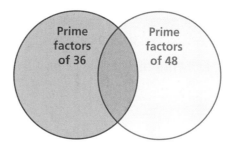

b. Repeat part (a) for the remaining number pairs in Activity 1.

c. **STRUCTURE** Compare the numbers in the overlap of the Venn diagrams to your results in Activity 1. What conjecture can you make about the relationship between these numbers and your results in Activity 1?

What Is Your Answer?

4. **IN YOUR OWN WORDS** How can you find the greatest common factor of two numbers? Give examples to support your explanation.

5. Can you think of another way to find the greatest common factor of two numbers? Explain.

Practice — Use what you learned about greatest common factors to complete Exercises 4–6 on page 34.

1.5 Lesson

Factors that are shared by two or more numbers are called **common factors**. The greatest of the common factors is called the **greatest common factor** (GCF). One way to find the GCF of two or more numbers is by listing factors.

EXAMPLE 1 Finding the GCF Using Lists of Factors

Find the GCF of 24 and 40.

List the factors of each number.

Factors of 24: ①,②, 3,④, 6,⑧, 12, 24   Circle the common factors.
Factors of 40: ①,②,④, 5,⑧, 10, 20, 40

The common factors of 24 and 40 are 1, 2, 4, and 8. The greatest of these common factors is 8.

∴ So, the GCF of 24 and 40 is 8.

Key Vocabulary
Venn diagram, *p. 30*
common factors, *p. 32*
greatest common factor, *p. 32*

Another way to find the GCF of two or more numbers is by using prime factors. The GCF is the product of the common prime factors of the numbers.

EXAMPLE 2 Finding the GCF Using Prime Factorizations

Find the GCF of 12 and 56.

Make a factor tree for each number.

Write the prime factorization of each number.

12 = ②·②· 3
56 = ②·②· 2 · 7   Circle the common prime factors.

2 · 2 = 4   Find the product of the common prime factors.

∴ So, the GCF of 12 and 56 is 4.

Study Tip
Examples 1 and 2 show two different methods for finding the GCF. After solving with one method, you can use the other method to check your answer.

On Your Own

Now You're Ready
Exercises 7–18

Find the GCF of the numbers using lists of factors.

1. 8, 36   2. 18, 72   3. 14, 28, 49

Find the GCF of the numbers using prime factorizations.

4. 20, 45   5. 32, 90   6. 45, 75, 120

EXAMPLE 3 Finding Two Numbers with a Given GCF

Which pair of numbers has a GCF of 15?

Ⓐ 10, 15   **Ⓑ** 30, 60   **Ⓒ** 21, 45   **Ⓓ** 45, 75

The number 15 cannot be a factor of the lesser number 10. So, you can eliminate Statement A.

The number 15 cannot be a factor of a number that does not have a 0 or 5 in the ones place. So, you can eliminate Statement C.

List the factors for Statements B and D. Then identify the GCF for each.

Choice B: **Factors of 30:** ①,②,③,⑤,⑥,⑩,⑮,㉚

Factors of 60: ①,②,③, 4,⑤,⑥,⑩, 12,⑮, 20,㉚, 60

The GCF of 30 and 60 is 30.

Choice D: **Factors of 45:** ①,③,⑤, 9,⑮, 45

Factors of 75: ①,③,⑤,⑮, 25, 75

The GCF of 45 and 75 is 15.

∴ The correct answer is **Ⓓ**.

EXAMPLE 4 Real-Life Application

* 18 bottles of nail polish
* 24 pairs of earrings
* 42 lollipops

You are filling piñatas for your sister's birthday party. The list shows the gifts you are putting into the piñatas. You want identical groups of gifts in each piñata with no gifts left over. What is the greatest number of piñatas you can make?

The GCF of the numbers of gifts represents the greatest number of identical groups of gifts you can make with no gifts left over. So, to find the number of piñatas, find the GCF.

$$18 = 2 \cdot 3 \cdot 3$$
$$24 = 2 \cdot 3 \cdot 2 \cdot 2$$
$$42 = 2 \cdot 3 \cdot 7$$

$2 \cdot 3 = 6$ Find the product of the common prime factors.

The GCF of 18, 24, and 42 is 6.

∴ So, you can make at most 6 piñatas.

On Your Own

Now You're Ready
Exercises 23–25

7. Write a pair of numbers whose greatest common factor is 10.

8. WHAT IF? In Example 4, you add 6 more pairs of earrings. Does this change your answer? Explain your reasoning.

1.5 Exercises

Vocabulary and Concept Check

1. **VOCABULARY** What is the greatest common factor (GCF) of two numbers?
2. **WRITING** Describe how to find the GCF of two numbers by using prime factorization.
3. **DIFFERENT WORDS, SAME QUESTION** Which is different? Find "both" answers.

> What is the greatest common factor of 24 and 32?
>
> What is the greatest common divisor of 24 and 32?
>
> What is the greatest prime factor of 24 and 32?
>
> What is the product of the common prime factors of 24 and 32?

Practice and Problem Solving

Use a Venn diagram to find the greatest common factor of the numbers.

4. 12, 30
5. 32, 54
6. 24, 108

Find the GCF of the numbers using lists of factors.

7. 6, 15
8. 14, 84
9. 45, 76
10. 39, 65
11. 51, 85
12. 40, 63

Find the GCF of the numbers using prime factorizations.

13. 45, 60
14. 27, 63
15. 36, 81
16. 72, 84
17. 61, 73
18. 189, 200

ERROR ANALYSIS Describe and correct the error in finding the GCF.

19.
$42 = 2 \cdot 3 \cdot 7$
$154 = 2 \cdot 7 \cdot 11$
The GCF is 7.

20.
$36 = 2^2 \cdot 3^2$
$60 = 2^2 \cdot 3 \cdot 5$
The GCF is $2 \cdot 3 = 6$.

21. **CLASSROOM** A teacher is making identical activity packets using 92 crayons and 23 sheets of paper. What is the greatest number of packets the teacher can make with no items left over?

22. **BALLOONS** You are making balloon arrangements for a birthday party. There are 16 white balloons and 24 red balloons. Each arrangement must be identical. What is the greatest number of arrangements you can make using every balloon?

34 Chapter 1 Numerical Expressions and Factors

Find the GCF of the numbers.

❹ 23. 35, 56, 63 **24.** 30, 60, 78 **25.** 42, 70, 84

26. OPEN-ENDED Write a set of three numbers that have a GCF of 16. What procedure did you use to find your answer?

27. REASONING You need to find the GCF of 256 and 400. Would you rather list their factors or use their prime factorizations? Explain.

CRITICAL THINKING Tell whether the statement is *always*, *sometimes*, or *never* true.

28. The GCF of two even numbers is 2.

29. The GCF of two prime numbers is 1.

30. When one number is a multiple of another, the GCF of the numbers is the greater of the numbers.

31. BOUQUETS A florist is making identical bouquets using 72 red roses, 60 pink roses, and 48 yellow roses. What is the greatest number of bouquets that the florist can make if no roses are left over? How many of each color are in each bouquet?

32. VENN DIAGRAM Consider the numbers 252, 270, and 300.

  a. Create a Venn diagram using the prime factors of the numbers.
  b. Use the Venn diagram to find the GCF of 252, 270, and 300.
  c. What is the GCF of 252 and 270? 252 and 300? Explain how you found your answer.

33. FRUIT BASKETS You are making fruit baskets using 54 apples, 36 oranges, and 73 bananas.

  a. Explain why you cannot make identical fruit baskets without leftover fruit.
  b. What is the greatest number of identical fruit baskets you can make with the least amount of fruit left over? Explain how you found your answer.

34. Problem Solving Two rectangular, adjacent rooms share a wall. One-foot-by-one-foot tiles cover the floor of each room. Describe how the greatest possible length of the adjoining wall is related to the total number of tiles in each room. Draw a diagram that represents one possibility.

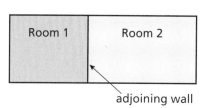

adjoining wall

Fair Game Review *What you learned in previous grades & lessons*

Tell which property is being illustrated. *(Skills Review Handbook)*

35. 13 + (29 + 7) = 13 + (7 + 29) **36.** 13 + (7 + 29) = (13 + 7) + 29

37. (6 × 37) × 5 = (37 × 6) × 5 **38.** (37 × 6) × 5 = 37 × (6 × 5)

39. MULTIPLE CHOICE In what order should you perform the operations in the expression 4 × 3 − 12 ÷ 2 + 5? *(Section 1.3)*

  Ⓐ ×, −, ÷, +   Ⓑ ×, ÷, −, +   Ⓒ ×, ÷, +, −   Ⓓ ×, +, −, ÷

1.6 Least Common Multiple

Essential Question How can you find the least common multiple of two numbers?

1 ACTIVITY: Identifying Common Multiples

Work with a partner. Using the first several multiples of each number, copy and complete the Venn diagram. Identify any *common multiples* of the two numbers.

a. 8 and 12

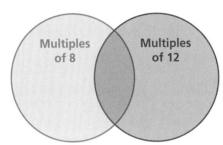

b. 4 and 14

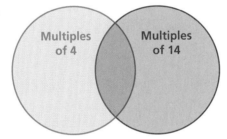

c. 10 and 15

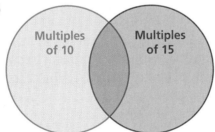

Common Multiples

In this lesson, you will
- use diagrams to identify common multiples.
- find least common multiples.

d. 20 and 35

e. Look at the Venn diagrams in parts (a)–(d). Explain how to identify the *least common multiple* of each pair of numbers. Then circle it in each diagram.

2 ACTIVITY: Interpreting a Venn Diagram of Prime Factors

Work with a partner.

a. Write the prime factorizations of 8 and 12. Use the results to complete the Venn diagram.

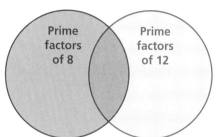

b. Repeat part (a) for the remaining number pairs in Activity 1.

c. **STRUCTURE** Compare the numbers from each section of the Venn diagrams to your results in Activity 1. What conjecture can you make about the relationship between these numbers and your results in Activity 1?

What Is Your Answer?

Math Practice

Construct Arguments

How can you use diagrams to support your explanation?

3. **IN YOUR OWN WORDS** How can you find the least common multiple of two numbers? Give examples to support your explanation.

4. The Venn diagram shows the prime factors of two numbers.

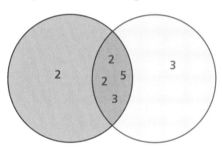

Use the diagram to do the following tasks.

a. Identify the two numbers.
b. Find the greatest common factor.
c. Find the least common multiple.

5. A student writes the prime factorizations of 8 and 12 in a table as shown. She claims she can use the table to find the greatest common factor and the least common multiple of 8 and 12. How is this possible?

8 =	2	2	2	
12 =	2	2		3

6. Can you think of another way to find the least common multiple of two or more numbers? Explain.

Practice — Use what you learned about least common multiples to complete Exercises 3–5 on page 40.

Section 1.6 Least Common Multiple 37

1.6 Lesson

Check It Out
Lesson Tutorials
BigIdeasMath.com

Multiples that are shared by two or more numbers are called **common multiples**. The least of the common multiples is called the **least common multiple** (LCM). You can find the LCM of two or more numbers by listing multiples or using prime factors.

EXAMPLE 1 — Finding the LCM Using Lists of Multiples

Key Vocabulary
common multiples, p. 38
least common multiple, p. 38

Find the LCM of 4 and 6.

List the multiples of each number.

Multiples of 4: 4, 8, ⑫, 16, 20, ㉔, 28, 32, ㊱, . . .   Circle the common multiples.
Multiples of 6: 6, ⑫, 18, ㉔, 30, ㊱, . . .

Some common multiples of 4 and 6 are 12, 24, and 36. The least of these common multiples is 12.

∴ So, the LCM of 4 and 6 is 12.

On Your Own

Find the LCM of the numbers using lists of multiples.

1. 3, 8
2. 9, 12
3. 6, 10

EXAMPLE 2 — Finding the LCM Using Prime Factorizations

Find the LCM of 16 and 20.

Make a factor tree for each number.

Write the prime factorization of each number. Circle each different factor where it appears the greater number of times.

16 = ②•②•②•②   2 appears more often here, so circle all 2s.
20 = 2 • 2 •⑤   5 appears once. Do not circle the 2s again.
2 • 2 • 2 • 2 • 5 = 80   Find the product of the circled factors.

∴ So, the LCM of 16 and 20 is 80.

On Your Own

Find the LCM of the numbers using prime factorizations.

4. 14, 18
5. 28, 36
6. 24, 90

EXAMPLE 3 Finding the LCM of Three Numbers

Find the LCM of 4, 15, and 18.

Write the prime factorization of each number. Circle each different factor where it appears the greatest number of times.

4 = ②•②   2 appears most often here, so circle both 2s.

15 = 3 •⑤   5 appears here only, so circle 5.

18 = 2 •③•③   3 appears most often here, so circle both 3s.

2 • 2 • 5 • 3 • 3 = 180   Find the product of the circled factors.

∴ So, the LCM of 4, 15, and 18 is 180.

On Your Own

Exercises 22–27

Find the LCM of the numbers.

7. 2, 5, 8   **8.** 6, 10, 12

9. Write a set of numbers whose least common multiple is 100.

EXAMPLE 4 Real-Life Application

A traffic light changes every 30 seconds. Another traffic light changes every 40 seconds. Both lights just changed. After how many minutes will both lights change at the same time again?

Find the LCM of 30 and 40 by listing multiples of each number. Circle the least common multiple.

Multiples of 30: 30, 60, 90, ⑫⓪, . . .

Multiples of 40: 40, 80, ⑫⓪, 160, . . .

The LCM is 120. So, both lights will change again after 120 seconds.

Because there are 60 seconds in 1 minute, there are 120 ÷ 60 = 2 minutes in 120 seconds.

∴ Both lights will change at the same time again after 2 minutes.

On Your Own

10. WHAT IF? In Example 4, the traffic light that changes every 40 seconds is adjusted to change every 45 seconds. Both lights just changed. After how many minutes will both lights change at the same time again?

1.6 Exercises

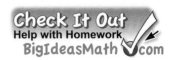

Vocabulary and Concept Check

1. **VOCABULARY** What is the least common multiple (LCM) of two numbers?
2. **WRITING** Describe how to find the LCM of two numbers by using prime factorization.

Practice and Problem Solving

Use a Venn diagram to find the least common multiple of the numbers.

3. 3, 7
4. 6, 8
5. 12, 15

Find the LCM of the numbers using lists of multiples.

6. 2, 9
7. 3, 4
8. 8, 9
9. 5, 8
10. 15, 20
11. 12, 18

Find the LCM of the numbers using prime factorizations.

12. 9, 21
13. 12, 27
14. 18, 45
15. 22, 33
16. 36, 60
17. 35, 50

18. **ERROR ANALYSIS** Describe and correct the error in finding the LCM.

$6 \times 9 = 54$
The LCM of 6 and 9 is 54.

19. **AQUATICS** You have diving lessons every fifth day and swimming lessons every third day. Today you have both lessons. In how many days will you have both lessons on the same day again?

20. **HOT DOGS** Hot dogs come in packs of 10, while buns come in packs of eight. What are the least numbers of packs you should buy in order to have the same numbers of hot dogs and buns?

21. **MODELING** Which model represents an LCM that is different from the other three? Explain your reasoning.

A.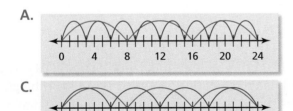

B.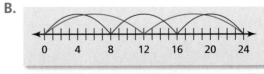

C.

D.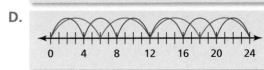

40   Chapter 1   Numerical Expressions and Factors

Find the LCM of the numbers.

22. 2, 3, 7   **23.** 3, 5, 11   **24.** 4, 9, 12

25. 6, 8, 15   **26.** 7, 18, 21   **27.** 9, 10, 28

28. REASONING You need to find the LCM of 13 and 14. Would you rather list their multiples or use their prime factorizations? Explain.

CRITICAL THINKING Tell whether the statement is *always*, *sometimes*, or *never* true.

29. The LCM of two different prime numbers is their product.

30. The LCM of a set of numbers is equal to one of the numbers in the set.

31. The GCF of two different numbers is the LCM of the numbers.

32. SUBWAY At Union Station, you notice that three subway lines just arrived at the same time. The table shows their arrival schedule. How long must you wait until all three lines arrive at Union Station at the same time again?

Subway Line	Arrival Time
A	every 10 min
B	every 12 min
C	every 15 min

33. RADIO CONTEST A radio station gives away $15 to every 15th caller, $25 to every 25th caller, and free concert tickets to every 100th caller. When will the station first give away *all* three prizes to one caller?

34. TREADMILL You and a friend are running on treadmills. You run 0.5 mile every 3 minutes, and your friend runs 2 miles every 14 minutes. You both start and stop running at the same time and run a whole number of miles. What is the least possible number of miles you and your friend can run?

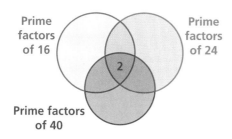

35. VENN DIAGRAM Refer to the Venn diagram.
  a. Copy and complete the Venn diagram.
  b. What is the LCM of 16, 24, and 40?
  c. What is the LCM of 16 and 40? 24 and 40?

36. When is the LCM of two numbers equal to their product?

Fair Game Review What you learned in previous grades & lessons

Write the product as a power. *(Section 1.2)*

37. 3×3 **38.** $5 \cdot 5 \cdot 5 \cdot 5$ **39.** $17 \times 17 \times 17 \times 17 \times 17$

40. MULTIPLE CHOICE Which two powers have the same value? *(Section 1.2)*

  Ⓐ 1^3 and 3^1 Ⓑ 2^4 and 4^2 Ⓒ 3^2 and 2^3 Ⓓ 4^3 and 3^4

Extension 1.6 Adding and Subtracting Fractions

Key Vocabulary
least common denominator, p. 42

Recall that you can add and subtract fractions with unlike denominators by writing equivalent fractions with a common denominator. One way to do this is by multiplying the numerator and the denominator of each fraction by the denominator of the other fraction.

EXAMPLE 1 — Adding Fractions Using a Common Denominator

Find $\dfrac{5}{8} + \dfrac{1}{6}$.

Rewrite the fractions with a common denominator. Use the product of the denominators as the common denominator.

$$\dfrac{5}{8} + \dfrac{1}{6} = \dfrac{5 \cdot 6}{8 \cdot 6} + \dfrac{1 \cdot 8}{6 \cdot 8}$$ Rewrite the fractions using a common denominator of $8 \cdot 6 = 48$.

$$= \dfrac{30}{48} + \dfrac{8}{48}$$ Multiply.

$$= \dfrac{38}{48}$$ Add the numerators.

$$= \dfrac{\cancel{2}^{1} \cdot 19}{\cancel{2}_{1} \cdot 24}$$ Divide out the common factor 2.

$$= \dfrac{19}{24}$$ Simplify.

Study Tip
A fraction is in *simplest form* when the numerator and the denominator have no common factors other than 1.

The **least common denominator** (LCD) of two or more fractions is the least common multiple (LCM) of the denominators. The LCD provides another method for adding and subtracting fractions with unlike denominators.

EXAMPLE 2 — Adding Fractions Using the LCD

Find $\dfrac{5}{8} + \dfrac{1}{6}$.

Find the LCM of the denominators.

Multiples of 8: 8, 16, ㉔, 32, 40, ㊽, . . .

Multiples of 6: 6, 12, 18, ㉔, 30, 36, 42, ㊽, . . .

The LCM of 8 and 6 is 24. So, the LCD is 24.

Common Multiples
In this extension, you will
• use least common multiples to add and subtract fractions.

$$\dfrac{5}{8} + \dfrac{1}{6} = \dfrac{5 \cdot 3}{8 \cdot 3} + \dfrac{1 \cdot 4}{6 \cdot 4}$$ Rewrite the fractions using the LCD, 24.

$$= \dfrac{15}{24} + \dfrac{4}{24}$$ Multiply.

$$= \dfrac{19}{24}$$ Add the numerators.

To add or subtract mixed numbers, first rewrite the numbers as improper fractions. Then find the common denominator.

EXAMPLE 3 Subtracting Mixed Numbers

Find $4\frac{3}{4} - 2\frac{3}{10}$.

Write the difference using improper fractions.

$$4\frac{3}{4} - 2\frac{3}{10} = \frac{19}{4} - \frac{23}{10}$$

Method 1: Use the product of the denominators as the common denominator.

$$\frac{19}{4} - \frac{23}{10} = \frac{19 \cdot 10}{4 \cdot 10} - \frac{23 \cdot 4}{10 \cdot 4}$$ Rewrite the fractions using a common denominator of $4 \cdot 10 = 40$.

$$= \frac{190}{40} - \frac{92}{40}$$ Multiply.

$$= \frac{98}{40}$$ Subtract the numerators.

$$= \frac{49}{20}, \text{ or } 2\frac{9}{20}$$ Simplify.

Study Tip

Notice that Method 1 uses the same procedure shown in Example 1. You can generalize the procedure using the rule $\frac{a}{b} \pm \frac{c}{d} = \frac{ad \pm bc}{bd}$.

Method 2: Use the LCD. The LCM of 4 and 10 is 20.

$$\frac{19}{4} - \frac{23}{10} = \frac{19 \cdot 5}{4 \cdot 5} - \frac{23 \cdot 2}{10 \cdot 2}$$ Rewrite the fractions using the LCD, 20.

$$= \frac{95}{20} - \frac{46}{20}$$ Multiply.

$$= \frac{49}{20}, \text{ or } 2\frac{9}{20}$$ Simplify.

Practice

Use the LCD to rewrite the fractions with the same denominator.

1. $\frac{1}{6}, \frac{3}{8}$
2. $\frac{4}{7}, \frac{3}{10}$
3. $\frac{5}{12}, \frac{2}{9}$
4. $\frac{3}{4}, \frac{5}{8}, \frac{1}{10}$

Copy and complete the statement using <, >, or =.

5. $\frac{4}{5}$ ___ $\frac{5}{6}$
6. $\frac{5}{14}$ ___ $\frac{3}{8}$
7. $2\frac{2}{5}$ ___ $\frac{24}{10}$
8. $4\frac{9}{25}$ ___ $4\frac{7}{20}$

Add or subtract. Write the answer in simplest form.

9. $\frac{2}{3} + \frac{3}{4}$
10. $\frac{6}{7} + \frac{1}{2}$
11. $\frac{7}{10} - \frac{5}{12}$
12. $\frac{13}{18} - \frac{5}{8}$
13. $2\frac{1}{6} + 3\frac{4}{9}$
14. $4\frac{3}{16} + 1\frac{1}{10}$
15. $1\frac{5}{6} - \frac{3}{4}$
16. $3\frac{2}{3} - 2\frac{4}{11}$

17. **COMPARING METHODS** List some advantages and disadvantages of each method shown in the examples. Which method do you prefer? Why?

Extension 1.6 Adding and Subtracting Fractions 43

1.4–1.6 Quiz

List the factor pairs of the number. *(Section 1.4)*

1. 48
2. 56

Write the prime factorization of the number. *(Section 1.4)*

3. 60
4. 72

Find the GCF of the numbers using lists of factors. *(Section 1.5)*

5. 18, 42
6. 24, 44, 52

Find the GCF of the numbers using prime factorizations. *(Section 1.5)*

7. 38, 68
8. 68, 76, 92

Find the LCM of the numbers using lists of multiples. *(Section 1.6)*

9. 8, 14
10. 3, 6, 16

Find the LCM of the numbers using prime factorizations. *(Section 1.6)*

11. 18, 30
12. 6, 24, 32

Add or subtract. Write the answer in simplest form. *(Section 1.6)*

13. $\frac{3}{5} + \frac{2}{3}$
14. $\frac{7}{8} - \frac{3}{4}$

15. **PICNIC BASKETS** You are creating identical picnic baskets using 30 sandwiches and 42 cookies. What is the greatest number of baskets that you can fill using all of the food? *(Section 1.5)*

16. **RIBBON** You have 52 inches of yellow ribbon and 64 inches of red ribbon. You want to cut the ribbons into pieces of equal length with no leftovers. What is the greatest length of the pieces that you can make? *(Section 1.5)*

17. **MUSIC LESSONS** You have piano lessons every fourth day and guitar lessons every sixth day. Today you have both lessons. In how many days will you have both lessons on the same day again? Explain. *(Section 1.6)*

18. **HAMBURGERS** Hamburgers come in packs of 20, while buns come in packs of 12. What is the least number of packs you should buy in order to have the same numbers of hamburgers and buns? *(Section 1.6)*

1 Chapter Review

Review Key Vocabulary

power, *p. 12*
base, *p. 12*
exponent, *p. 12*
perfect square, *p. 13*
numerical expression, *p. 18*
evaluate, *p. 18*
order of operations, *p. 18*

factor pair, *p. 26*
prime factorization, *p. 26*
factor tree, *p. 26*
Venn diagram, *p. 30*
common factors, *p. 32*
greatest common factor (GCF), *p. 32*

common multiples, *p. 38*
least common multiple (LCM), *p. 38*
least common denominator (LCD), *p. 42*

Review Examples and Exercises

1.1 Whole Number Operations (pp. 2–9)

Use the tens place because 203 is less than 508.

$$203 \overline{)5081}$$
$$\underline{-406}$$
$$102$$
quotient: 2

Divide 508 by 203: There are two groups of 203 in 508.
Multiply 2 and 203.
Subtract 406 from 508.

Next, bring down the 1 and divide the ones.

25 R6
$$203 \overline{)5081}$$
$$\underline{-406}\downarrow$$
$$1021$$
$$\underline{-1015}$$
$$6$$

Divide 1021 by 203: There are five groups of 203 in 1021.
Multiply 5 and 203.
Subtract 1015 from 1021.

∴ The quotient of 5081 and 203 is $25\frac{6}{203}$.

Exercises

Find the value of the expression. Use estimation to check your answer.

1. $4382 + 2899$
2. $8724 - 3568$
3. 192×38
4. $216 \div 31$

1.2 Powers and Exponents (pp. 10–15)

Evaluate 6^2.

$6^2 = 6 \cdot 6 = 36$ Write as repeated multiplication and simplify.

Exercises

Find the value of the power.

5. 7^3
6. 2^6
7. 4^4

1.3 Order of Operations (pp. 16–21)

Evaluate $4^3 - 15 \div 5$.

$4^3 - 15 \div 5 = 64 - 15 \div 5$ Evaluate 4^3.

$ = 64 - 3$ Divide 15 by 5.

$ = 61$ Subtract 3 from 64.

Exercises

Evaluate the expression.

8. $3 \times 6 - 12 \div 6$
9. $20 \times (3^2 - 4) \div 50$
10. $5 + (4^2 + 2) \div 6$

1.4 Prime Factorization (pp. 24–29)

Write the prime factorization of 18.

Find a factor pair and draw "branches."

Circle the prime factors as you find them.

Continue until each branch ends at a prime factor.

The prime factorization of 18 is $2 \cdot 3 \cdot 3$, or $2 \cdot 3^2$.

Exercises

List the factor pairs of the number.

11. 28
12. 44
13. 63

Write the prime factorization of the number.

14. 42
15. 50
16. 66

1.5 Greatest Common Factor (pp. 30–35)

a. Find the GCF of 32 and 76.

Factors of 32: ①,②,④, 8, 16, 32
Factors of 76: ①,②,④, 19, 38, 76

The greatest of the common factors is 4.

So, the GCF of 32 and 76 is 4.

b. Find the GCF of 45 and 63.

$45 = ③ \cdot ③ \cdot 5$
$63 = ③ \cdot ③ \cdot 7$
$3 \cdot 3 = 9$

So, the GCF of 45 and 63 is 9.

Exercises

Find the GCF of the numbers using lists of factors.

17. 27, 45   **18.** 30, 48   **19.** 28, 48, 64

Find the GCF of the numbers using prime factorizations.

20. 24, 90   **21.** 52, 68   **22.** 32, 56, 96

1.6 Least Common Multiple (pp. 36–43)

a. Find the LCM of 8 and 12.

Make a factor tree for each number.

Write the prime factorization of each number. Circle each different factor where it appears the greater number of times.

$8 = ②\cdot②\cdot②$ 2 appears more often here, so circle all 2s.

$12 = 2 \cdot 2 \cdot ③$ 3 appears once. Do not circle the 2s again.

$2 \cdot 2 \cdot 2 \cdot 3 = 24$ Find the product of the circled factors.

∴ So, the LCM of 8 and 12 is 24.

b. Find $\dfrac{1}{2} + \dfrac{1}{3}$.

The LCM of 2 and 3 is 6. So, the LCD is 6.

$$\dfrac{1}{2} + \dfrac{1}{3} = \dfrac{1 \cdot 3}{2 \cdot 3} + \dfrac{1 \cdot 2}{3 \cdot 2} = \dfrac{3}{6} + \dfrac{2}{6} = \dfrac{5}{6}$$

Exercises

Find the LCM of the numbers using lists of multiples.

23. 4, 14   **24.** 6, 20   **25.** 12, 28

Find the LCM of the numbers using prime factorizations.

26. 6, 45   **27.** 10, 12   **28.** 18, 27

Add or subtract. Write the answer in simplest form.

29. $\dfrac{2}{7} + \dfrac{1}{4}$ **30.** $\dfrac{5}{9} + \dfrac{3}{8}$ **31.** $3\dfrac{5}{6} - 2\dfrac{7}{15}$

32. WATER PITCHER A water pitcher contains $\dfrac{2}{3}$ gallon of water. You add $\dfrac{5}{7}$ gallon of water to the pitcher. How much water does the pitcher contain?

1 Chapter Test

Check It Out
Test Practice
BigIdeasMath.com

Find the value of the expression. Use estimation to check your answer.

1. $3963 + 2379$
2. $6184 - 2348$
3. 184×26
4. $207 \div 23$

Find the value of the power.

5. 2^3
6. 15^2
7. 5^4

Evaluate the expression.

8. $11 \times 8 - 6 \div 2$
9. $5 + 2^3 \div 4 - 2$
10. $6 + 4(11 - 2) \div 3^2$

List the factor pairs of the number.

11. 52
12. 66

Write the prime factorization of the number.

13. 46
14. 28

Find the GCF of the numbers using lists of factors.

15. 24, 54
16. 16, 32, 72

Find the GCF of the numbers using prime factorizations.

17. 52, 65
18. 18, 45, 63

Find the LCM of the numbers using lists of multiples.

19. 14, 21
20. 9, 24

Find the LCM of the numbers using prime factorizations.

21. 26, 39
22. 6, 12, 14

23. **BRACELETS** You have 16 yellow beads, 20 red beads, and 24 orange beads to make identical bracelets. What is the greatest number of bracelets that you can make using all the beads?

24. **MARBLES** A bag contains equal numbers of green and blue marbles. You can divide all the green marbles into groups of 12 and all the blue marbles into groups of 16. What is the least number of each color of marble that can be in the bag?

25. **SCALE** You place a $3\frac{3}{8}$-pound weight on the left side of a balance scale and a $1\frac{1}{5}$-pound weight on the right side. How much weight do you need to add to the right side to balance the scale?

48 Chapter 1 Numerical Expressions and Factors

1 Cumulative Assessment

1. You are making identical bagel platters using 40 plain bagels, 30 raisin bagels, and 24 blueberry bagels. What is the greatest number of platters that you can make if there are no leftover bagels?

   A. 2
   B. 6
   C. 8
   D. 10

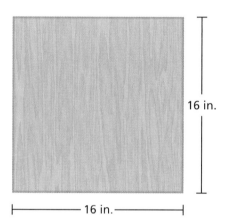

Test-Taking Strategy
Solve Directly or Eliminate Choices

How many hyenas are $5-2^2-1$ hyenas?
Ⓐ 0   Ⓑ 2   Ⓒ 8   Ⓓ 10

Survival strategy: A

"Which strategy would you use on this one: solve directly or eliminate choices?"

2. The top of an end table is a square with a side length of 16 inches. What is the area of the tabletop?

   16 in.

   16 in.

   F. 16 in.^{2}
   G. 32 in.^{2}
   H. 64 in.^{2}
   I. 256 in.^{2}

3. Which number is equivalent to the expression below?

   $$3 \cdot 2^3 - 8 \div 4$$

   A. 0
   B. 4
   C. 22
   D. 214

4. What is the least common multiple of 14 and 49?

5. Which number is equivalent to the expression 7059 ÷ 301?

- **F.** 23
- **G.** $23\frac{136}{7059}$
- **H.** $23\frac{136}{301}$
- **I.** 136

6. You are building identical displays for the school fair using 65 blue boxes and 91 yellow boxes. What is the greatest number of displays you can build using all the boxes?

- **A.** 13
- **B.** 35
- **C.** 91
- **D.** 156

7. You hang the two strands of decorative lights shown below.

Strand 1: changes between red and blue every 15 seconds

Strand 2: changes between green and gold every 18 seconds

Both strands just changed color. After how many seconds will the strands change color at the same time again?

- **F.** 3 seconds
- **G.** 30 seconds
- **H.** 90 seconds
- **I.** 270 seconds

8. Which expression is equivalent to $\frac{29}{63}$?

- **A.** $\frac{28}{60} + \frac{1}{3}$
- **B.** $\frac{4}{27} + \frac{25}{36}$
- **C.** $\frac{5}{21} + \frac{2}{9}$
- **D.** $\frac{22}{47} + \frac{7}{16}$

9. Which expression is *not* equivalent to 32?

- **F.** $6^2 - 8 \div 2$
- **G.** $30 \div 2 + 5^2 - 8$
- **H.** $30 + 4^2 \div (2 + 6)$
- **I.** $8^2 \div 4 - 2$

10. Which number is equivalent to the expression 148 × 27?

- **A.** 3696
- **B.** 3896
- **C.** 3946
- **D.** 3996

11. You have 60 nickels, 48 dimes, and 42 quarters. You want to divide the coins into identical groups with no coins left over. What is the greatest number of groups that you can make?

12. Erica was evaluating the expression in the box below.

$$56 \div (2^3 - 1) \times 4 = 56 \div (8 - 1) \times 4$$
$$= 56 \div 7 \times 4$$
$$= 56 \div 28$$
$$= 2$$

What should Erica do to correct the error that she made?

F. Divide 56 by 8 because operations are performed left to right.

G. Multiply 1 by 4 because multiplication is done before subtraction.

H. Divide 56 by 7 because operations are performed left to right.

I. Divide 56 by 8 and multiply 1 by 4 because division and multiplication are performed before subtraction.

13. Find the greatest common factor for each pair of numbers.

10 and 15   10 and 21   15 and 21

What can you conclude about the greatest common factor of 10, 15, and 21? Explain your reasoning.

14. Which number is *not* a perfect square?

A. 64

B. 81

C. 96

D. 100

15. Which number pair has a least common multiple of 48?

F. 4, 12

G. 6, 8

H. 8, 24

I. 16, 24

16. Which number is equivalent to the expression below?

$$\frac{3(6 + 2^2) + 2}{8}$$

A. 3

B. 4

C. 7

D. $24\frac{1}{4}$

2 Fractions and Decimals

2.1 **Multiplying Fractions**

2.2 **Dividing Fractions**

2.3 **Dividing Mixed Numbers**

2.4 **Adding and Subtracting Decimals**

2.5 **Multiplying Decimals**

2.6 **Dividing Decimals**

"Dear Sir: You say that MOST humans use only a fraction of their brain power."

"But, $\frac{3}{2}$ is a fraction. So, does that mean that SOME humans use one and a half of their brain power?"

"One of my homework problems is 'How many halves are in five halves?'"

What You Learned Before

"On a scale from 1 to 10, how do you like my painting?"

● Estimating Whole Number Products and Quotients

Example 1 Estimate 32×88.

32 is close to 30.

$32 \times 88 \approx 30 \times 90 = 2700$

88 is close to 90.

Example 2 Estimate $176 \div 57$.

176 is close to 180.

$176 \div 57 \approx 180 \div 60 = 3$

57 is close to 60.

Try It Yourself
Estimate the product or the quotient.

1. 9×23
2. 19×22
3. 49×21
4. 38×61
5. $38 \div 9$
6. $63 \div 22$
7. $118 \div 19$
8. $245 \div 62$

● Multiplying and Dividing Whole Numbers

Example 3 Find 356×21.

$$\begin{array}{r} \overset{1}{}\overset{1}{}\\ 356 \\ \times\ \ 21 \\ \hline 356 \\ +\ 7120 \\ \hline 7476 \end{array}$$

Example 4 Find $765 \div 3$.

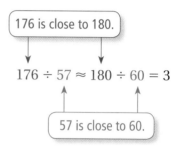

Try It Yourself
Find the product or the quotient.

9. 425×9
10. 721×18
11. 599×29
12. 503×12
13. $7\overline{)280}$
14. $4\overline{)428}$
15. $14\overline{)532}$
16. $23\overline{)8303}$

2.1 Multiplying Fractions

Essential Question What does it mean to multiply fractions?

1 ACTIVITY: Multiplying Fractions

Work with a partner. A bottle of water is $\frac{1}{2}$ full. You drink $\frac{2}{3}$ of the water. How much of the bottle of water do you drink?

THINK ABOUT THE QUESTION: To help you think about this question, rewrite the question.

Words: What is $\frac{2}{3}$ of $\frac{1}{2}$?   **Numbers:** $\frac{2}{3} \times \frac{1}{2} = ?$

Here is one way to get the answer.

- **Draw** a length of $\frac{1}{2}$.

Because you want to find $\frac{2}{3}$ of the length, divide it into 3 equal sections.

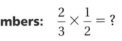

Now, you need to think of a way to divide $\frac{1}{2}$ into 3 equal parts.

- **Rewrite** $\frac{1}{2}$ as a fraction whose numerator is divisible by 3.

Because the length is divided into 3 equal sections, multiply the numerator and denominator by 3.

$\frac{1}{2} = \frac{1 \times 3}{2 \times 3} = \frac{3}{6}$

In this form, you see that $\frac{3}{6}$ can be divided into 3 equal parts of $\frac{1}{6}$.

- Each part is $\frac{1}{6}$ of the bottle of water, and you drank two of them. Written as multiplication, you have

$\frac{2}{3} \times \frac{1}{2} = \frac{2}{\boxed{}} = \frac{\boxed{}}{\boxed{}}$.

So, you drank $\frac{\boxed{}}{\boxed{}}$ of the bottle of water.

Dividing Fractions

In this lesson, you will
- use models to multiply fractions.
- multiply fractions by fractions.

54   Chapter 2   Fractions and Decimals

2 ACTIVITY: Multiplying Fractions

Work with a partner. A park has a playground that is $\frac{3}{4}$ of its width and $\frac{4}{5}$ of its length. What fraction of the park is covered by the playground?

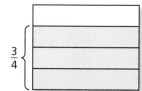

Fold a piece of paper horizontally into fourths and shade three of the fourths to represent $\frac{3}{4}$.

Fold the paper vertically into fifths and shade $\frac{4}{5}$ of the paper another color.

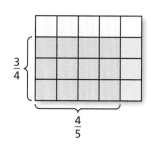

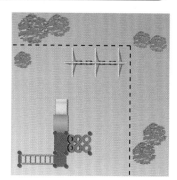

Count the total number of squares. This number is the denominator. The numerator is the number of squares shaded with both colors.

$\frac{3}{4} \times \frac{4}{5} = \frac{\square}{\square} = \frac{\square}{\square}$. So, $\frac{\square}{\square}$ of the park is covered by the playground.

Inductive Reasoning

Work with a partner. Complete the table by using a model or folding paper.

	Exercise	Verbal Expression	Answer
1	3. $\frac{2}{3} \times \frac{1}{2}$	$\frac{2}{3}$ of $\frac{1}{2}$	
2	4. $\frac{3}{4} \times \frac{4}{5}$	$\frac{3}{4}$ of $\frac{4}{5}$	
	5. $\frac{2}{3} \times \frac{5}{6}$		
	6. $\frac{1}{6} \times \frac{1}{4}$		
	7. $\frac{2}{5} \times \frac{1}{2}$		
	8. $\frac{5}{8} \times \frac{4}{5}$		

Math Practice

Consider Similar Problems

What are the similarities in constructing the models for each problem? What are the differences?

What Is Your Answer?

9. **IN YOUR OWN WORDS** What does it mean to multiply fractions?

10. **STRUCTURE** Write a general rule for multiplying fractions.

Practice — Use what you learned about multiplying fractions to complete Exercises 4–11 on page 59.

Section 2.1 Multiplying Fractions 55

2.1 Lesson

🔑 Key Idea

Multiplying Fractions

Words Multiply the numerators and multiply the denominators.

Numbers $\dfrac{3}{7} \times \dfrac{1}{2} = \dfrac{3 \times 1}{7 \times 2} = \dfrac{3}{14}$

Algebra $\dfrac{a}{b} \cdot \dfrac{c}{d} = \dfrac{a \cdot c}{b \cdot d}$, where $b, d \neq 0$

EXAMPLE 1 — Multiplying Fractions

Find $\dfrac{1}{5} \times \dfrac{1}{3}$.

$\dfrac{1}{5} \times \dfrac{1}{3} = \dfrac{1 \times 1}{5 \times 3}$ ← Multiply the numerators.
← Multiply the denominators.

$= \dfrac{1}{15}$ Simplify.

EXAMPLE 2 — Multiplying Fractions with Common Factors

Find $\dfrac{8}{9} \times \dfrac{3}{4}$. Estimate $1 \times \dfrac{3}{4} = \dfrac{3}{4}$

$\dfrac{8}{9} \times \dfrac{3}{4} = \dfrac{8 \times 3}{9 \times 4}$ ← Multiply the numerators.
← Multiply the denominators.

$= \dfrac{\overset{2}{\cancel{8}} \times \overset{1}{\cancel{3}}}{\underset{3}{\cancel{9}} \times \underset{1}{\cancel{4}}}$ Divide out common factors.

$= \dfrac{2}{3}$ Simplify.

∴ The product is $\dfrac{2}{3}$. Reasonable? $\dfrac{2}{3} \approx \dfrac{3}{4}$ ✓

Study Tip

When the numerator of one fraction is the same as the denominator of another fraction, you can use mental math to multiply. For example, $\dfrac{4}{5} \times \dfrac{5}{9} = \dfrac{4}{9}$ because you can divide out the common factor 5.

● On Your Own

Now You're Ready
Exercises 4–19

Multiply. Write the answer in simplest form.

1. $\dfrac{1}{2} \times \dfrac{5}{6}$
2. $\dfrac{7}{8} \times \dfrac{1}{4}$
3. $\dfrac{3}{7} \times \dfrac{2}{3}$
4. $\dfrac{4}{9} \times \dfrac{3}{10}$

EXAMPLE 3 **Real-Life Application**

You have $\frac{2}{3}$ of a bag of flour. You use $\frac{3}{4}$ of the flour to make empanada dough. How much of the entire bag do you use to make the dough?

Method 1: Use a model. Six of the 12 squares have both types of shading.

∴ So, you use $\frac{6}{12} = \frac{1}{2}$ of the entire bag.

Method 2: To find $\frac{3}{4}$ of $\frac{2}{3}$, multiply.

$$\frac{3}{4} \times \frac{2}{3} = \frac{\overset{1}{\cancel{3}} \times \overset{1}{\cancel{2}}}{\underset{2}{\cancel{4}} \times \underset{1}{\cancel{3}}}$$ Multiply the numerators and the denominators. Divide out common factors.

$$= \frac{1}{2}$$ Simplify.

∴ So, you use $\frac{1}{2}$ of the entire bag.

On Your Own

5. **WHAT IF?** In Example 3, you use $\frac{1}{4}$ of the flour to make the dough. How much of the entire bag do you use to make the dough?

Key Idea

Multiplying Mixed Numbers

Write each mixed number as an improper fraction. Then multiply as you would with fractions.

EXAMPLE 4 **Multiplying a Fraction and a Mixed Number**

Find $\frac{1}{2} \times 2\frac{3}{4}$.   **Estimate** $\frac{1}{2} \times 3 = 1\frac{1}{2}$

$$\frac{1}{2} \times 2\frac{3}{4} = \frac{1}{2} \times \frac{11}{4}$$ Write $2\frac{3}{4}$ as the improper fraction $\frac{11}{4}$.

$$= \frac{1 \times 11}{2 \times 4}$$ Multiply the numerators and the denominators.

$$= \frac{11}{8}, \text{ or } 1\frac{3}{8}$$ Simplify.

∴ The product is $1\frac{3}{8}$.   **Reasonable?** $1\frac{3}{8} \approx 1\frac{1}{2}$ ✓

Section 2.1   Multiplying Fractions   **57**

Multiply. Write the answer in simplest form.

26. $1\frac{1}{3} \times \frac{2}{3}$

27. $6\frac{2}{3} \times \frac{3}{10}$

28. $2\frac{1}{2} \times \frac{4}{5}$

29. $\frac{3}{5} \times 3\frac{1}{3}$

30. $7\frac{1}{2} \times \frac{2}{3}$

31. $\frac{5}{9} \times 3\frac{3}{5}$

32. $\frac{3}{4} \times 1\frac{1}{3}$

33. $3\frac{3}{4} \times \frac{2}{5}$

34. $4\frac{3}{8} \times \frac{4}{5}$

35. $\frac{3}{7} \times 2\frac{5}{6}$

36. $1\frac{3}{10} \times 18$

37. $15 \times 2\frac{4}{9}$

38. $1\frac{1}{6} \times 6\frac{3}{4}$

39. $2\frac{5}{12} \times 2\frac{2}{3}$

40. $5\frac{5}{7} \times 3\frac{1}{8}$

41. $2\frac{4}{5} \times 4\frac{1}{16}$

ERROR ANALYSIS Describe and correct the error in finding the product.

42. ✗ $4 \times 3\frac{7}{10} = 12\frac{7}{10}$

43. ✗ $2\frac{1}{2} \times 7\frac{4}{5} = (2 \times 7) + \left(\frac{1}{2} \times \frac{4}{5}\right)$
$= 14 + \frac{2}{5} = 14\frac{2}{5}$

44. VITAMIN C A vitamin C tablet contains $\frac{1}{40}$ of a gram of vitamin C. You take $1\frac{1}{2}$ tablets every day. How many grams of vitamin C do you take every day?

45. SCHOOL BANNER You make a banner for a football rally.

   a. What is the area of the banner?

   b. You add a $\frac{1}{4}$-foot border on each side. What is the new area of the banner?

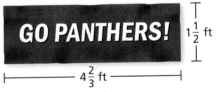

46. NUMBER SENSE Without calculating, is $1\frac{1}{6} \cdot \frac{4}{5}$ less than or greater than $1\frac{1}{6}$? Is the product less than or greater than $\frac{4}{5}$? Explain your reasoning.

Multiply. Write the answer in simplest form.

47. $\frac{1}{2} \times \frac{3}{5} \times \frac{4}{9}$

48. $\frac{4}{7} \times 4\frac{3}{8} \times \frac{5}{6}$

49. $1\frac{1}{15} \times 5\frac{2}{5} \times 4\frac{7}{12}$

50. $\left(\frac{3}{5}\right)^3$

51. $\left(\frac{4}{5}\right)^2 \times \left(\frac{3}{4}\right)^2$

52. $\left(\frac{5}{6}\right)^2 \times \left(1\frac{1}{10}\right)^2$

53. PICTURES Three pictures hang side by side on a wall. What is the total area of the wall that the pictures cover?

54. OPEN-ENDED Find a fraction that, when multiplied by $\frac{1}{2}$, is less than $\frac{1}{4}$.

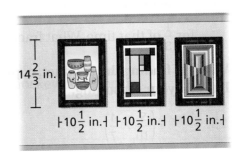

60   Chapter 2   Fractions and Decimals

55. **DISTANCES** You are in a bike race. When you get to the first checkpoint, you are $\frac{2}{5}$ of the distance to the second checkpoint. When you get to the second checkpoint, you are $\frac{1}{4}$ of the distance to the finish. What is the distance from the start to the first checkpoint?

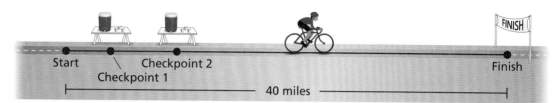

56. **NUMBER SENSE** Is the product of two positive mixed numbers ever less than 1? Explain.

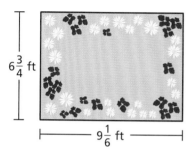
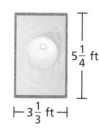

57. **MODELING** You plan to add a fountain to your garden.
    a. Draw a diagram of the fountain in the garden. Label the dimensions.
    b. Describe two methods for finding the area of the garden that surrounds the fountain.
    c. Find the area. Which method did you use, and why?

58. **COOKING** The cooking time for a ham is $\frac{2}{5}$ of an hour for each pound.
    a. How long should you cook a ham that weighs $12\frac{3}{4}$ pounds?
    b. Dinner time is 4:45 P.M. What time should you start cooking the ham?

59. **PETS** You ask 150 people about their pets. The results show that $\frac{9}{25}$ of the people own a dog. Of the people who own a dog, $\frac{1}{6}$ of them also own a cat.
    a. What fraction of the people own a dog and a cat?
    b. **Reasoning** How many people own a dog but not a cat? Explain.

Fair Game Review What you learned in previous grades & lessons

Find the prime factorization of the number. *(Section 1.4)*

60. 24   61. 45   62. 53   63. 60

64. **MULTIPLE CHOICE** A science experiment calls for $\frac{3}{4}$ cup of baking powder. You have $\frac{1}{3}$ cup of baking powder. How much more baking powder do you need? *(Section 1.6)*

    Ⓐ $\frac{1}{4}$ cup   Ⓑ $\frac{5}{12}$ cup   Ⓒ $\frac{4}{7}$ cup   Ⓓ $1\frac{1}{12}$ cups

2.2 Dividing Fractions

Essential Question How can you divide by a fraction?

1 ACTIVITY: Dividing by a Fraction

Work with a partner. Write the division problem and solve it using a model.

a. How many two-thirds are in three?

The division problem is ☐ ÷ ☐/☐.

How many groups of $\frac{2}{3}$ are in 3? ☐

The remaining piece represents ☐/☐ of $\frac{2}{3}$.

So, there are ☐ ☐/☐ groups of $\frac{2}{3}$ in 3.

∴ So, ☐ ÷ ☐/☐ = ☐ ☐/☐.

Dividing Fractions

In this lesson, you will
- write reciprocals of numbers.
- use models to divide fractions.
- divide fractions by fractions.
- solve real-life problems.

b. How many halves are in five halves?

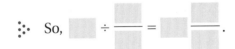

c. How many four-fifths are in eight?

d. How many one-thirds are in seven halves?

e. How many three-fourths are in five halves?

62 Chapter 2 Fractions and Decimals

2 ACTIVITY: Using Tables to Recognize a Pattern

Work with a partner.

a. Complete each table.

Division Table

$8 \div 16$	$\frac{1}{2}$
$8 \div 8$	1
$8 \div 4$	2
$8 \div 2$	4
$8 \div 1$	8
$8 \div \frac{1}{2}$	
$8 \div \frac{1}{4}$	
$8 \div \frac{1}{8}$	

Multiplication Table

$8 \times \frac{1}{16}$	$\frac{1}{2}$
$8 \times \frac{1}{8}$	1
$8 \times \frac{1}{4}$	2
$8 \times \frac{1}{2}$	4
8×1	8
8×2	
8×4	
8×8	

Math Practice

Analyze Relationships
How is multiplying numbers similar to dividing numbers?

b. Describe the relationship between the red numbers in the division table and the red numbers in the multiplication table.

c. Describe the relationship between the blue numbers in the division table and the blue numbers in the multiplication table.

d. **STRUCTURE** Make a conjecture about how you can use multiplication to divide by a fraction.

e. Test your conjecture using the problems in Activity 1.

What Is Your Answer?

3. **IN YOUR OWN WORDS** How can you divide by a fraction? Give an example.

4. How many halves are in a fourth? Explain how you found your answer.

Practice — Use what you learned about dividing fractions to complete Exercises 11–18 on page 67.

Section 2.2 Dividing Fractions

2.2 Lesson

Check It Out
Lesson Tutorials
BigIdeasMath.com

Key Vocabulary
reciprocals, p. 64

Two numbers whose product is 1 are **reciprocals**. To write the reciprocal of a number, write the number as a fraction. Then invert the fraction. So, the reciprocal of a fraction $\frac{a}{b}$ is $\frac{b}{a}$, where a and $b \neq 0$.

The Meaning of a Word • Invert

When you **invert** a glass, you turn it over.

Study Tip
The product of a nonzero number and its reciprocal is 1.
$$\frac{a}{b} \cdot \frac{b}{a} = 1$$
This is called the *Multiplicative Inverse Property*. You will learn more about this property in Chapter 7.

EXAMPLE 1 Writing Reciprocals

	Original Number	Fraction	Reciprocal	Check
a.	$\frac{3}{5}$	$\frac{3}{5}$	$\frac{5}{3}$	$\frac{3}{5} \times \frac{5}{3} = 1$
b.	$\frac{9}{5}$	$\frac{9}{5}$	$\frac{5}{9}$	$\frac{9}{5} \times \frac{5}{9} = 1$
c.	2	$\frac{2}{1}$	$\frac{1}{2}$	$\frac{2}{1} \times \frac{1}{2} = 1$

Study Tip
When any number is multiplied by 0, the product is 0. So, the number 0 does not have a reciprocal.

On Your Own

Now You're Ready
Exercises 7–10

Write the reciprocal of the number.

1. $\frac{3}{4}$ 2. 5   3. $\frac{7}{2}$ 4. $\frac{4}{9}$

🔑 Key Idea

Dividing Fractions

Words To divide a number by a fraction, multiply the number by the reciprocal of the fraction.

Numbers $\dfrac{1}{5} \div \dfrac{3}{4} = \dfrac{1}{5} \times \dfrac{4}{3} = \dfrac{1 \times 4}{5 \times 3}$

Algebra $\dfrac{a}{b} \div \dfrac{c}{d} = \dfrac{a}{b} \cdot \dfrac{d}{c} = \dfrac{a \cdot d}{b \cdot c}$, where b, c, and $d \neq 0$

EXAMPLE 2 Dividing a Fraction by a Fraction

Find $\dfrac{1}{6} \div \dfrac{2}{3}$.

$\dfrac{1}{6} \div \dfrac{2}{3} = \dfrac{1}{6} \times \dfrac{3}{2}$ Multiply by the reciprocal of $\dfrac{2}{3}$, which is $\dfrac{3}{2}$.

$= \dfrac{1 \times \cancel{3}^{1}}{\cancel{6}_{2} \times 2}$ Multiply fractions. Divide out the common factor 3.

$= \dfrac{1}{4}$ Simplify.

EXAMPLE 3 Dividing a Whole Number by a Fraction

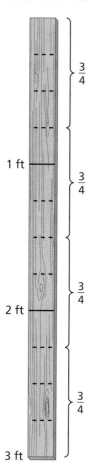

A piece of wood is 3 feet long. How many $\dfrac{3}{4}$-foot pieces can you cut from the piece of wood?

Method 1: Draw a diagram. Mark each foot on the diagram. Then divide each foot into $\dfrac{1}{4}$-foot sections.

Count the number of $\dfrac{3}{4}$-foot pieces of wood. There are four.

∴ So, you can cut four $\dfrac{3}{4}$-foot pieces from the piece of wood.

Method 2: Divide 3 by $\dfrac{3}{4}$ to find the number of $\dfrac{3}{4}$-foot pieces.

$3 \div \dfrac{3}{4} = 3 \times \dfrac{4}{3}$ Multiply by the reciprocal of $\dfrac{3}{4}$, which is $\dfrac{4}{3}$.

$= \dfrac{\cancel{3}^{1} \times 4}{\cancel{3}_{1}}$ Multiply. Divide out the common factor 3.

$= 4$ Simplify.

∴ So, you can cut four $\dfrac{3}{4}$-foot pieces from the piece of wood.

On Your Own

Divide. Write the answer in simplest form.

5. $\dfrac{2}{7} \div \dfrac{1}{3}$ 6. $\dfrac{1}{2} \div \dfrac{1}{8}$ 7. $\dfrac{3}{8} \div \dfrac{1}{4}$ 8. $\dfrac{2}{5} \div \dfrac{3}{10}$

9. How many $\dfrac{1}{2}$-foot pieces can you cut from a 7-foot piece of wood?

EXAMPLE 4 **Dividing a Fraction by a Whole Number**

Find $\dfrac{4}{5} \div 2$.

$\dfrac{4}{5} \div 2 = \dfrac{4}{5} \div \dfrac{2}{1}$ Write 2 as an improper fraction.

$= \dfrac{4}{5} \times \dfrac{1}{2}$ Multiply by the reciprocal of $\dfrac{2}{1}$, which is $\dfrac{1}{2}$.

$= \dfrac{\overset{2}{\cancel{4}} \times 1}{5 \times \underset{1}{\cancel{2}}}$ Multiply fractions. Divide out the common factor 2.

$= \dfrac{2}{5}$ Simplify.

On Your Own

Now You're Ready
Exercises 11–26

Divide. Write the answer in simplest form.

10. $\dfrac{1}{2} \div 3$ 11. $\dfrac{2}{3} \div 10$

12. $\dfrac{5}{8} \div 4$ 13. $\dfrac{6}{7} \div 4$

EXAMPLE 5 **Using Order of Operations**

Evaluate $\dfrac{3}{8} + \dfrac{5}{6} \div 5$.

$\dfrac{3}{8} + \dfrac{5}{6} \div 5 = \dfrac{3}{8} + \dfrac{5}{6} \times \dfrac{1}{5}$ Multiply by the reciprocal of 5, which is $\dfrac{1}{5}$.

$= \dfrac{3}{8} + \dfrac{\overset{1}{\cancel{5}} \times 1}{6 \times \underset{1}{\cancel{5}}}$ Multiply $\dfrac{5}{6}$ and $\dfrac{1}{5}$. Divide out the common factor 5.

$= \dfrac{3}{8} + \dfrac{1}{6}$ Simplify.

$= \dfrac{18}{48} + \dfrac{8}{48}$ Rewrite fractions using a common denominator.

$= \dfrac{26}{48}$, or $\dfrac{13}{24}$ Simplify.

Study Tip
You can use the LCD, 24, to add the fractions in Example 5.
$\dfrac{3}{8} + \dfrac{1}{6} = \dfrac{9}{24} + \dfrac{4}{24} = \dfrac{13}{24}$

On Your Own

Now You're Ready
Exercises 43–51

Evaluate the expression. Write the answer in simplest form.

14. $\dfrac{4}{5} + \dfrac{2}{5} \div 4$ 15. $\dfrac{3}{8} \div \dfrac{3}{4} - \dfrac{1}{6}$ 16. $\dfrac{8}{9} \div 2 \div 8$

Chapter 2   Fractions and Decimals

2.2 Exercises

Vocabulary and Concept Check

1. **OPEN-ENDED** Write a fraction and its reciprocal.

2. **WHICH ONE DOESN'T BELONG?** Which of the following does *not* belong with the other three? Explain your reasoning.

$$\frac{1}{3} \quad \frac{1}{6} \quad \frac{2}{9} \quad \frac{1}{8}$$

MATCHING Match the expression with its value.

3. $\frac{2}{5} \div \frac{8}{15}$
4. $\frac{8}{15} \div \frac{2}{5}$
5. $\frac{2}{15} \div \frac{8}{5}$
6. $\frac{8}{5} \div \frac{2}{15}$

A. $\frac{1}{12}$
B. $\frac{3}{4}$
C. 12
D. $1\frac{1}{3}$

Practice and Problem Solving

Write the reciprocal of the number.

 7. 8
8. $\frac{6}{7}$
9. $\frac{2}{5}$
10. $\frac{8}{11}$

Divide. Write the answer in simplest form.

 11. $\frac{1}{8} \div \frac{1}{4}$
12. $\frac{5}{6} \div \frac{2}{7}$
13. $12 \div \frac{3}{4}$
14. $8 \div \frac{2}{5}$

15. $\frac{3}{7} \div 6$
16. $\frac{12}{25} \div 4$
17. $\frac{2}{9} \div \frac{2}{3}$
18. $\frac{8}{15} \div \frac{4}{5}$

19. $\frac{1}{3} \div \frac{1}{9}$
20. $\frac{7}{10} \div \frac{3}{8}$
21. $\frac{14}{27} \div 7$
22. $\frac{5}{8} \div 15$

23. $\frac{27}{32} \div \frac{7}{8}$
24. $\frac{4}{15} \div \frac{10}{13}$
25. $9 \div \frac{4}{9}$
26. $10 \div \frac{5}{12}$

ERROR ANALYSIS Describe and correct the error in finding the quotient.

27.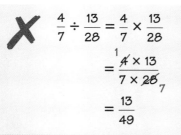

28. $$\frac{2}{5} \div \frac{8}{9} = \frac{5}{2} \times \frac{8}{9}$$
$$= \frac{5 \times \overset{4}{\cancel{8}}}{\underset{1}{\cancel{2}} \times 9}$$
$$= \frac{20}{9}$$

29. **REASONING** How can you use estimation to show that the quotient in Exercise 28 is incorrect?

Section 2.2 Dividing Fractions

30. APPLE PIE You have $\frac{3}{5}$ of an apple pie. You divide the remaining pie into 5 equal slices. What fraction of the original pie is each slice?

31. ANIMALS How many times longer is the baby alligator than the baby gecko?

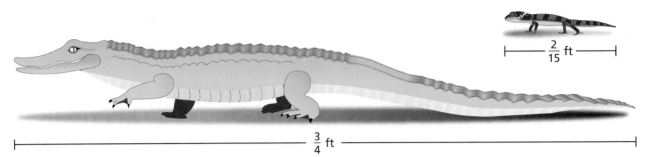

Determine whether the numbers are reciprocals. If not, write the reciprocal of each number.

32. $9, \frac{1}{9}$

33. $\frac{4}{5}, \frac{10}{8}$

34. $\frac{5}{6}, \frac{15}{18}$

35. $\frac{6}{5}, \frac{5}{6}$

Copy and complete the statement.

36. $\frac{5}{12} \times \underline{} = 1$

37. $3 \times \underline{} = 1$

38. $7 \div \underline{} = 56$

Without finding the quotient, copy and complete the statement using <, >, or =. Explain your reasoning.

39. $5 \div \frac{7}{9} \; \underline{} \; 5$

40. $\frac{3}{7} \div 1 \; \underline{} \; \frac{3}{7}$

41. $8 \div \frac{3}{4} \; \underline{} \; 8$

42. $\frac{5}{6} \div \frac{7}{8} \; \underline{} \; \frac{5}{6}$

Evaluate the expression. Write the answer in simplest form.

⑤ 43. $\frac{1}{6} \div 6 \div 6$

44. $\frac{7}{12} \div 14 \div 6$

45. $\frac{3}{5} \div \frac{4}{7} \div \frac{9}{10}$

46. $4 \div \frac{8}{9} - \frac{1}{2}$

47. $\frac{3}{4} + \frac{5}{6} \div \frac{2}{3}$

48. $\frac{7}{8} - \frac{3}{8} \div 9$

49. $\frac{9}{16} \div \frac{3}{4} \cdot \frac{2}{13}$

50. $\frac{3}{14} \cdot \frac{2}{5} \div \frac{6}{7}$

51. $\frac{10}{27} \cdot \left(\frac{3}{8} \div \frac{5}{24} \right)$

52. REASONING Use a model to evaluate the quotient $\frac{1}{2} \div \frac{1}{6}$. Explain.

53. VIDEO CHATTING You use $\frac{1}{8}$ of your battery for every $\frac{2}{5}$ of an hour that you video chat. You use $\frac{3}{4}$ of your battery video chatting. How long did you video chat?

54. NUMBER SENSE When is the reciprocal of a fraction a whole number? Explain.

55. BUDGETS The table shows the portions of a family budget that are spent on several expenses.

Expense	Portion of Budget
Housing	$\frac{1}{4}$
Food	$\frac{1}{12}$
Automobiles	$\frac{1}{15}$
Recreation	$\frac{1}{40}$

  a. How many times more is the expense for housing than for automobiles?
  b. How many times more is the expense for food than for recreation?
  c. The expense for automobile fuel is $\frac{1}{60}$ of the total expenses. What fraction of the automobile expense is spent on fuel?

56. PROBLEM SOLVING You have 6 pints of glaze. It takes $\frac{7}{8}$ of a pint to glaze a bowl and $\frac{9}{16}$ of a pint to glaze a plate.

  a. How many bowls could you glaze? How many plates could you glaze?
  b. You want to glaze 5 bowls, and then use the rest for plates. How many plates can you glaze? How much glaze will be left over?
  c. How many of each object could you glaze so that there is no glaze left over? Explain how you found your answer.

57. Reasoning A water tank is $\frac{1}{8}$ full. The tank is $\frac{3}{4}$ full when 42 gallons of water are added to the tank.

  a. How much water can the tank hold?
  b. How much water was originally in the tank?
  c. How much water is in the tank when it is $\frac{1}{2}$ full?

Fair Game Review *What you learned in previous grades & lessons*

Find the GCF of the numbers. *(Section 1.5)*

58. 8, 16 **59.** 24, 66 **60.** 48, 80 **61.** 15, 45, 100

62. MULTIPLE CHOICE How many inches are in $5\frac{1}{2}$ yards? *(Skills Review Handbook)*

  Ⓐ $15\frac{1}{2}$ Ⓑ $16\frac{1}{2}$ Ⓒ 66   Ⓓ 198

2.3 Lesson

🔑 Key Idea

Dividing Mixed Numbers

Write each mixed number as an improper fraction. Then divide as you would with proper fractions.

EXAMPLE 1 **Dividing a Mixed Number by a Fraction**

Find $2\frac{1}{4} \div \frac{3}{8}$.

$2\frac{1}{4} \div \frac{3}{8} = \frac{9}{4} \div \frac{3}{8}$ Write $2\frac{1}{4}$ as the improper fraction $\frac{9}{4}$.

$= \frac{9}{4} \times \frac{8}{3}$ Multiply by the reciprocal of $\frac{3}{8}$, which is $\frac{8}{3}$.

$= \frac{\overset{3}{\cancel{9}} \times \overset{2}{\cancel{8}}}{\underset{1}{\cancel{4}} \times \underset{1}{\cancel{3}}}$ Multiply fractions. Divide out common factors.

$= 6$ Simplify.

Check

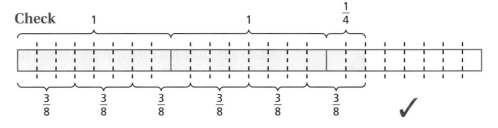

EXAMPLE 2 **Dividing Mixed Numbers**

Find $3\frac{5}{6} \div 1\frac{2}{3}$. **Estimate** $4 \div 2 = 2$

$3\frac{5}{6} \div 1\frac{2}{3} = \frac{23}{6} \div \frac{5}{3}$ Write each mixed number as an improper fraction.

$= \frac{23}{6} \times \frac{3}{5}$ Multiply by the reciprocal of $\frac{5}{3}$, which is $\frac{3}{5}$.

$= \frac{23 \times \overset{1}{\cancel{3}}}{\underset{2}{\cancel{6}} \times 5}$ Multiply fractions. Divide out common factors.

$= \frac{23}{10}$, or $2\frac{3}{10}$ Simplify.

∴ So, the quotient is $2\frac{3}{10}$. **Reasonable?** $2\frac{3}{10} \approx 2$ ✓

On Your Own

Divide. Write the answer in simplest form.

Now You're Ready
Exercises 5–20

1. $1\frac{3}{7} \div \frac{2}{3}$
2. $2\frac{1}{6} \div \frac{3}{4}$
3. $8\frac{1}{4} \div 1\frac{1}{2}$
4. $6\frac{4}{5} \div 2\frac{1}{8}$

72 Chapter 2 Fractions and Decimals

EXAMPLE 3 Using Order of Operations

Evaluate $5\frac{1}{4} \div 1\frac{1}{8} - \frac{2}{3}$.

$$5\frac{1}{4} \div 1\frac{1}{8} - \frac{2}{3} = \frac{21}{4} \div \frac{9}{8} - \frac{2}{3}$$ Write each mixed number as an improper fraction.

$$= \frac{21}{4} \times \frac{8}{9} - \frac{2}{3}$$ Multiply by the reciprocal of $\frac{9}{8}$, which is $\frac{8}{9}$.

$$= \frac{\overset{7}{\cancel{21}} \times \overset{2}{\cancel{8}}}{\underset{1}{\cancel{4}} \times \underset{3}{\cancel{9}}} - \frac{2}{3}$$ Multiply $\frac{21}{4}$ and $\frac{8}{9}$. Divide out common factors.

$$= \frac{14}{3} - \frac{2}{3}$$ Simplify.

$$= \frac{12}{3}, \text{ or } 4$$ Subtract.

Remember

Be sure to check your answers whenever possible. In Example 3, you can use estimation to check that your answer is reasonable.

$5\frac{1}{4} \div 1\frac{1}{8} - \frac{2}{3}$
$\approx 5 \div 1 - 1$
$= 5 - 1$
$= 4$ ✓

EXAMPLE 4 Real-Life Application

One serving of tortilla soup is $1\frac{2}{3}$ cups. A restaurant cook makes 50 cups of soup. Is there enough to serve 35 people? Explain.

Divide 50 by $1\frac{2}{3}$ to find the number of available servings.

$$50 \div 1\frac{2}{3} = \frac{50}{1} \div \frac{5}{3}$$ Rewrite each number as an improper fraction.

$$= \frac{50}{1} \cdot \frac{3}{5}$$ Multiply by the reciprocal of $\frac{5}{3}$, which is $\frac{3}{5}$.

$$= \frac{\overset{10}{\cancel{50}} \cdot 3}{1 \cdot \underset{1}{\cancel{5}}}$$ Multiply fractions. Divide out common factors.

$$= 30$$ Simplify.

∴ No. Because 30 is less than 35, there is not enough soup to serve 35 people.

On Your Own

Now You're Ready
Exercises 26–37

Evaluate the expression. Write the answer in simplest form.

5. $1\frac{1}{2} \div \frac{1}{6} - \frac{7}{8}$

6. $3\frac{1}{3} \div \frac{5}{6} + \frac{8}{9}$

7. $\frac{2}{5} + 2\frac{4}{5} \div 1\frac{3}{4}$

8. $\frac{2}{3} - 1\frac{4}{7} \div 4\frac{5}{7}$

9. In Example 4, can 30 cups of tortilla soup serve 15 people? Explain.

Section 2.3 Dividing Mixed Numbers 73

2.3 Exercises

Vocabulary and Concept Check

1. **VOCABULARY** What is the reciprocal of $7\frac{1}{3}$?

2. **NUMBER SENSE** Is $5\frac{1}{4} \div 3\frac{1}{2}$ the same as $3\frac{1}{2} \div 5\frac{1}{4}$? Explain.

3. **NUMBER SENSE** Is the reciprocal of an improper fraction *sometimes*, *always*, or *never* a proper fraction? Explain.

4. **DIFFERENT WORDS, SAME QUESTION** Which is different? Find "both" answers.

What is $5\frac{1}{2}$ divided by $\frac{1}{8}$?	Find the quotient of $5\frac{1}{2}$ and $\frac{1}{8}$.
What is $5\frac{1}{2}$ times 8?	Find the product of $5\frac{1}{2}$ and $\frac{1}{8}$.

Practice and Problem Solving

Divide. Write the answer in simplest form.

5. $2\frac{1}{4} \div \frac{3}{4}$
6. $3\frac{4}{5} \div \frac{2}{5}$
7. $8\frac{1}{8} \div \frac{5}{6}$
8. $7\frac{5}{9} \div \frac{4}{7}$

9. $7\frac{1}{2} \div 1\frac{9}{10}$
10. $3\frac{3}{4} \div 2\frac{1}{12}$
11. $7\frac{1}{5} \div 8$
12. $8\frac{4}{7} \div 15$

13. $8\frac{1}{3} \div \frac{2}{3}$
14. $9\frac{1}{6} \div \frac{5}{6}$
15. $13 \div 10\frac{5}{6}$
16. $12 \div 5\frac{9}{11}$

17. $\frac{7}{8} \div 3\frac{1}{16}$
18. $\frac{4}{9} \div 1\frac{7}{15}$
19. $4\frac{5}{16} \div 3\frac{3}{8}$
20. $6\frac{2}{9} \div 5\frac{5}{6}$

21. **ERROR ANALYSIS** Describe and correct the error in finding the quotient.

    $$3\frac{1}{2} \div 1\frac{2}{3} = 3\frac{1}{2} \times 1\frac{3}{2} = \frac{7}{2} \times \frac{5}{2} = \frac{35}{4} = 8\frac{3}{4}$$

22. **DOG FOOD** A bag contains 42 cups of dog food. Your dog eats $2\frac{1}{3}$ cups of dog food each day. How many days does the bag of dog food last?

23. **HAMBURGERS** How many $\frac{1}{4}$-pound hamburgers can you make from $3\frac{1}{2}$ pounds of ground beef?

24. **BOOKS** How many $1\frac{3}{5}$-inch-thick books can fit on a $14\frac{1}{2}$-inch-long bookshelf?

25. **LOGIC** Alexei uses the model shown to state that $2\frac{1}{2} \div 1\frac{1}{6} = 2\frac{1}{6}$. Is Alexei correct? Justify your answer using the model.

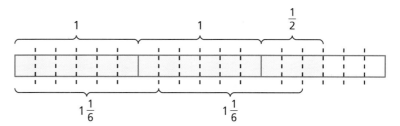

Evaluate the expression. Write the answer in simplest form.

26. $3 \div 1\frac{1}{5} + \frac{1}{2}$

27. $4\frac{2}{3} - 1\frac{1}{3} \div 2$

28. $\frac{2}{5} + 2\frac{1}{6} \div \frac{5}{6}$

29. $5\frac{5}{6} \div 3\frac{3}{4} - \frac{2}{9}$

30. $6\frac{1}{2} - \frac{7}{8} \div 5\frac{11}{16}$

31. $9\frac{1}{6} \div 5 + 3\frac{1}{3}$

32. $3\frac{3}{5} + 4\frac{4}{15} \div \frac{4}{9}$

33. $\frac{3}{5} \times \frac{7}{12} \div 2\frac{7}{10}$

34. $4\frac{3}{8} \div \frac{3}{4} \times \frac{4}{7}$

35. $1\frac{9}{11} \times 4\frac{7}{12} \div \frac{2}{3}$

36. $3\frac{4}{15} \div \left(8 \times 6\frac{3}{10}\right)$

37. $2\frac{5}{14} \div \left(2\frac{5}{8} \times 1\frac{3}{7}\right)$

38. **TRAIL MIX** You have 12 cups of granola and $8\frac{1}{2}$ cups of peanuts to make trail mix. What is the greatest number of full batches of trail mix you can make? Explain how you found your answer.

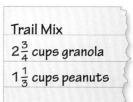

Trail Mix
$2\frac{3}{4}$ cups granola
$1\frac{1}{3}$ cups peanuts

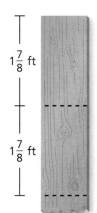

39. **RAMPS** You make skateboard ramps by cutting pieces from a board that is $12\frac{1}{2}$ feet long.

   a. Estimate how many ramps you can cut from the board. Is your estimate reasonable? Explain.

   b. How many ramps can you cut from the board? How much wood is left over?

40. **Reasoning** At a track meet, the longest shot-put throw by a boy is 25 feet 8 inches. The longest shot-put throw by a girl is 19 feet 3 inches. How many times greater is the longest shot-put throw by the boy than by the girl?

Fair Game Review *What you learned in previous grades & lessons*

Write the number as a decimal. *(Skills Review Handbook)*

41. forty-three hundredths

42. thirteen thousandths

43. three and eight tenths

44. seven and nine thousandths

45. **MULTIPLE CHOICE** The winner in a vote for class president received $\frac{3}{4}$ of the 240 votes. How many votes did the winner receive? *(Skills Review Handbook)*

  Ⓐ 60   Ⓑ 150   Ⓒ 180   Ⓓ 320

2 Study Help

Check It Out
Graphic Organizer
BigIdeasMath.com

You can use a **notetaking organizer** to write notes, vocabulary, and questions about a topic. Here is an example of a notetaking organizer for dividing fractions.

Write important vocabulary or formulas in this space.

Dividing fractions

$\dfrac{a}{b} \div \dfrac{c}{d} = \dfrac{a}{b} \cdot \dfrac{d}{c}$

$= \dfrac{a \cdot d}{b \cdot c}$

(where b, c, and $d \ne 0$)

To divide a number by a fraction, multiply the number by the reciprocal of the fraction.

Example:

$\dfrac{1}{5} \div \dfrac{3}{4} = \dfrac{1}{5} \times \dfrac{4}{3} = \dfrac{1 \times 4}{5 \times 3} = \dfrac{4}{15}$

Write your notes about the topic in this space.

Write your questions about the topic in this space.

How do you divide a mixed number by a fraction?

On Your Own

Make notetaking organizers to help you study these topics.

1. multiplying fractions
2. multiplying mixed numbers
3. dividing mixed numbers

After you complete this chapter, make notetaking organizers for the following topics.

4. adding and subtracting decimals
5. multiplying decimals by whole numbers
6. multiplying decimals by decimals
7. dividing decimals by whole numbers
8. dividing decimals by decimals

"The notetaking organizer in my math class gave me an idea of how to organize my doggy biscuits."

2.1–2.3 Quiz

Multiply. Write the answer in simplest form. *(Section 2.1)*

1. $\dfrac{3}{7} \times \dfrac{1}{4}$

2. $\dfrac{9}{10} \times \dfrac{2}{3}$

3. $1\dfrac{1}{6} \times \dfrac{2}{5}$

4. $3\dfrac{1}{2} \times 5\dfrac{7}{10}$

Divide. Write the answer in simplest form. *(Section 2.2 and Section 2.3)*

5. $\dfrac{1}{9} \div \dfrac{1}{3}$

6. $7 \div \dfrac{5}{8}$

7. $4\dfrac{7}{8} \div \dfrac{1}{8}$

8. $7\dfrac{2}{3} \div 1\dfrac{1}{9}$

Evaluate the expression. Write the answer in simplest form. *(Section 2.2 and Section 2.3)*

9. $6 \div \dfrac{2}{3} + \dfrac{1}{2}$

10. $\dfrac{7}{12} \div \dfrac{1}{4} \times \dfrac{9}{14}$

11. $3\dfrac{1}{3} \times 3\dfrac{3}{4} \div \dfrac{5}{6}$

12. $6\dfrac{2}{9} \div \left(4 \times 1\dfrac{1}{6} \right)$

13. **MALL** In a mall, $\dfrac{1}{15}$ of the stores sell shoes. There are 180 stores in the mall. How many of the stores sell shoes? *(Section 2.1)*

14. **CONCERT FLOOR** The floor of a concert venue is $100\dfrac{3}{4}$ feet by $75\dfrac{1}{2}$ feet. What is the area of the floor? *(Section 2.1)*

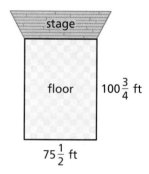

15. **BAND** Band members make $\dfrac{2}{3}$ of their profit from selling concert tickets. They make $\dfrac{1}{5}$ of their profit from selling band merchandise at the concerts. How many times more profit do they make from ticket sales than from merchandise sales? *(Section 2.2)*

16. **SKATEBOARDS** You are cutting as many $32\dfrac{1}{4}$-inch sections as you can out of the board to make skateboards. How many skateboards can you make? *(Section 2.3)*

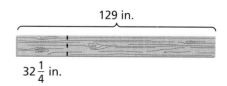

2.4 Adding and Subtracting Decimals

Essential Question How can you add and subtract decimals?

Base ten blocks can be used to model numbers.

1 one   1 tenth   1 hundredth

1 ACTIVITY: Modeling a Sum

Work with a partner. Use base ten blocks to find the sum.

a. 1.23 + 0.87

Which base ten blocks do you need to model the numbers in the sum? How many of each do you need?

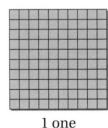 +

1.23 + 0.87

How many of each base ten block do you have when you combine the blocks?

☐ ones   ☐ tenths   ☐ hundredths

How many of each base ten block do you have when you trade the blocks?

☐ ones   ☐ tenths   ☐ hundredths

So, 1.23 + 0.87 = ☐.

b. 1.25 + 1.35   c. 2.14 + 0.92   d. 0.73 + 0.86

2 ACTIVITY: Modeling a Difference

Adding and Subtracting Decimals

In this lesson, you will
- use models to add and subtract decimals.
- add and subtract decimals.

Work with a partner. Use base ten blocks to find the difference.

a. 2.43 − 0.73

Which number is shown by the model? ☐

Circle the portion of the model that represents 0.73.

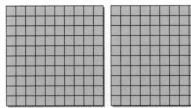 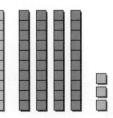

So, 2.43 − 0.73 = ☐.

b. 1.86 − 1.26   c. 3.72 − 0.5   d. 1.58 − 0.09

78 Chapter 2 Fractions and Decimals

3 ACTIVITY: Making a Conjecture

Work with a partner.

a. Find each sum or difference.

| 123 + 87 | 125 + 135 | 214 + 92 | 73 + 86 |
| 243 − 73 | 186 − 126 | 372 − 50 | 158 − 9 |

b. How are the numerical expressions in part (a) related to the numerical expressions in Activities 1 and 2? How are the sums and differences related?

c. **STRUCTURE** There is a relationship between adding and subtracting decimals and adding and subtracting whole numbers. What conjecture can you make about this relationship?

4 ACTIVITY: Using a Place Value Chart

Work with a partner. Use the place value chart to find the sum or difference.

Place Value Chart

Columns: millions | hundred thousands | ten thousands | thousands | hundreds | tens | ones | and | tenths | hundredths | thousandths | ten-thousandths | hundred-thousandths | millionths

Math Practice

Analyze Conjectures
How can the conjecture you wrote in Activity 3 help you to solve these problems?

a. 16.05 + 2.94
b. 7.421 + 92.55
c. 38.72 − 8.61
d. 64.968 − 51.167

What Is Your Answer?

5. **MODELING** Describe two real-life examples of when you would need to add and subtract decimals.

6. **IN YOUR OWN WORDS** How can you add and subtract decimals?

Practice

Use what you learned about adding and subtracting decimals to complete Exercises 3–4 on page 82.

2.4 Lesson

🔑 Key Idea

Adding and Subtracting Decimals
To add or subtract decimals, write the numbers vertically and line up the decimal points. Then bring down the decimal point and add or subtract as you would with whole numbers.

EXAMPLE 1 Adding Decimals

a. Add 8.13 + 2.76.   Estimate 8.13 + 2.76 ≈ 8 + 3 = 11

Line up the decimal points.

```
   8.13
 + 2.76
  -----
  10.89
```

Add as you would with whole numbers.

Reasonable? 10.89 ≈ 11 ✓

Study Tip
Be sure to add or subtract only digits that have the same place value.

b. Add 1.459 + 23.7.

```
     1
   1.459
 +23.700
  ------
  25.159
```

Insert zeros so that both numbers have the same number of decimal places.

EXAMPLE 2 Subtracting Decimals

a. Subtract 5.508 − 3.174.   Estimate 5.508 − 3.174 ≈ 6 − 3 = 3

Line up the decimal points.

```
       4 10
    5.5̸0̸ 8
  − 3.1 7 4
    ------
    2.3 3 4
```

Subtract as you would with whole numbers.

Reasonable? 2.334 ≈ 3 ✓

b. Subtract 21.9 − 1.605.

```
         9
     8 10 10
   21.9̸0̸0̸
  − 1.6 0 5
    -------
   20.2 9 5
```

Insert zeros so that both numbers have the same number of decimal places.

● On Your Own

Now You're Ready
Exercises 5–16

Add or subtract.

1. 4.206 + 10.85
2. 15.5 + 8.229
3. 78.41 + 90.99
4. 6.34 − 5.33
5. 27.9 − 0.905
6. 18.626 − 13.88

EXAMPLE 3 Real-Life Application

Your meal at the school cafeteria costs $3.45. Your friend's meal costs $3.90. You pay for both meals with a $10 bill. How much change do you receive?

Use a verbal model to solve the problem.

$$\boxed{\text{amount of change}} = \boxed{\text{amount given}} - \left(\boxed{\text{cost of your meal}} + \boxed{\text{cost of friend's meal}}\right)$$

$= \boxed{10.00} - (\boxed{3.45} + \boxed{3.90})$ Substitute values.

$= 10.00 - 7.35$ Add inside parentheses.

$= 2.65$ Subtract.

∴ So, you receive $2.65.

EXAMPLE 4 Real-Life Application

The Lincoln Memorial Reflecting Pool is approximately rectangular. Its width is 50.9 meters, and its length is 618.44 meters. You walk the perimeter of the pool. About how many meters do you walk?

Draw a diagram and label the dimensions.

Find the sum of the side lengths.

```
  1 1 2
  618.44
   50.90
  618.44
+  50.90
--------
 1338.68
```

∴ So, you walk about 1339 meters.

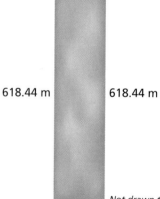

On Your Own

7. **WHAT IF?** In Example 3, your meal costs $4.10 and your friend's meal costs $3.65. You pay for both meals with a $20 bill. How much change do you receive?

8. Find the perimeter of the triangle.

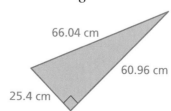

Section 2.4 Adding and Subtracting Decimals

2.4 Exercises

Vocabulary and Concept Check

1. **CHOOSE TOOLS** Why is it helpful to estimate the answer before adding or subtracting decimals?

2. **WRITING** When adding or subtracting decimals, how can you be sure to add or subtract only digits that have the same place value?

Practice and Problem Solving

Write and evaluate the numerical expression modeled by the base ten blocks.

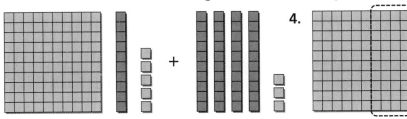

Add.

5. $7.82 + 3.209$
6. $3.7 + 2.774$
7. $12.829 + 10.07$
8. $20.35 + 13.748$
9. $17.440 + 12.497$
10. $15.255 + 19.058$

Subtract.

11. $4.58 - 3.12$
12. $8.629 - 5.309$
13. $6.98 - 2.614$
14. $15.131 - 11.57$
15. $13.5 - 10.856$
16. $25.82 - 22.936$

ERROR ANALYSIS Describe and correct the error in the solution.

17.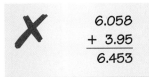

18.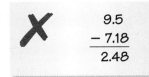

Breakfast Menu
7:30 A.M. to 11:00 A.M.

2 Eggs (any style)	$2.95	Bacon & Eggs	$3.95
Steak & Eggs	$6.25	Cheese Omelet	$3.55
Ham & Eggs	$3.95	Ham Omelet	$4.35
Sausage & Eggs	$3.95	Ham & Cheese Omelet	$4.95
Salami & Eggs	$3.95		

19. **BREAKFAST** You order the sausage and eggs breakfast, and your friend orders the ham omelet. How much is the bill before taxes and tip?

20. **HAM & CHEESE** How much more does the ham and cheese omelet cost than the cheese omelet?

Evaluate the expression.

21. $6.105 + 10.4 + 3.075$

22. $22.6 - 12.286 - 3.542$

23. $15.35 + 7.604 - 12.954$

24. $16.5 - 13.45 + 7.293$

25. $25.92 - 18.478 + 8.164$

26. $23.45 + 17.75 - 19.618$

27. STRUCTURE When is the sum of two decimals equal to a whole number? When is the difference of two decimals equal to a whole number?

28. OPEN-ENDED Write three decimals that have a sum of 27.905.

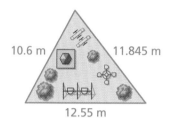

29. DAY CARE A day-care center is building a new outdoor play area. The diagram shows the dimensions in meters. How much fencing is needed to enclose the play area?

10.6 m 11.845 m 12.55 m

30. HOMEWORK You work 1.15 hours on English homework and 1.75 hours on math homework. Your science homework takes 1.05 hours less than your math homework. How many hours do you work on homework?

ASTRONOMY An astronomical unit (AU) is the average distance of Earth from the Sun. In Exercises 31–34, use the table that shows the average distance of each planet in our solar system from the Sun.

31. How much farther is Jupiter from the Sun than Mercury?

32. How much farther is Neptune from the Sun than Mars?

33. Estimate the greatest distance between Earth and Uranus.

34. Estimate the greatest distance between Venus and Saturn.

35. The length of a rectangle is twice the width. The perimeter of the rectangle can be expressed as $3 \cdot 13.7$. What is the width?

Planet	Average Distance from the Sun (AU)
Mercury	0.387
Venus	0.723
Earth	1.000
Mars	1.524
Jupiter	5.203
Saturn	9.537
Uranus	19.189
Neptune	30.07

Fair Game Review *What you learned in previous grades & lessons*

Multiply. Write the answer in simplest form. *(Section 2.1)*

36. $\dfrac{7}{10} \times \dfrac{5}{7}$

37. $\dfrac{5}{6} \times \dfrac{3}{10}$

38. $\dfrac{3}{4} \times \dfrac{2}{9}$

39. $\dfrac{2}{5} \times \dfrac{1}{8}$

40. MULTIPLE CHOICE What is the LCM of 6, 12, and 18? *(Section 1.6)*

　Ⓐ 6　　　Ⓑ 18　　　Ⓒ 36　　　Ⓓ 72

Section 2.4　Adding and Subtracting Decimals

2.5 Multiplying Decimals

Essential Question How can you multiply decimals?

1 ACTIVITY: Multiplying Decimals Using a Rectangle

Work with a partner. Use a rectangle to find the product.

a. 2.7 • 1.3

Arrange base ten blocks to form a rectangle of length 2.7 units and width 1.3 units.

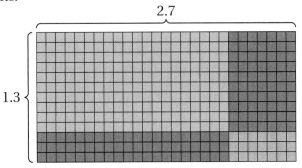

The area of the rectangle represents the product.
Find the total area represented by each grouping of base ten blocks.

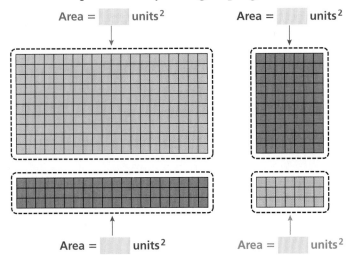

The area of the rectangle is:

$$\begin{array}{r} \square \\ \square \\ \square \\ +\square \\ \hline \square \text{ units}^2 \end{array}$$

So, 2.7 • 1.3 = 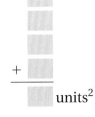.

b. 1.8 • 1.1   c. 4.6 • 1.2   d. 3.2 • 2.4

Multiplying Decimals

In this lesson, you will
- use models to multiply decimals.
- multiply decimals.

84   Chapter 2   Fractions and Decimals

② ACTIVITY: Multiplying Decimals Using an Area Model

Work with a partner. Use an area model to find the product. Explain your reasoning.

a. 0.8 • 0.5

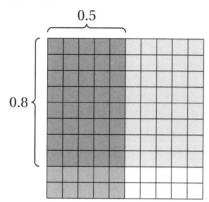

Math Practice

View as Components

How can you use an area model to find the product?

Because ▭ hundredths are shaded with both colors, the product is

$\dfrac{\boxed{}}{100} = \boxed{}$.

So, 0.8 • 0.5 = ▭.

b. 0.3 • 0.5   c. 0.7 • 0.6   d. 0.2 • 0.9

③ ACTIVITY: Making a Conjecture

Work with a partner.

a. Find each product.

| 27 • 13 | 18 • 11 | 46 • 12 | 32 • 24 |
| 8 • 5 | 3 • 5 | 7 • 6 | 2 • 9 |

b. How are the numerical expressions in part (a) related to the numerical expressions in Activities 1 and 2? How are the products related?

c. **STRUCTURE** What conjecture can you make about the relationship between multiplying decimals and multiplying whole numbers?

What Is Your Answer?

4. **IN YOUR OWN WORDS** How can you multiply decimals?

Practice — Use what you learned about multiplying decimals to complete Exercises 9–12 on page 89.

2.5 Lesson

Check It Out
Lesson Tutorials
BigIdeasMath.com

 Key Idea

Multiplying Decimals by Whole Numbers

Words Multiply as you would with whole numbers. Then count the number of decimal places in the decimal factor. The product has the same number of decimal places.

Numbers
$$\begin{array}{r} 13.91 \\ \times\ \ \ 7 \\ \hline 97.37 \end{array}$$ 2 decimal places

$$\begin{array}{r} 6.218 \\ \times\ \ \ 4 \\ \hline 24.872 \end{array}$$ 3 decimal places

EXAMPLE 1 — Multiplying Decimals and Whole Numbers

a. Find 6×3.91.

Estimate $6 \times 4 = 24$

$$\begin{array}{r} \overset{5}{3.91} \\ \times\ \ \ 6 \\ \hline 23.46 \end{array}$$

← 2 decimal places

← Count 2 decimal places from right to left.

∴ So, $6 \times 3.91 = 23.46$.

Reasonable? $23.46 \approx 24$ ✓

b. Find 3×0.016.

Estimate $3 \times 0 = 0$

$$\begin{array}{r} \overset{1}{0.016} \\ \times\ \ \ 3 \\ \hline 0.048 \end{array}$$

← 3 decimal places

← To have 3 decimal places, insert zeros to the left of 48.

∴ So, $3 \times 0.016 = 0.048$.

Reasonable? $0.048 \approx 0$ ✓

EXAMPLE 2 — Use Mental Math

How high is a stack of 100 dimes?

1.35 millimeters

Method 1: Multiply 1.35 by 100.

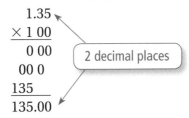

2 decimal places

Method 2: You are multiplying by a power of 10. Use mental math.

There are two zeros in 100. So, move the decimal point in 1.35 two places to the right.

$1.35 \times 100 = 135.\ = 135$

∴ So, a stack of 100 dimes is 135 millimeters high.

On Your Own

Now You're Ready
Exercises 13–24

Multiply. Use estimation to check your answer.

1. 12.3×8
2. 5×14.51
3. 0.88×9
4. 0.003×10

5. A quarter is 1.75 millimeters thick. How high is a stack of 1000 quarters? Solve using both methods.

86 Chapter 2 Fractions and Decimals

The rule for multiplying two decimals is similar to the rule for multiplying a decimal by a whole number.

 **Key Idea**

Multiplying Decimals by Decimals

Words Multiply as you would with whole numbers. Then add the number of decimal places in the factors. The sum is the number of decimal places in the product.

Numbers
```
    4.716  ←   3 decimal places
  ×   0.2  ← + 1 decimal place
    0.9432 ←   4 decimal places
```

EXAMPLE 3 Multiplying Decimals

a. **Multiply 4.8 × 7.2.**   **Estimate** $5 \times 7 = 35$

```
    4.8   ←—— 1 decimal place
  × 7.2   ←—— + 1 decimal place
     96
    336
   34.56  ←—— 2 decimal places
```

∴ So, 4.8 × 7.2 = 34.56.   **Reasonable?** $34.56 \approx 35$ ✓

b. **Multiply 3.1 × 0.05.**   **Estimate** $3 \times 0 = 0$

```
    3.1    ←—— 1 decimal place
  × 0.05   ←—— + 2 decimal places
    0.155  ←—— 3 decimal places
```

∴ So, 3.1 × 0.05 = 0.155.   **Reasonable?** $0.155 \approx 0$ ✓

On Your Own

Exercises 30–45

Multiply. Use estimation to check your answer.

6. 8.1 × 5.6
7. 2.7 × 9.04
8. 6.32 × 0.09
9. 1.785 × 0.2

EXAMPLE 4 Evaluating an Expression

What is the value of 2.44(4.5 − 3.175)?

Ⓐ 3.233 **Ⓑ** 3.599 **Ⓒ** 7.805 **Ⓓ** 32.33

Step 1: Subtract first because the minus sign is in parentheses.

```
    9
  4 10̸10
  4.5̸0̸0̸
− 3.1 7 5
  1.3 2 5
```

So, 2.44(4.5 − 3.175) = 2.44(1.325).

Step 2: Multiply the result from Step 1 by 2.44.

```
    1.3 2 5
  ×   2.4 4
    5 3 0 0
    5 3 0 0
  2 6 5 0
  3.2 3 3 0 0
```

∴ The correct answer is **Ⓐ**.

● **On Your Own**

Now You're Ready
Exercises 52–60

Evaluate the expression.

10. 12.67 + 8.2 • 1.9   **11.** 6.4(1.8 • 7.5)

EXAMPLE 5 Real-Life Application

You buy 2.75 pounds of tomatoes. You hand the cashier a $10 bill. How much change will you receive?

Step 1: Find the cost of the tomatoes. Multiply 1.89 by 2.75.

```
    1.8 9   ← 2 decimal places
  × 2.7 5   ← + 2 decimal places
    9 4 5
  1 3 2 3
  3 7 8
  5.1 9 7 5  ← 4 decimal places
```

The cost of 2.75 pounds of tomatoes is $5.20.

Step 2: Subtract the cost of the tomatoes from the amount of money you hand the cashier.

10.00 − 5.20 = $4.80

∴ So, you will receive $4.80 in change.

● **On Your Own**

12. WHAT IF? You buy 2.25 pounds of grapes. You hand the cashier a $5 bill. How much change will you receive?

2.5 Exercises

Vocabulary and Concept Check

1. **NUMBER SENSE** If you know 12 × 24 = 288, how can you find 1.2 × 2.4?

2. **NUMBER SENSE** Is the product 1.23 × 8 greater than or less than 8? Explain.

Copy the problem and place the decimal point in the product.

3. 1.7 8
    × 4.9
    ‾‾‾‾‾‾
    8 7 2 2

4. 9.2 4
    × 0.6 8
    ‾‾‾‾‾‾‾
    6 2 8 3 2

5. 3.7 5
    × 5.2 2
    ‾‾‾‾‾‾‾
    1 9 5 7 5 0

How many decimal places are in the product?

6. 6.17 × 8.2
7. 1.684 × 10.2
8. 0.053 × 2.78

Practice and Problem Solving

Use base ten blocks or an area model to find the product.

9. 2.1
    × 1.5

10. 0.6
    × 0.4

11. 0.7
    × 0.3

12. 2.7
    × 2.3

Multiply. Use estimation to check your answer.

13. 4.8
    × 7

14. 6.3
    × 5

15. 7.19
    × 16

16. 0.87
    × 21

17. 1.95
    × 11

18. 5.89
    × 5

19. 3.472
    × 4

20. 8.188
    × 12

21. 100 × 0.024
22. 19 × 0.004
23. 0.0038 × 9
24. 10 × 0.0093

ERROR ANALYSIS Describe and correct the error in the solution.

25.

26.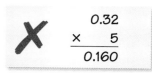

27. **MOON** The weight of an object on the Moon is about 0.167 of its weight on Earth. How much does a 180-pound astronaut weigh on the Moon?

28. **BAMBOO** A bamboo plant grows about 1.25 feet each day. Find the growth in one week.

29. **NAILS** A fingernail grows about 0.1 millimeter each day. How much does a fingernail grow in 30 days? 90 days?

Section 2.5 Multiplying Decimals 89

Multiply.

30. 0.7 × 0.2

31. 0.08 × 0.3

32. 0.007 × 0.03

33. 0.0008 × 0.09

34. 0.004 × 0.9

35. 0.06 × 0.5

36. 0.0008 × 0.004

37. 0.0002 × 0.06

38. 12.4 × 0.2

39. 18.6 × 5.9

40. 7.91 × 0.72

41. 1.16 × 3.35

42. 6.478 × 18.21

43. 1.9 × 7.216

44. 0.0021 × 18.2

45. 6.109 × 8.4

46. ERROR ANALYSIS Describe and correct the error in the solution.

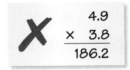

47. TAKEOUT A Chinese restaurant offers buffet takeout for $4.99 per pound. How much does your takeout meal cost?

48. CROPLAND Alabama has about 2.51 million acres of cropland. Florida has about 1.15 times as much cropland as Alabama. How much cropland does Florida have?

49. GOLD On a tour of an old gold mine, you find a nugget containing 0.82 ounce of gold. Gold is worth $1566.80 per ounce. How much is your nugget worth?

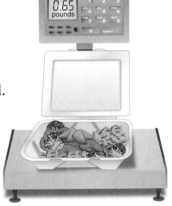

50. BUILDING HEIGHTS One meter is approximately 3.28 feet. Find the height of each building in feet by multiplying its height in meters by 3.28.

Continent	Tallest Building	Height (meters)
Africa	Carlton Centre Office Tower	223
Asia	Burj Khalifa	828
Australia	Q1 Tower	323
Europe	The Shard	310
North America	Willis Tower	442
South America	Gran Torre	300

51. REASONING Show how to evaluate 7.12 × 8.22 × 100 without multiplying the two decimals.

ORDER OF OPERATIONS Evaluate the expression.

52. 2.4 × 16 + 7

53. 6.85 × 2 × 10

54. 1.047 × 5 − 0.88

55. 4.32(3.7 + 1.65)

56. 23.98 − 1.7^2 · 7.6

57. 12 · 5.16 + 10.064

58. 0.9(8.2 · 20.35)

59. 7.5^2(6.084 − 5.44)

60. 6.8 · 2.18 · 3.95

61. REASONING Without multiplying, how many decimal places does 3.4^2 have? 3.4^3? 3.4^4? Explain your reasoning.

REPEATED REASONING Describe the pattern. Find the next three numbers.

62. 1, 0.6, 0.36, 0.216, . . .

63. 15, 1.5, 0.15, 0.015, . . .

64. 0.04, 0.02, 0.01, 0.005, . . .

65. 5, 7.5, 11.25, 16.875, . . .

66. FOOD You buy 2.6 pounds of apples and 1.475 pounds of peaches. You hand the cashier a $20 bill. How much change will you receive?

Apples $1.23/pound   Peaches $1.88/pound

67. MILEAGE A car can travel 22.36 miles on one gallon of gasoline.

   a. How far can the car travel on 8.5 gallons of gasoline?

   b. A hybrid car can travel 33.1 miles on one gallon of gasoline. How much farther can the hybrid car travel on 8.5 gallons of gasoline?

68. OPEN-ENDED You and four friends have dinner at a restaurant.

   a. Draw a restaurant menu that has main items, desserts, and beverages, with their prices.

   b. Write a guest check that shows what each of you ate. Find the subtotal.

   c. Multiply by 0.07 to find the tax. Then find the total.

   d. Round the total to the nearest whole number. Multiply by 0.20 to estimate a tip. Including the tip, how much did you spend?

GUEST CHECK

69. **Geometry** A rectangular painting has an area of 9.52 square feet.

   a. Draw three different ways in which this can happen.

   b. The cost of a frame depends on the perimeter of the painting. Which of your drawings from part (a) is the least expensive to frame? Explain your reasoning.

   c. The thin, black framing costs $1 per foot. The fancy framing costs $5 per foot. Will the fancy framing cost five times as much as the black framing? Explain why or why not.

   d. Suppose the cost of a frame depends on the outside perimeter of the frame. Does this change your answer to part (c)? Explain why or why not.

Fair Game Review What you learned in previous grades & lessons

Divide. *(Skills Review Handbook)*

70. 78 ÷ 3   **71.** 65 ÷ 13   **72.** 57 ÷ 19   **73.** 84 ÷ 12

74. MULTIPLE CHOICE How many edges does the rectangular prism at the right have? *(Skills Review Handbook)*

   Ⓐ 4   Ⓑ 6

   Ⓒ 8   Ⓓ 12

2.6 Dividing Decimals

Essential Question How can you use base ten blocks to model decimal division?

1 ACTIVITY: Dividing Decimals

Work with a partner. Use base ten blocks to model the division. Then find the quotient.

a. $2.4 \div 0.6$

Begin by modeling 2.4.

2.4

How many of each base ten block did you use?

☐ ones

☐ tenths

☐ hundredths

Next, think of the division problem $2.4 \div 0.6$ as the question,

"How can you divide 2.4 into groups of 0.6?"

Rearrange the model for 2.4 into groups of 0.6. There are ☐ groups of 0.6.

So, $2.4 \div 0.6 =$ ☐.

b. $1.8 \div 2$ c. $3.9 \div 3$ d. $2.8 \div 0.7$ e. $3.2 \div 0.4$

f. Write and solve the division problem represented by the model.

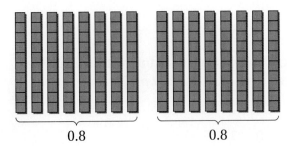

0.8   0.8

Dividing Decimals

In this lesson, you will
- use models to divide decimals.
- divide decimals.

2 ACTIVITY: Dividing Decimals

Work with a partner. Use base ten blocks to model the division. Then find the quotient.

Math Practice

Evaluate Results
What can you do to check the reasonableness of your answer?

a. $0.3 \div 0.06$

Model 0.3.   Replace tenths with hundredths.   How many 0.06s are in 0.3? Divide hundredths into groups of 0.06.

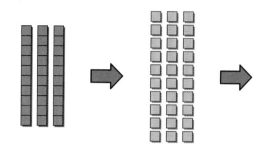

There are ____ groups of 0.06. So, $0.3 \div 0.06 =$ ____.

b. $0.2 \div 0.04$
c. $0.6 \div 0.01$
d. $0.16 \div 0.08$
e. $0.28 \div 0.07$

What Is Your Answer?

3. **IN YOUR OWN WORDS** How can you use base ten blocks to model decimal division? Use examples from Activity 1 and Activity 2 as part of your answer.

4. **WRITING** Newton's poem is about dividing fractions. Write a poem about dividing decimals.

"When you must divide a fraction, do this very simple action: Flip what you're dividing BY, and then it's easy—multiply!"

5. Think of your own cartoon about dividing decimals. Draw your cartoon.

Practice — Use what you learned about dividing decimals to complete Exercises 8–11 on page 97.

Section 2.6   Dividing Decimals   93

2.6 Lesson

Key Idea

Dividing Decimals by Whole Numbers

Words Place the decimal point in the quotient above the decimal point in the dividend. Then divide as you would with whole numbers. Continue until there is no remainder.

Numbers

$$4\overline{)7.32}^{\,1.83}$$

Place the decimal point in the quotient above the decimal point in the dividend.

EXAMPLE 1 Dividing Decimals by Whole Numbers

a. Find $7.6 \div 4$.   **Estimate** $8 \div 4 = 2$

$$\begin{array}{r} 1.9 \\ 4\overline{)7.6} \\ -\,4 \\ \hline 3\,6 \\ -\,3\,6 \\ \hline 0 \end{array}$$

Place the decimal point in the quotient above the decimal point in the dividend.

∴ So, $7.6 \div 4 = 1.9$.   **Reasonable?** $1.9 \approx 2$ ✓

b. Find $4.38 \div 12$.

$$\begin{array}{r} 0.365 \\ 12\overline{)4.380} \\ -\,3\,6 \\ \hline 78 \\ -\,72 \\ \hline 60 \\ -\,60 \\ \hline 0 \end{array}$$

Place the decimal point in the quotient above the decimal point in the dividend.

Insert a zero and continue to divide.

∴ So, $4.38 \div 12 = 0.365$.   **Check** $0.365 \times 12 = 4.38$ ✓

On Your Own

Now You're Ready
Exercises 12–23

Divide. Use estimation to check your answer.

1. $36.4 \div 2$ 2. $22.2 \div 6$ 3. $59.64 \div 7$
4. $43.26 \div 14$ 5. $6.2 \div 4$ 6. $3.12 \div 16$

94   Chapter 2   Fractions and Decimals

 Key Idea

Dividing Decimals by Decimals

Words Multiply the divisor *and* the dividend by a power of 10 to make the divisor a whole number. Then place the decimal point in the quotient and divide as you would with whole numbers. Continue until there is no remainder.

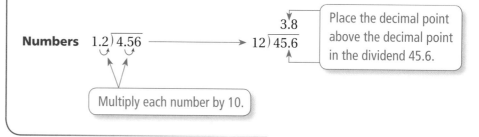

EXAMPLE 2 Dividing Decimals

a. **Find 18.2 ÷ 1.4.**

Study Tip

Multiplying the divisor and the dividend by a power of 10 does not change the quotient.

For example:
18.2 ÷ 1.4 = 13
182 ÷ 14 = 13
1820 ÷ 140 = 13

So, 18.2 ÷ 1.4 = 13.   **Check** 13 × 1.4 = 18.2 ✓

b. **Find 0.273 ÷ 0.39.**

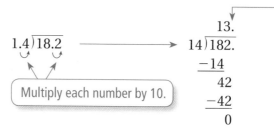

So, 0.273 ÷ 0.39 = 0.7.   **Check** 0.7 × 0.39 = 0.273 ✓

On Your Own

Exercises 36–39

Divide. Check your answer.

7. $1.2\overline{)9.6}$

8. $3.4\overline{)57.8}$

9. 21.643 ÷ 2.3

10. 0.459 ÷ 0.51

EXAMPLE 3 Inserting Zeros in the Dividend and the Quotient

Divide 2.45 ÷ 0.007.

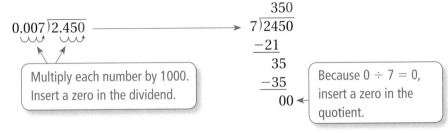

Study Tip
Remember to check your answer by multiplying the quotient by the divisor.

Multiply each number by 1000. Insert a zero in the dividend.

Because 0 ÷ 7 = 0, insert a zero in the quotient.

So, 2.45 ÷ 0.007 = 350.

On Your Own

Now You're Ready
Exercises 40–43

Divide. Check your answer.

11. 3.8 ÷ 0.16
12. 15.6 ÷ 0.78
13. 7.2 ÷ 0.048
14. 0.18 ÷ 0.003

EXAMPLE 4 Real-Life Application

How many times more cellular phone subscribers were there in 2011 than in 1991? Round to the nearest whole number.

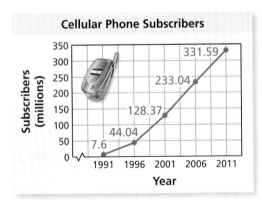

Cellular Phone Subscribers

From the graph, there were 331.59 million subscribers in 2011 and 7.6 million in 1991. So, divide 331.59 by 7.6.

Estimate 320 ÷ 8 = 40

```
                43.6   ← Rounds to 44.
7.6)331.59  →  76)3315.9
                −304
                 275
                −228
                  47 9
                 −45 6
                   2 3
```

So, there were about 44 times more subscribers in 2011 than in 1991.

Reasonable? 44 ≈ 40 ✓

On Your Own

15. How many times more subscribers were there in 2006 than in 1996? Round to the nearest whole number.

2.6 Exercises

Vocabulary and Concept Check

1. **NUMBER SENSE** Fix the one that is not correct.

    $4\overline{)24.4} = 6.1$ $4\overline{)244} = 61$ $4\overline{)2.44} = 6.1$

    Copy the problem and place the decimal point in the correct location.

    2. $18.6 \div 4 = 465$ 3. $6.38 \div 11 = 58$ 4. $88.27 \div 7 = 1261$

    Rewrite the problem so that the divisor is a whole number.

    5. $4.7\overline{)13.6}$ 6. $0.21\overline{)17.66}$ 7. $2.16\overline{)18.5}$

Practice and Problem Solving

Use base ten blocks to find the quotient.

8. $3.6 \div 0.3$ 9. $2.6 \div 0.2$ 10. $0.72 \div 0.06$ 11. $0.36 \div 0.04$

Divide. Use estimation to check your answer.

12. $6\overline{)25.2}$ 13. $5\overline{)33.5}$ 14. $7\overline{)3.5}$ 15. $8\overline{)10.4}$

16. $38.7 \div 9$ 17. $37.6 \div 4$ 18. $43.4 \div 7$ 19. $25.6 \div 8$

20. $44.64 \div 8$ 21. $0.294 \div 3$ 22. $3.6 \div 24$ 23. $64.26 \div 18$

ERROR ANALYSIS Describe and correct the error in finding the quotient.

24.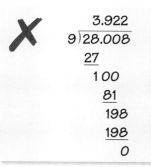

25.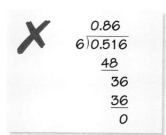

26. **TEXT MESSAGING** You send 40 text messages in one month. The total cost is $4.80. How much does each text message cost?

27. **SUNBLOCK** Of the two bottles of sunblock shown, which is the better buy? Explain.

4-ounce bottle $8.49

5-ounce bottle $10.29

Section 2.6 Dividing Decimals 97

ORDER OF OPERATIONS Evaluate the expression.

28. 7.68 + 3.18 ÷ 12

29. 10.56 ÷ 3 − 1.9

30. 19.6 ÷ 7 × 9

31. 5.5 × 16.56 ÷ 9

32. 35.25 ÷ 5 ÷ 3

33. 13.41 × (5.4 ÷ 9)

34. FRUIT PUNCH Which pack of fruit punch is the best buy? Explain.

35. SALE You buy 3 pairs of jeans for $35.95 each and get a fourth pair for free. What is your cost per pair of jeans?

Divide. Check your answer.

② **36.** 2.1)25.2

37. 3.8)34.2

38. 36.47 ÷ 0.7

39. 0.984 ÷ 12.3

③ **40.** 4.23 ÷ 0.012

41. 0.52 ÷ 0.0013

42. 95.04 ÷ 0.0132

43. 32.2 ÷ 0.07

Divide. Round to the nearest hundredth if necessary.

44. 80.88 ÷ 8.425

45. 0.8 ÷ 0.6

46. 38.9 ÷ 6.44

47. 11.6 ÷ 0.95

48. ERROR ANALYSIS Describe and correct the error in rewriting the problem.

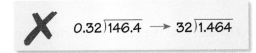

49. TICKETS Tickets to the school musical cost $6.25. The amount received from ticket sales is $706.25. How many tickets were sold?

50. HEIGHT A person's running stride is about 1.14 times the person's height. Your friend's stride is 5.472 feet. How tall is your friend?

51. MP3 PLAYER You have 3.4 gigabytes available on your MP3 player. Each song is about 0.004 gigabyte. How many more songs can you download onto your MP3 player?

52. SWIMMING The table shows the top three times in a swimming event at the Summer Olympics. The event consists of a team of four women swimming 100 meters each.

Women's 4 × 100 Freestyle Relay		
Medal	Country	Time (seconds)
Gold	Australia	215.94
Silver	United States	216.39
Bronze	Netherlands	217.59

  a. Suppose the times of all four swimmers on each team were the same. For each team, how much time does it take a swimmer to swim 100 meters?

  b. Suppose each U.S. swimmer completed 100 meters a quarter second faster. Would the U.S. team have won the gold medal? Explain your reasoning.

Without finding the quotient, copy and complete the statement using <, >, or =.

53. 6.66 ÷ 0.74 ▢ 66.6 ÷ 7.4   **54.** 32.2 ÷ 0.7 ▢ 3.22 ÷ 7

55. 160.72 ÷ 16.4 ▢ 160.72 ÷ 1.64   **56.** 75.6 ÷ 63 ▢ 7.56 ÷ 0.63

57. BEES To approximate the number of bees in a hive, multiply the number of bees that leave the hive in one minute by 3 and divide by 0.014. You count 25 bees leaving a hive in one minute. How many bees are in the hive?

58. PROBLEM SOLVING You are saving money to buy a new bicycle that costs $155.75. You have $30 and plan to save $5 each week. Your aunt decides to give you an additional $10 each week.

  a. How many weeks will you have to save until you have enough money to buy the bicycle?

  b. How many more weeks would you have to save to buy a new bicycle that costs $203.89? Explain how you found your answer.

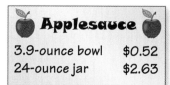

59. PRECISION A store sells applesauce in two sizes.

  a. How many *bowls* of applesauce fit in a *jar*? Round your answer to the nearest hundredth.

  b. Explain two ways to find the better buy.

  c. What is the better buy?

60. Geometry The large rectangle's dimensions are three times the dimensions of the small rectangle.

  a. How many times greater is the perimeter of the large rectangle compared to the perimeter of the small rectangle?

  b. How many times greater is the area of the large rectangle compared to the area of the small rectangle?

  c. Are the answers to parts (a) and (b) the same? *Explain* why or why not.

  d. What happens in parts (a) and (b) if the dimensions of the large rectangle are two times the dimensions of the small rectangle?

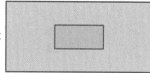

Fair Game Review *What you learned in previous grades & lessons*

Add or subtract. Write your answer in simplest form. *(Section 1.6)*

61. $\frac{1}{2} + \frac{2}{3}$ **62.** $\frac{2}{5} + \frac{3}{4}$ **63.** $\frac{3}{10} - \frac{1}{4}$ **64.** $\frac{11}{12} - \frac{7}{8}$

65. MULTIPLE CHOICE Melissa earns $7.40 an hour working at a grocery store. She works 14.25 hours this week. How much does she earn? *(Section 2.5)*

  Ⓐ $83.13   **Ⓑ** $105.45   **Ⓒ** $156.75   **Ⓓ** $1054.50

2.4–2.6 Quiz

Add or subtract. *(Section 2.4)*

1. $6.329 + 14.38$
2. $43.56 + 41.82$
3. $85.8 - 2.354$
4. $26.782 - 14.96$

Multiply. Use estimation to check your answer. *(Section 2.5)*

5. 7.6×5
6. 0.62×17
7. 0.54×0.9
8. 4.16×0.7

Divide. Use estimation to check your answer. *(Section 2.6)*

9. $5 \overline{)8.4}$
10. $6 \overline{)6.48}$
11. $5.6 \div 0.7$
12. $1.8 \div 0.03$

13. **FIELD HOCKEY** A field hockey field is rectangular. Its width is 54.88 meters, and its length is 91.46 meters. Find the perimeter of the field. *(Section 2.4)*

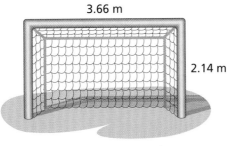

3.66 m
2.14 m

14. **GEOMETRY** Find the area of the mouth of the field hockey goal. *(Section 2.5)*

15. **BROADWAY** The bar graph shows the yearly attendance at traveling Broadway shows. *(Section 2.6)*

    a. Suppose the attendance was the same each month in 2008. How many people attended each month?

    b. How many times more people attended shows in 2006 than in 2009? Round your answer to the nearest tenth.

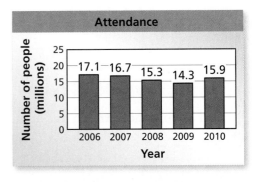

2 Chapter Review

Review Key Vocabulary

reciprocals, *p. 64*

Review Examples and Exercises

2.1 Multiplying Fractions (pp. 54–61)

a. Find $\dfrac{1}{4} \times \dfrac{3}{5}$.

$$\dfrac{1}{4} \times \dfrac{3}{5} = \dfrac{1 \times 3}{4 \times 5} = \dfrac{3}{20}$$ Multiply the numerators and the denominators.

b. Find $\dfrac{3}{5} \times 1\dfrac{1}{8}$.

$$\dfrac{3}{5} \times 1\dfrac{1}{8} = \dfrac{3}{5} \times \dfrac{9}{8}$$ Write $1\dfrac{1}{8}$ as the improper fraction $\dfrac{9}{8}$.

$$= \dfrac{3 \times 9}{5 \times 8} = \dfrac{27}{40}$$ Multiply the numerators and the denominators.

Exercises

Multiply. Write the answer in simplest form.

1. $\dfrac{1}{8} \times \dfrac{5}{7}$
2. $\dfrac{3}{5} \times \dfrac{1}{2}$
3. $\dfrac{2}{9} \times \dfrac{3}{4}$
4. $\dfrac{3}{10} \times \dfrac{4}{5}$
5. $2\dfrac{2}{3} \times \dfrac{4}{5}$
6. $\dfrac{2}{7} \times 4\dfrac{4}{9}$
7. $1\dfrac{5}{6} \times 2\dfrac{3}{8}$
8. $2\dfrac{3}{10} \times 5\dfrac{1}{3}$

2.2 Dividing Fractions (pp. 62–69)

Find $\dfrac{3}{7} \div \dfrac{5}{8}$.

$$\dfrac{3}{7} \div \dfrac{5}{8} = \dfrac{3}{7} \times \dfrac{8}{5}$$ Multiply by the reciprocal of $\dfrac{5}{8}$, which is $\dfrac{8}{5}$.

$$= \dfrac{3 \times 8}{7 \times 5} = \dfrac{24}{35}$$ Multiply fractions and simplify.

Exercises

Divide. Write the answer in simplest form.

9. $\dfrac{1}{9} \div \dfrac{2}{5}$
10. $\dfrac{3}{4} \div \dfrac{5}{6}$
11. $5 \div \dfrac{1}{3}$
12. $\dfrac{8}{9} \div \dfrac{3}{10}$

2.3 Dividing Mixed Numbers (pp. 70–75)

Find $3\frac{3}{4} \div 1\frac{1}{2}$.

$3\frac{3}{4} \div 1\frac{1}{2} = \frac{15}{4} \div \frac{3}{2}$ Write each mixed number as an improper fraction.

$= \frac{15}{4} \times \frac{2}{3}$ Multiply by the reciprocal of $\frac{3}{2}$, which is $\frac{2}{3}$.

$= \frac{\overset{5}{\cancel{15}} \times \overset{1}{\cancel{2}}}{\underset{2}{\cancel{4}} \times \underset{1}{\cancel{3}}}$ Multiply fractions. Divide out common factors.

$= \frac{5}{2}$, or $2\frac{1}{2}$ Simplify.

Exercises

Divide. Write the answer in simplest form.

13. $1\frac{2}{5} \div \frac{4}{7}$ **14.** $2\frac{3}{8} \div \frac{3}{5}$ **15.** $4\frac{1}{8} \div 2\frac{1}{4}$ **16.** $5\frac{5}{8} \div 1\frac{2}{9}$

17. PANCAKES A box contains 10 cups of pancake mix. You use $\frac{2}{3}$ cup each time you make pancakes. How many times can you make pancakes?

2.4 Adding and Subtracting Decimals (pp. 78–83)

a. Add 7.36 + 2.22.

Line up the decimal points.

```
  7.36
+ 2.22
------
  9.58
```
Add as you would with whole numbers.

b. Subtract 5.467 − 2.736.

Line up the decimal points.

```
  5.467
- 2.736
-------
  2.731
```
Subtract as you would with whole numbers.

Exercises

Add or subtract.

18. $3.78 + 8.94$ **19.** $19.89 + 4.372$

20. $7.638 - 2.365$ **21.** $14.21 - 4.103$

2.5 Multiplying Decimals (pp. 84–91)

Find 7.5 × 5.3.

```
    7.5  ←——— 1 decimal place
  × 5.3  ←——— + 1 decimal place
    2 2 5
  + 3 7 5
  3 9.7 5  ←——— 2 decimal places
```

So, 7.5 × 5.3 = 39.75.

Exercises

Multiply. Use estimation to check your answer.

22. 5.3 × 8   **23.** 6.1 × 7   **24.** 4.68 × 3

25. 9.475 × 8.03   **26.** 0.27 × 4.42   **27.** 0.051 × 0.244

28. AREA Find the area of the computer screen.

2.6 Dividing Decimals (pp. 92–99)

Find 22.8 ÷ 1.2.

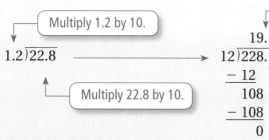

So, 22.8 ÷ 1.2 = 19.

Exercises

Divide. Use estimation to check your answer.

29. 6.8 ÷ 4   **30.** 13.2 ÷ 6 + 4   **31.** 49.7 ÷ 7

32. 0.12)3.6   **33.** 2.5)0.125   **34.** 3.9)22.23

2 Chapter Test

Multiply. Write the answer in simplest form.

1. $\dfrac{9}{16} \times \dfrac{2}{3}$
2. $\dfrac{1}{10} \times \dfrac{5}{6}$
3. $1\dfrac{3}{7} \times 6\dfrac{7}{10}$

Divide. Write the answer in simplest form.

4. $\dfrac{1}{6} \div \dfrac{1}{3}$
5. $10 \div \dfrac{2}{5}$
6. $8\dfrac{3}{4} \div 2\dfrac{7}{8}$

Add or subtract.

7. $4.92 + 3.79$
8. $5.138 + 2.624$
9. $5.316 - 1.942$

Multiply. Use estimation to check your answer.

10. 6.7×8
11. 0.4×0.7
12. 4.87×7.23

Divide. Use estimation to check your answer.

13. $5.6 \div 7$
14. $2.6 \div 0.02$
15. $4\overline{)9.32}$
16. $0.25\overline{)5.46}$

17. **DVD SALE** Which deal is the better buy?

18. **BLOG** You spend $2\dfrac{1}{2}$ hours online. You spend $\dfrac{1}{5}$ of that time writing a blog. How long do you spend writing your blog?

19. **GRAPES** A grocery store sells grapes for $1.99 per pound. You buy 2.34 pounds of the grapes. How much do you pay?

20. **PHOTOGRAPHY** A motocross rider is in the air for 2.5 seconds. Your camera can take a picture every 0.125 second. Your friend's camera can take a picture every 0.15 second.

    a. How many times faster is your camera than your friend's camera?

    b. How many more pictures can you take while the rider is in the air?

2 Cumulative Assessment

1. At a party, 10 people equally shared $2\frac{1}{2}$ gallons of ice cream. How much ice cream did each person eat?

   A. $\frac{1}{5}$ gal C. $\frac{2}{5}$ gal

   B. $\frac{1}{4}$ gal D. $\frac{3}{4}$ gal

2. What is the value of the expression below?

   $$4.643 + 11.02 \div 2.32$$

3. Which number is equivalent to the expression below?

   $$2 \cdot 4^2 + 3(6 \div 2)$$

   F. 25   H. 73

   G. 41   I. 105

Test-Taking Strategy
Estimate the Answer

"Using estimation you can see that the answer is about 3. So, you should choose B."

4. Your friend divided two decimal numbers. Her work is shown in the box below. What should your friend change in order to divide the two decimal numbers correctly?

   $$0.07\overline{)14.56} \rightarrow 7\overline{)14.56}^{\,2.08}$$

   A. Rewrite the problem as $0.07\overline{)0.1456}$.   C. Rewrite the problem as $7\overline{)0.1456}$.

   B. Rewrite the problem as $0.07\overline{)1456}$.   D. Rewrite the problem as $7\overline{)1456}$.

5. You bought some grapes at a farm stand. You paid $2.48 per pound.

   What was the total amount that you paid for the grapes?

Cumulative Assessment 105

6. The steps your friend took to divide two mixed numbers are shown below.

$$4\frac{2}{3} \div 2\frac{1}{4} = \frac{14}{3} \times \frac{9}{4}$$
$$= \frac{21}{2}$$
$$= 10\frac{1}{2}$$

What should your friend change in order to divide the two mixed numbers correctly?

F. Find a common denominator of 3 and 4.

G. Multiply by the reciprocal of $\frac{14}{3}$.

H. Multiply by the reciprocal of $\frac{9}{4}$.

I. Rename $4\frac{2}{3}$ as $3\frac{5}{3}$.

7. Which pair of numbers does *not* have a least common multiple less than 100?

 A. 10, 15
 B. 12, 16
 C. 16, 18
 D. 18, 24

8. You are making identical snack bags. You have 18 fruit-chew snacks and 24 granola snacks. What is the greatest number of snack bags that you can make with no snacks left over?

 F. 1
 G. 2
 H. 3
 I. 6

9. Which expression is *not* equivalent to $\frac{2}{3}$?

 A. $\frac{1}{4} + \frac{1}{3} \div \frac{4}{5}$
 B. $\frac{13}{30} + \frac{1}{5} \div \frac{6}{7}$
 C. $\frac{5}{6} - \frac{1}{8} \div \frac{1}{2}$
 D. $\frac{13}{18} - \frac{1}{26} \div \frac{9}{13}$

10. Which number is equivalent to $5.139 - 2.64$?

   F. 2.499
   G. 2.599
   H. 3.519
   I. 3.599

11. Which expression is equivalent to $\frac{4}{9} \div \frac{5}{7}$?

   A. $\frac{20}{63}$
   B. $\frac{28}{45}$
   C. $\frac{45}{28}$
   D. $\frac{63}{20}$

12. Which of the following expressions is equivalent to a perfect square?

   F. $3 + 2^2 \times 7$
   G. $34 + 18 \div 3^2$
   H. $(80 + 4) \div 4$
   I. $3^2 + 6 \times 5 \div 3$

13. You are filling baskets using 18 green eggs, 36 red eggs, and 54 blue eggs. What is the greatest number of baskets that you can fill so that the baskets are identical and there are no eggs left over?

   A. 3
   B. 6
   C. 9
   D. 18

14. A walkway was built using identical concrete blocks.

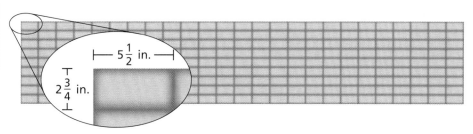

Part A How much longer, in inches, is the length of the walkway than the width of the walkway? Show your work and explain your reasoning.

Part B How many times longer is the length of the walkway than the width of the walkway? Show your work and explain your reasoning.

3 Algebraic Expressions and Properties

- **3.1** Algebraic Expressions
- **3.2** Writing Expressions
- **3.3** Properties of Addition and Multiplication
- **3.4** The Distributive Property

"Did you know that 5 × 6 = 6 × 5, but 5 ÷ 6 ≠ 6 ÷ 5?"

"Only certain operations like addition and multiplication preserve equality when you switch the numbers around."

"Descartes, evaluate this expression when x = 2 to determine the number of cat treats you are going to eat today."

"Remember that you evaluate an algebraic expression by substituting the value of x into the expression."

What You Learned Before

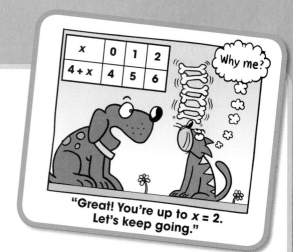

"Great! You're up to x = 2. Let's keep going."

● Interpreting Numerical Expressions

Example 1 Write a sentence interpreting the expression 3 × (19,762 + 418).

∴ 3 × (19,762 + 418) is 3 times as large as 19,762 + 418.

Example 2 Write a sentence interpreting the expression (316 + 43,449) + 5.

∴ (316 + 43,449) + 5 is 5 more than 316 + 43,449.

Example 3 Write a sentence interpreting the expression (20,008 − 752) ÷ 2.

∴ (20,008 − 752) ÷ 2 is half as large as 20,008 − 752.

Try It Yourself
Write a sentence interpreting the expression.

1. 3 × (372 + 20,967)
2. 2 × (432 + 346,322)
3. 4 × (6722 + 4086)
4. (115 + 36,372) + 6
5. (392 + 75,325) + 78
6. (352 + 46,795) + 100
7. (30,929 + 425) ÷ 2
8. (58,742 − 721) ÷ 2
9. (96,792 + 564) ÷ 3

● Using Order of Operations

Example 4 Simplify $4^2 \div 2 + 3(9 - 5)$.

First:	Parentheses	$4^2 \div 2 + 3(9 - 5) = 4^2 \div 2 + 3 \cdot 4$
Second:	Exponents	$= 16 \div 2 + 3 \cdot 4$
Third:	Multiplication and Division (from left to right)	$= 8 + 12$
Fourth:	Addition and Subtraction (from left to right)	$= 20$

Try It Yourself
Simplify the expression.

10. $3^2 + 5(4 - 2)$
11. $3 + 4 \div 2$
12. $10 \div 5 \cdot 3$
13. $4(3^3 - 8) \div 2$
14. $3 \cdot 6 - 4 \div 2$
15. $12 + 7 \cdot 3 - 24$

3.1 Algebraic Expressions

Essential Question How can you write and evaluate an expression that represents a real-life problem?

1 ACTIVITY: Reading and Re-Reading

Work with a partner.

a. You babysit for 3 hours. You receive $12. What is your hourly wage?
   - Write the problem. Underline the important numbers and units you need to solve the problem.
   - Read the problem carefully a second time. Circle the key word for the question.

   > You babysit for 3 hours. You receive $12.
   > What is your hourly wage?

   - Write each important number or word, with its units, on a piece of paper. Write +, −, ×, ÷, and = on five other pieces of paper.

   - Arrange the pieces of paper to answer the key word question, "What is your hourly wage?"

   - Evaluate the expression that represents the hourly wage.

   hourly wage = ▢ ÷ ▢   Write.

   = ▢   Evaluate.

   ∴ So, your hourly wage is $ ▢ per hour.

b. How can you use your hourly wage to find how much you will receive for any number of hours worked?

Algebraic Expressions
In this lesson, you will
- use order of operations to evaluate algebraic expressions.
- solve real-life problems.

110   Chapter 3   Algebraic Expressions and Properties

2 ACTIVITY: Reading and Re-Reading

Math Practice

Make Sense of Quantities

What are the units in the problem? How does this help you write an expression?

Work with a partner. Use the strategy shown in Activity 1 to write an expression for each problem. After you have written the expression, evaluate it using mental math or some other method.

a. You wash cars for 2 hours. You receive $6. How much do you earn per hour?

b. You have $60. You buy a pair of jeans and a shirt. The pair of jeans costs $27. You come home with $15. How much did you spend on the shirt?

c. For lunch, you buy 5 sandwiches that cost $3 each. How much do you spend?

d. You are running a 4500-foot race. How much farther do you have to go after running 2000 feet?

e. A young rattlesnake grows at a rate of about 20 centimeters per year. How much does a young rattlesnake grow in 2 years?

What Is Your Answer?

3. **IN YOUR OWN WORDS** How can you write and evaluate an expression that represents a real-life problem? Give one example with addition, one with subtraction, one with multiplication, and one with division.

Use what you learned about evaluating expressions to complete Exercises 4–7 on page 115.

Section 3.1   Algebraic Expressions   111

3.1 Lesson

Key Vocabulary
algebraic expression, *p. 112*
terms, *p. 112*
variable, *p. 112*
coefficient, *p. 112*
constant, *p. 112*

An **algebraic expression** is an expression that may contain numbers, operations, and one or more symbols. Parts of an algebraic expression are called **terms**.

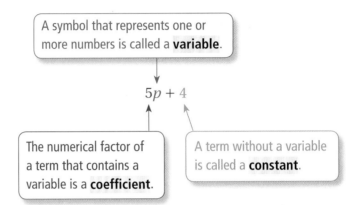

EXAMPLE 1 Identifying Parts of an Algebraic Expression

Identify the terms, coefficients, and constants in each expression.

a. $5x + 13$

Terms: $5x$, 13

Coefficient: 5

Constant: 13

b. $2z^2 + y + 3$

Terms: $2z^2$, $1y$, 3

Coefficients: 2, 1

Constant: 3

Study Tip
A variable by itself has a coefficient of 1. So, the term y in Example 1(b) has a coefficient of 1.

On Your Own

Exercises 8–13

Identify the terms, coefficients, and constants in the expression.

1. $12 + 10c$
2. $15 + 3w + \dfrac{1}{2}$
3. $z^2 + 9z$

EXAMPLE 2 Writing Algebraic Expressions Using Exponents

Write each expression using exponents.

a. $d \cdot d \cdot d \cdot d$

Because d is used as a factor 4 times, its exponent is 4.

∴ So, $d \cdot d \cdot d \cdot d = d^4$.

b. $1.5 \cdot h \cdot h \cdot h$

Because h is used as a factor 3 times, its exponent is 3.

∴ So, $1.5 \cdot h \cdot h \cdot h = 1.5h^3$.

Chapter 3 Algebraic Expressions and Properties

On Your Own

Now You're Ready
Exercises 16–21

Write the expression using exponents.

4. $j \cdot j \cdot j \cdot j \cdot j \cdot j$

5. $9 \cdot k \cdot k \cdot k \cdot k \cdot k$

To evaluate an algebraic expression, substitute a number for each variable. Then use the order of operations to find the value of the numerical expression.

EXAMPLE 3 Evaluating Algebraic Expressions

a. Evaluate $k + 10$ when $k = 25$.

$$k + 10 = 25 + 10 \qquad \text{Substitute 25 for } k.$$
$$= 35 \qquad \text{Add 25 and 10.}$$

Study Tip
You can write the product of 4 and n in several ways.
$4 \cdot n$
$4n$
$4(n)$

b. Evaluate $4 \cdot n$ when $n = 12$.

$$4 \cdot n = 4 \cdot 12 \qquad \text{Substitute 12 for } n.$$
$$= 48 \qquad \text{Multiply 4 and 12.}$$

On Your Own

Now You're Ready
Exercises 25–32

6. Evaluate $24 + c$ when $c = 9$.

7. Evaluate $d - 17$ when $d = 30$.

EXAMPLE 4 Evaluating an Expression with Two Variables

Evaluate $a \div b$ when $a = 16$ and $b = \dfrac{2}{3}$.

$$a \div b = 16 \div \dfrac{2}{3} \qquad \text{Substitute 16 for } a \text{ and } \dfrac{2}{3} \text{ for } b.$$
$$= 16 \cdot \dfrac{3}{2} \qquad \text{Multiply by the reciprocal of } \dfrac{2}{3}, \text{ which is } \dfrac{3}{2}.$$
$$= 24 \qquad \text{Multiply.}$$

On Your Own

Now You're Ready
Exercises 33–36

Evaluate the expression when $p = 24$ and $q = 8$.

8. $p \div q$

9. $q + p$

10. $p - q$

11. pq

EXAMPLE 5 Evaluating Expressions with Two Operations

a. Evaluate $3x - 14$ when $x = 5$.

$3x - 14 = 3(5) - 14$ Substitute 5 for x.

$= 15 - 14$ Using order of operations, multiply 3 and 5.

$= 1$ Subtract 14 from 15.

b. Evaluate $z^2 + 8.5$ when $z = 2$.

$z^2 + 8.5 = 2^2 + 8.5$ Substitute 2 for z.

$= 4 + 8.5$ Using order of operations, evaluate 2^2.

$= 12.5$ Add 4 and 8.5.

On Your Own

Now You're Ready
Exercises 43–51

Evaluate the expression when $y = 6$.

12. $5y + 1$ 13. $30 - 24 \div y$ 14. $y^2 - 7$ 15. $1.5 + y^2$

EXAMPLE 6 Real-Life Application

You are saving money to buy a skateboard. You begin with $45 and you save $3 each week. The expression $45 + 3w$ gives the amount of money you save after w weeks.

a. How much will you have after 4 weeks, 10 weeks, and 20 weeks?

b. After 20 weeks, can you buy the skateboard? Explain.

a. Substitute the given number of weeks for w.

Number of Weeks, w	$45 + 3w$	Amount Saved
4	$45 + 3(4)$	$45 + 12 = \$57$
10	$45 + 3(10)$	$45 + 30 = \$75$
20	$45 + 3(20)$	$45 + 60 = \$105$

b. After 20 weeks, you have $105. So, you cannot buy the $125 skateboard.

On Your Own

16. **WHAT IF?** In Example 6, the expression for how much money you have after w weeks is $45 + 4w$. Can you buy the skateboard after 20 weeks? Explain.

114   Chapter 3   Algebraic Expressions and Properties

3.1 Exercises

Vocabulary and Concept Check

1. **WHICH ONE DOESN'T BELONG?** Which expression does *not* belong with the other three? Explain your reasoning.

   | $2x + 1$ | $5w \cdot c$ | $3(4) + 5$ | $y \div z$ |

2. **NUMBER SENSE** Which step in the order of operations is first? second? third? fourth?

   | Add or subtract from left to right. | Multiply or divide from left to right. |
   | Evaluate terms with exponents. | Perform operations in parentheses. |

3. **NUMBER SENSE** Will the value of the expression $20 - x$ *increase*, *decrease*, or *stay the same* as x increases? Explain.

Practice and Problem Solving

Write and evaluate an expression for the problem.

4. You receive $8 for raking leaves for 2 hours. What is your hourly wage?

5. Music lessons cost $20 per week. How much do 6 weeks of lessons cost?

6. The scores on your first two history tests were 82 and 95. By how many points did you improve on your second test?

7. You buy a hat for $12 and give the cashier a $20 bill. How much change do you receive?

Identify the terms, coefficients, and constants in the expression.

8. $7h + 3$

9. $g + 12 + 9g$

10. $5c^2 + 7d$

11. $2m^2 + 15 + 2p^2$

12. $6 + n^2 + \dfrac{1}{2}d$

13. $8x + \dfrac{x^2}{3}$

Terms: 2, x^2, y
Coefficient: 2
Constant: none

14. **ERROR ANALYSIS** Describe and correct the error in identifying the terms, coefficients, and constants in the algebraic expression $2x^2y$.

15. **PERIMETER** You can use the expression $2\ell + 2w$ to find the perimeter of a rectangle where ℓ is the length and w is the width.

    a. Identify the terms, coefficients, and constants in the expression.

    b. Interpret the coefficients of the terms.

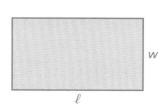

Section 3.1 Algebraic Expressions 115

Write each expression using exponents.

16. $b \cdot b \cdot b$
17. $g \cdot g \cdot g \cdot g \cdot g$
18. $8 \cdot w \cdot w \cdot w \cdot w$
19. $5.2 \cdot y \cdot y \cdot y$
20. $a \cdot a \cdot c \cdot c$
21. $2.1 \cdot x \cdot z \cdot z \cdot z \cdot z$

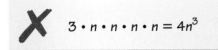

22. **ERROR ANALYSIS** Describe and correct the error in writing the product using exponents.

23. **AREA** Write an expression using exponents that represents the area of the square.

24. **ST. IVES** Suppose the man in the St. Ives poem has x wives, each wife has x sacks, each sack has x cats, and each cat has x kits. Write an expression using exponents that represents the total number of kits, cats, sacks, and wives going to St. Ives.

> As I was going to St. Ives
> I met a man with seven wives
> Each wife had seven sacks
> Each sack had seven cats
> Each cat had seven kits
> Kits, cats, sacks, wives
> How many were going to St. Ives?

ALGEBRA Evaluate the expression when $a = 3$, $b = 2$, and $c = 12$.

25. $6 + a$
26. $b \cdot 5$
27. $c - 1$
28. $27 \div a$
29. $12 - b$
30. $c + 5$
31. $2a$
32. $c \div 6$
33. $a + b$
34. $c - a$
35. $\dfrac{c}{a}$
36. $b \cdot c$

37. **ERROR ANALYSIS** Describe and correct the error in evaluating the expression when $m = 8$.

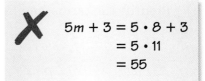

38. **LAWNS** You earn $15n$ dollars for mowing n lawns. How much do you earn for mowing one lawn? seven lawns?

39. **PLANT** After m months, the height of a plant is $10 + 3m$ millimeters. How tall is the plant after eight months? three years?

Copy and complete the table.

40.

x	3	6	9
x · 8			

41.

x	2	4	8
64 ÷ x			

42. **FALLING OBJECT** An object falls $16t^2$ feet in t seconds. You drop a rock from a bridge that is 75 feet above the water. Will the rock hit the water in 2 seconds? Explain.

ALGEBRA Evaluate the expression when $a = 10$, $b = 9$, and $c = 4$.

⑤ **43.** $2a + 3$ **44.** $4c - 7.8$ **45.** $\dfrac{a}{4} + \dfrac{1}{3}$

46. $\dfrac{24}{b} + 8$ **47.** $c^2 + 6$ **48.** $a^2 - 18$

49. $a + 9c$ **50.** $bc + 12.3$ **51.** $3a + 2b - 6c$

Standard Rentals $3

New Releases $4

52. MOVIES You rent x new releases and y standard rentals. Which expression tells you how much money you will need?

$3x + 4y$ $4x + 3y$ $7(x + y)$

53. WATER PARK You float 2000 feet along a "Lazy River" water ride. The ride takes less than 10 minutes. Give two examples of possible times and speeds. Illustrate the water ride with a drawing.

54. SCIENCE CENTER The expression $20a + 13c$ is the cost (in dollars) for a adults and c students to enter a science center.

  a. How much does it cost for an adult? a student? Explain your reasoning.

  b. Find the total cost for 4 adults and 24 students.

  c. You find the cost for a group. Then the numbers of adults and students in the group both double. Does the cost double? Explain your answer using an example.

  d. In part (b), the number of adults is cut in half, but the number of students doubles. Is the cost the same? Explain your answer.

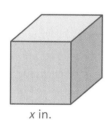

x in.

55. Reasoning The volume of the cube is equal to four times the area of one of its faces. What is the volume of the cube?

 Fair Game Review What you learned in previous grades & lessons

Find the value of the power. *(Section 1.2)*

56. 3^5 **57.** 8^3 **58.** 7^4 **59.** 2^8

60. MULTIPLE CHOICE Which numbers have a least common multiple of 24? *(Section 1.6)*

  Ⓐ 4, 6   Ⓑ 2, 22   Ⓒ 3, 8   Ⓓ 6, 12

3.2 Writing Expressions

Essential Question How can you write an expression that represents an unknown quantity?

1 ACTIVITY: Ordering Lunch

Work with a partner. You use a $20 bill to buy lunch at a café. You order a sandwich from the menu board shown.

a. Complete the table. In the last column, write a numerical expression for the amount of change received.

b. **REPEATED REASONING** Write an expression for the amount of change you receive when you order any sandwich from the menu board.

Sandwich	Price (dollars)	Change Received (dollars)
Reuben		
BLT		
Egg salad		
Roast beef		

Algebraic Expressions
In this lesson, you will
- use variables to represent numbers in algebraic expressions.
- write algebraic expressions.

c. Compare the expression you wrote in part (b) with the expressions in the last column of the table in part (a).

d. The café offers several side dishes, each at the same price. You order a chicken salad sandwich and two side dishes. Write an expression for the total amount of money you spend. Explain how you wrote your expression.

e. The expression $20 - 4.65s$ represents the amount of change one customer receives after ordering from the menu board. Explain what each part of the expression represents. Do you know what the customer ordered? Explain your reasoning.

2 ACTIVITY: Words That Imply Addition or Subtraction

Math Practice

Use Expressions
How do the key words in the phrase help you write the given relationship as an expression?

Work with a partner.

a. Complete the table.

Variable	Phrase	Expression
n	4 **more than** a number	
m	the **difference** of a number and 3	
x	the **sum** of a number and 8	
p	10 **less than** a number	
n	7 units **farther** away	
t	8 minutes **sooner**	
w	12 minutes **later**	
y	a number **increased** by 9	

b. Here is a word problem that uses one of the expressions in the table.

You arrive at the café 8 minutes sooner than your friend. Your friend arrives at 6:42 P.M. When did you arrive?

Which expression from the table can you use to solve the problem?

c. Write a problem that uses a different expression from the table.

3 ACTIVITY: Words That Imply Multiplication or Division

Work with a partner. Match each phrase with an expression.

the product of a number and 3 $n \div 3$

the quotient of 3 and a number $4p$

4 times a number $n \cdot 3$

a number divided by 3 $2m$

twice a number $3 \div n$

What Is Your Answer?

4. **IN YOUR OWN WORDS** How can you write an expression that represents an unknown quantity? Give examples to support your explanation.

Practice

Use what you learned about writing expressions to complete Exercises 9–12 on page 122.

3.2 Lesson

Some words imply math operations.

Operation	Addition	Subtraction	Multiplication	Division
Key Words and Phrases	added to plus sum of more than increased by total of and	subtracted from minus difference of less than decreased by fewer than take away	multiplied by times product of twice of	divided by quotient of

EXAMPLE 1 Writing Numerical Expressions

Write the phrase as an expression.

a. 8 fewer than 21

$21 - 8$ The phrase *fewer than* means *subtraction*.

b. the product of 30 and 9

30×9, or $30 \cdot 9$ The phrase *product of* means *multiplication*.

EXAMPLE 2 Writing Algebraic Expressions

Write the phrase as an expression.

a. 14 more than a number x

$x + 14$ The phrase *more than* means *addition*.

b. a number y minus 75

$y - 75$ The word *minus* means *subtraction*.

c. the quotient of 3 and a number z

$3 \div z$, or $\dfrac{3}{z}$ The phrase *quotient of* means *division*.

Common Error

When writing expressions involving subtraction or division, order is important. For example, the quotient of a number x and 2 means
$x \div 2$, not $2 \div x$.

On Your Own

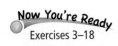
Exercises 3–18

Write the phrase as an expression.

1. the sum of 18 and 35
2. 6 times 50
3. 25 less than a number b
4. a number x divided by 4
5. the total of a number t and 11
6. 100 decreased by a number k

EXAMPLE 3 **Writing an Algebraic Expression**

The length of Interstate 90 from the West Coast to the East Coast is 153.5 miles more than 2 times the length of Interstate 15 from southern California to northern Montana. Let m be the length of Interstate 15. Which expression can you use to represent the length of Interstate 90?

(A) $2m + 153.5$ **(B)** $2m - 153.5$ **(C)** $153.5 - 2m$ **(D)** $153.5m + 2$

The word *times* means multiplication. So, multiply 2 and m.

The phrase *more than* means addition. So, add $2m$ and 153.5.

$$2m + 153.5$$

∴ The correct answer is **(A)**.

EXAMPLE 4 **Real-Life Application**

You plant a cypress tree that is 10 inches tall. Each year, its height increases by 15 inches.

a. Make a table that shows the height of the tree for 4 years. Then write an expression for the height after t years.

b. What is the height after 9 years?

a. The height is *increasing*, so *add* 15 each year as shown in the table.

Year, t	Height (inches)
0	10
1	$10 + 15(1) = 25$
2	$10 + 15(2) = 40$
3	$10 + 15(3) = 55$
4	$10 + 15(4) = 70$

When t is 0, the height is 10 inches.

You can see that an expression is $10 + 15t$.

∴ So, the height after year t is $10 + 15t$.

b. Evaluate $10 + 15t$ when $t = 9$.

$$10 + 15t = 10 + 15(9) = 145$$

∴ After 9 years, the height of the tree is 145 inches.

Study Tip

Sometimes, like in Example 3, a variable represents a single value. Other times, like in Example 4, a variable can represent more than one value.

On Your Own

Now You're Ready
Exercises 27–30

7. Your friend has 5 more than twice as many game tokens as your sister. Let t be the number of game tokens your sister has. Write an expression for the number of game tokens your friend has.

8. **WHAT IF?** In Example 4, what is the height of the cypress tree after 16 years?

Section 3.2   Writing Expressions

3.2 Exercises

Vocabulary and Concept Check

1. **DIFFERENT WORDS, SAME QUESTION** Which is different? Write "both" expressions.

   | 12 more than x | x increased by 12 | x take away 12 | the sum of x and 12 |

2. **REASONING** You pay $0.25p$ dollars to print p photos. What does the coefficient represent?

Practice and Problem Solving

Write the phrase as an expression.

3. 5 less than 8
4. the product of 3 and 12
5. 28 divided by 7
6. the total of 6 and 10
7. 3 fewer than 18
8. 17 added to 15
9. 13 subtracted from a number x
10. 5 times a number d
11. the quotient of 18 and a number a
12. the difference of a number s and 6
13. 7 increased by a number w
14. a number b squared
15. the sum of a number y and 4
16. the difference of 12 and a number x
17. twice a number z
18. a number t cubed

ERROR ANALYSIS Describe and correct the error in writing the phrase as an expression.

19. the quotient of 8 and a number y

    ✗ $\dfrac{y}{8}$

20. 16 decreased by a number x

    ✗ $x - 16$

21. **DINNER** Five friends share the cost of a dinner equally.
    a. Write an expression for the cost per person.
    b. Make up a total cost and test your expression. Is the result reasonable?

22. **TV SHOW** A television show has 19 episodes per season.
    a. Copy and complete the table.
    b. Write an expression for the number of episodes in n seasons.

Seasons	1	2	3	4	5
Episodes					

Give two ways to write the expression as a phrase.

23. $n + 6$
24. $4w$
25. $15 - b$
26. $14 - 3z$

Write the phrase as an expression. Then evaluate when $x = 5$ and $y = 20$.

27. 3 less than the quotient of a number y and 4
28. the sum of a number x and 4, all divided by 3
29. 6 more than the product of 8 and a number x
30. the quotient of 40 and the difference of a number y and 16

31. **MODELING** It costs $3 to bowl a game and $2 for shoe rental.
    a. Make a table for the cost of up to 5 games.
    b. Write an expression for the cost of g games.
    c. Use your expression to find the cost of 8 games.

32. **PUZZLE** Florida has 8 less than 5 times the number of counties in Arizona. Georgia has 25 more than twice the number of counties in Florida.
    a. Write an expression for the number of counties in Florida.
    b. Write an expression for the number of counties in Georgia.
    c. Arizona has 15 counties. How many do Florida and Georgia have?

33. **PATTERNS** There are 140 people in a singing competition. The graph shows the results for the first five rounds.
    a. Write an expression for the number of people after each round.
    b. How many people compete in the ninth round? Explain your reasoning.

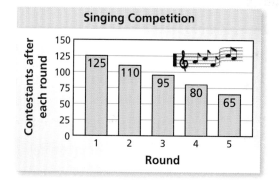

34. **NUMBER SENSE** The difference between two numbers is 8. The lesser number is a. Write an expression for the greater number.

35. **Reasoning** One number is four times another. The greater number is x. Write an expression for the lesser number.

Fair Game Review *What you learned in previous grades & lessons*

Evaluate the expression. *(Skills Review Handbook)*

36. $8 + (22 + 15)$
37. $(13 + 9) + 37$
38. $(13 \times 6) \times 5$
39. $4 \times (7 \times 5)$

40. **MULTIPLE CHOICE** A grocery store is making fruit baskets using 144 apples, 108 oranges, and 90 pears. Each basket will be identical. What is the greatest number of fruit baskets the store can make using all the fruit? *(Section 1.5)*

   Ⓐ 6   Ⓑ 9   Ⓒ 16   Ⓓ 18

3 Study Help

You can use an **information wheel** to organize information about a topic. Here is an example of an information wheel for identifying parts of an algebraic expression.

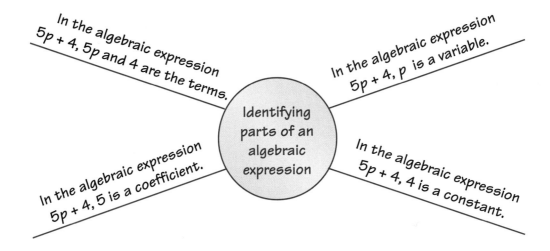

On Your Own

Make information wheels to help you study these topics.

1. evaluating algebraic expressions
2. writing algebraic expressions

After you complete this chapter, make information wheels for the following topics.

3. Commutative Properties of Addition and Multiplication
4. Associative Properties of Addition and Multiplication
5. Addition Property of Zero
6. Multiplication Properties of Zero and One
7. Distributive Property
8. factoring expressions

"My information wheel for Fluffy has matching adjectives and nouns."

124   Chapter 3   Algebraic Expressions and Properties

3.1–3.2 Quiz

Identify the terms, coefficients, and constants of the expression. *(Section 3.1)*

1. $6q + 1$
2. $3r^2 + 4r + 8$

Write the expression using exponents. *(Section 3.1)*

3. $s \cdot s \cdot s \cdot s$
4. $2 \cdot t \cdot t \cdot t \cdot t \cdot t$

Evaluate the expression when $a = 8$ and $b = 2$. *(Section 3.1)*

5. $a + 5$
6. ab
7. $a^2 - 6$

Copy and complete the table. *(Section 3.1)*

8.

x	x + 6
1	
2	
3	

9.

x	3x − 5
3	
6	
9	

Write the phrase as an expression. *(Section 3.2)*

10. the sum of 28 and 35
11. a number x divided by 2
12. the product of a number m and 23
13. 10 less than a number a

14. **COUPON** The expression $p - 15$ is the amount you pay after using the coupon on a purchase of p dollars. How much do you pay for a purchase of $83? *(Section 3.1)*

Coupon
Good for $15 off any purchase of $75 or more

15. **AMUSEMENT PARK** The expression $15a + 12c$ is the cost (in dollars) of admission at an amusement park for a adults and c children. Find the total cost for 5 adults and 10 children. *(Section 3.1)*

16. **MOVING TRUCK** To rent a moving truck for the day, it costs $33 plus $1 for each mile driven. *(Section 3.2)*

   a. Write an expression for the cost to rent the truck.
   b. You drive the truck 300 miles. How much do you pay?

3.3 Lesson

Check It Out
Lesson Tutorials
BigIdeasMath.com

Key Vocabulary
equivalent expressions, p. 128

Expressions with the same value, like 12 + 7 and 7 + 12, are **equivalent expressions**. You can use the Commutative and Associative Properties to write equivalent expressions.

Key Ideas

Commutative Properties

Words Changing the order of addends or factors does not change the sum or product.

Numbers $5 + 8 = 8 + 5$ **Algebra** $a + b = b + a$
$5 \cdot 8 = 8 \cdot 5$ $a \cdot b = b \cdot a$

Associative Properties

Words Changing the grouping of addends or factors does not change the sum or product.

Numbers $(7 + 4) + 2 = 7 + (4 + 2)$
$(7 \cdot 4) \cdot 2 = 7 \cdot (4 \cdot 2)$

Algebra $(a + b) + c = a + (b + c)$
$(a \cdot b) \cdot c = a \cdot (b \cdot c)$

EXAMPLE 1 Using Properties to Write Equivalent Expressions

a. Simplify the expression $7 + (12 + x)$.

$7 + (12 + x) = (7 + 12) + x$ Associative Property of Addition
$= 19 + x$ Add 7 and 12.

b. Simplify the expression $(6.1 + x) + 8.4$.

$(6.1 + x) + 8.4 = (x + 6.1) + 8.4$ Commutative Property of Addition
$= x + (6.1 + 8.4)$ Associative Property of Addition
$= x + 14.5$ Add 6.1 and 8.4.

c. Simplify the expression $5(11y)$.

$5(11y) = (5 \cdot 11)y$ Associative Property of Multiplication
$= 55y$ Multiply 5 and 11.

Study Tip
One way to check whether expressions are equivalent is to evaluate each expression for any value of the variable. In Example 1(a), use $x = 2$.
$7 + (12 + x) = 19 + x$
$7 + (12 + 2) \stackrel{?}{=} 19 + 2$
$21 = 21$ ✓

On Your Own

Now You're Ready
Exercises 5–8

Simplify the expression. Explain each step.

1. $10 + (a + 9)$
2. $\left(c + \dfrac{2}{3}\right) + \dfrac{1}{2}$
3. $5(4n)$

128   Chapter 3   Algebraic Expressions and Properties   Multi-Language Glossary at BigIdeasMath.com

Key Ideas

Addition Property of Zero

Words The sum of any number and 0 is that number.

Numbers $7 + 0 = 7$ **Algebra** $a + 0 = a$

Multiplication Properties of Zero and One

Words The product of any number and 0 is 0.

The product of any number and 1 is that number.

Numbers $9 \cdot 0 = 0$ **Algebra** $a \cdot 0 = 0$
$4 \cdot 1 = 4$ $a \cdot 1 = a$

EXAMPLE 2 Using Properties to Write Equivalent Expressions

a. **Simplify the expression $9 \cdot 0 \cdot p$.**

$9 \cdot 0 \cdot p = (9 \cdot 0) \cdot p$ Associative Property of Multiplication

$= 0 \cdot p = 0$ Multiplication Property of Zero

b. **Simplify the expression $4.5 \cdot r \cdot 1$.**

$4.5 \cdot r \cdot 1 = 4.5 \cdot (r \cdot 1)$ Associative Property of Multiplication

$= 4.5 \cdot r$ Multiplication Property of One

$= 4.5r$

EXAMPLE 3 Real-Life Application

You and six friends play on a basketball team. A sponsor paid $100 for the league fee, x dollars for each player's T-shirt, and $68.25 for trophies. Write an expression for the total amount the sponsor paid.

Common Error

You and six friends are on the team, so use the expression $7x$, not $6x$, to represent the cost of the T-shirts.

Add the league fee, the cost of the T-shirts, and the cost of the trophies.

$100 + 7x + 68.25 = 7x + 100 + 68.25$ Commutative Property of Addition

$= 7x + 168.25$ Add 100 and 68.25.

∴ An expression for the total amount is $7x + 168.25$.

On Your Own

Now You're Ready
Exercises 9–23

Simplify the expression. Explain each step.

4. $12 \cdot b \cdot 0$ **5.** $1 \cdot m \cdot 24$ **6.** $(t + 15) + 0$

7. WHAT IF? In Example 3, your sponsor paid $54.75 for trophies. Write an expression for the total amount the sponsor paid.

3.3 Exercises

Vocabulary and Concept Check

1. **NUMBER SENSE** Write an example of a sum of fractions. Show that the Commutative Property of Addition is true for the sum.

2. **OPEN-ENDED** Write an algebraic expression that can be simplified using the Associative Property of Addition.

3. **OPEN-ENDED** Write an algebraic expression that can be simplified using the Associative Property of Multiplication and the Multiplication Property of One.

4. **WHICH ONE DOESN'T BELONG?** Which statement does *not* belong with the other three? Explain your reasoning.

$7 + (x + 4) = 7 + (4 + x)$	$(3 + b) + 2 = (b + 3) + 2$
$9 + (7 + w) = (9 + 7) + w$	$(4 + n) + 6 = (n + 4) + 6$

Practice and Problem Solving

Tell which property the statement illustrates.

5. $5 \cdot p = p \cdot 5$

6. $2 + (12 + r) = (2 + 12) + r$

7. $4 \cdot (x \cdot 10) = (4 \cdot x) \cdot 10$

8. $x + 7.5 = 7.5 + x$

9. $(c + 2) + 0 = c + 2$

10. $a \cdot 1 = a$

11. **ERROR ANALYSIS** Describe and correct the error in stating the property that the statement illustrates.

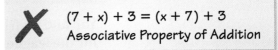

$(7 + x) + 3 = (x + 7) + 3$
Associative Property of Addition

Simplify the expression. Explain each step.

12. $6 + (5 + x)$

13. $(14 + y) + 3$

14. $6(2b)$

15. $7(9w)$

16. $3.2 + (x + 5.1)$

17. $(0 + a) + 8$

18. $9 \cdot c \cdot 4$

19. $(18.6 \cdot d) \cdot 1$

20. $\left(3k + 4\frac{1}{5}\right) + 8\frac{3}{5}$

21. $(2.4 + 4n) + 9$

22. $(3s) \cdot 8$

23. $z \cdot 0 \cdot 12$

24. **GEOMETRY** The expression $12 + x + 4$ represents the perimeter of a triangle. Simplify the expression.

25. **SCOUT COOKIES** A case of Scout cookies has 10 cartons. A carton has 12 boxes. The amount you earn on a whole case is $10(12x)$ dollars.

    a. What does x represent?
    b. Simplify the expression.

26. **STRUCTURE** The volume of the rectangular prism is 12.5 • x • 1.
    a. Simplify the expression.
    b. Match x = 0.25, 12.5, and 144 with the object. Explain.
       A. siding for a house   B. ruler   C. square floor tile

Write the phrase as an expression. Then simplify the expression.

27. 7 plus the sum of a number x and 5

28. the product of 8 and a number y multiplied by 9

Copy and complete the statement using the specified property.

	Property	Statement
29.	Associative Property of Multiplication	7(2y) =
30.	Commutative Property of Multiplication	13.2 • (x • 1) =
31.	Associative Property of Addition	17 + (6 + 2x) =
32.	Addition Property of Zero	2 + (c + 0) =
33.	Multiplication Property of One	1 • w • 16 =

34. **HATS** You and a friend sell hats at a fair booth. You sell 16 hats on the first shift and 21 hats on the third shift. Your friend sells x hats on the second shift.

    a. Write an expression for the number of hats sold.
    b. The expression 37(14) + 10x represents the amount that you both earned. How can you tell that your friend was selling the hats for a discounted price?
    c. **Reasoning** You earned more money than your friend. What can you say about the value of x?

Fair Game Review What you learned in previous grades & lessons

Evaluate the expression. *(Section 1.3)*

35. 7(10 + 4)   36. 12(10 − 1)   37. 6(5 + 10)   38. 8(30 − 5)

Find the prime factorization of the number. *(Section 1.4)*

39. 37   40. 144   41. 147   42. 205

43. **MULTIPLE CHOICE** A bag has 16 blue, 20 red, and 24 green marbles. What fraction of the marbles in the bag are blue? *(Skills Review Handbook)*

    Ⓐ $\frac{1}{5}$ Ⓑ $\frac{4}{15}$ Ⓒ $\frac{4}{11}$ Ⓓ $\frac{11}{15}$

Section 3.3   Properties of Addition and Multiplication

3.4 The Distributive Property

Essential Question How do you use mental math to multiply two numbers?

The Meaning of a Word • Distribute

When you **distribute** something to each person in a group,

you give that thing to each person in the group.

1 ACTIVITY: Modeling a Property

Work with a partner.

a. **MODELING** Draw two rectangles of the same width but with different lengths on a piece of grid paper. Label the dimensions.

b. Write an expression for the total area of the rectangles.

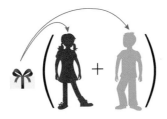

c. Rearrange the rectangles by aligning the shortest sides to form one rectangle. Label the dimensions. Write an expression for the area.

d. Can the expressions from parts (b) and (c) be set equal to each other? Explain.

e. **REPEATED REASONING** Repeat this activity using different rectangles. Explain how this illustrates the Distributive Property. Write a rule for the Distributive Property.

Equivalent Expressions

In this lesson, you will
- use the Distributive Property to find products.
- use the Distributive Property to simplify algebraic expressions.

2 ACTIVITY: Using Mental Math

Math Practice

Find Entry Points

How can you rewrite the larger number as the sum of two numbers so that you can use mental math?

Work with a partner. Use the method shown to find the product.

a. **Sample:** 23×6

$$
\begin{array}{r}
23 \\
\times\ 6 \\
\hline
120 \\
+\ 18 \\
\hline
138
\end{array}
$$

23 is $20 + 3$.

Multiply 20 and 6.
Multiply 3 and 6.
Add.

So, $23 \times 6 = 138$.

b. 33×7 c. 47×9
d. 28×5 e. 17×4

3 ACTIVITY: Using Mental Math

Work with a partner. Use the Distributive Property and mental math to find the product.

Hmmm. Which method is easier?

a. **Sample:** 6×23

$6 \times 23 = 6 \times (20 + 3)$ Write 23 as the sum of 20 and 3.
$ = (6 \times 20) + (6 \times 3)$ Distribute the 6 over the sum.
$ = 120 + 18$ Find the products.
$ = 138$ Add.

So, $6 \times 23 = 138$.

b. 5×17 c. 8×26
d. 20×19 e. 40×29
f. 25×39 g. 15×47

What Is Your Answer?

4. Compare the methods in Activities 2 and 3.

5. **IN YOUR OWN WORDS** How do you use mental math to multiply two numbers? Give examples to support your explanation.

Practice

Use what you learned about the Distributive Property to complete Exercises 5–8 on page 137.

Section 3.4   The Distributive Property

3.4 Lesson

Key Vocabulary
like terms, p. 136

Key Idea

Distributive Property

Words To multiply a sum or difference by a number, multiply each number in the sum or difference by the number outside the parentheses. Then evaluate.

Numbers $3(7 + 2) = 3 \times 7 + 3 \times 2$ **Algebra** $a(b + c) = ab + ac$

$3(7 - 2) = 3 \times 7 - 3 \times 2$ $a(b - c) = ab - ac$

EXAMPLE 1 · Using Mental Math

Use the Distributive Property and mental math to find 8×53.

$8 \times 53 = 8(50 + 3)$ Write 53 as $50 + 3$.

$= 8(50) + 8(3)$ Distributive Property

$= 400 + 24$ Multiply.

$= 424$ Add.

EXAMPLE 2 · Using the Distributive Property

Use the Distributive Property to find $\frac{1}{2} \times 2\frac{3}{4}$.

$\frac{1}{2} \times 2\frac{3}{4} = \frac{1}{2} \times \left(2 + \frac{3}{4}\right)$ Rewrite $2\frac{3}{4}$ as the sum $2 + \frac{3}{4}$.

$= \left(\frac{1}{2} \times 2\right) + \left(\frac{1}{2} \times \frac{3}{4}\right)$ Distributive Property

$= 1 + \frac{3}{8}$ Multiply.

$= 1\frac{3}{8}$ Add.

On Your Own

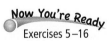
Exercises 5–16

Use the Distributive Property to find the product.

1. 5×41
2. 9×19
3. $6(37)$
4. $\frac{2}{3} \times 1\frac{1}{2}$
5. $\frac{1}{4} \times 4\frac{1}{5}$
6. $\frac{2}{7} \times 3\frac{3}{4}$

EXAMPLE 3 Simplifying Algebraic Expressions

Use the Distributive Property to simplify the expression.

a. $4(n + 5)$

$4(n + 5) = 4(n) + 4(5)$ Distributive Property

$ = 4n + 20$ Multiply.

b. $12(2y - 3)$

$12(2y - 3) = 12(2y) - 12(3)$ Distributive Property

$ = 24y - 36$ Multiply.

c. $9(6 + x + 2)$

$9(6 + x + 2) = 9(6) + 9(x) + 9(2)$ Distributive Property

$ = 54 + 9x + 18$ Multiply.

$ = 9x + 54 + 18$ Commutative Property of Addition

$ = 9x + 72$ Add 54 and 18.

Study Tip

You can use the Distributive Property when there are more than two terms in the sum or difference.

On Your Own

Use the Distributive Property to simplify the expression.

7. $7(a + 2)$ **8.** $3(d - 11)$ **9.** $7(2 + 6 - 4d)$

Now You're Ready
Exercises 17–32

EXAMPLE 4 Real-Life Application

José is x years old. His brother, Felipe, is 2 years older than José. Their aunt, Maria, is three times as old as Felipe. Write and simplify an expression that represents Maria's age in years.

Name	Description	Expression
José	He is x years old.	x
Felipe	He is 2 years *older* than José. So, *add* 2 to x.	$x + 2$
Maria	She is three *times* as old as Felipe. So, *multiply* 3 and $(x + 2)$.	$3(x + 2)$

$3(x + 2) = 3(x) + 3(2)$ Distributive Property

$ = 3x + 6$ Multiply.

∴ Maria's age in years is represented by the expression $3x + 6$.

On Your Own

10. Alexis is x years old. Her sister, Gloria, is 7 years older than Alexis. Their grandfather is five times as old as Gloria. Write and simplify an expression that represents their grandfather's age in years.

In an algebraic expression, **like terms** are terms that have the same variables raised to the same exponents. Constant terms are also like terms.

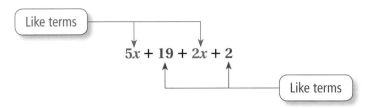

Use the Distributive Property to *combine* like terms.

EXAMPLE 5 Combining Like Terms

Simplify each expression.

a. $3x + 9 + 2x - 5$

$3x + 9 + 2x - 5 = 3x + 2x + 9 - 5$	Commutative Property of Addition
$= (3 + 2)x + 9 - 5$	Distributive Property
$= 5x + 4$	Simplify.

b. $y + y + y$

$y + y + y = 1y + 1y + 1y$	Multiplication Property of One
$= (1 + 1 + 1)y$	Distributive Property
$= 3y$	Add coefficients.

c. $7z + 2(z - 5y)$

$7z + 2(z - 5y) = 7z + 2(z) - 2(5y)$	Distributive Property
$= 7z + 2z - 10y$	Multiply.
$= (7 + 2)z - 10y$	Distributive Property
$= 9z - 10y$	Add coefficients.

On Your Own

Exercises 39–53

Simplify the expression.

11. $8 + 3z - z$ **12.** $3(b + 5) + b + 2$

3.4 Exercises

Vocabulary and Concept Check

1. **WRITING** One meaning of the word *distribute* is "to give something to each member of a group." How can this help you remember the Distributive Property?

2. **OPEN-ENDED** Write an algebraic expression in which you use the Distributive Property and then the Associative Property of Addition to simplify.

3. **WHICH ONE DOESN'T BELONG?** Which expression does *not* belong with the other three? Explain your reasoning.

   $2(x + 2)$ $5(x - 8)$ $4 + (x \cdot 4)$ $8(9 - x)$

4. Identify the like terms in the expression $8x + 1 + 7x + 4$.

Practice and Problem Solving

Use the Distributive Property and mental math to find the product.

 5. 3×21 6. 9×76 7. $12(43)$ 8. $5(88)$

9. 18×52 10. 8×27 11. $8(63)$ 12. $7(28)$

Use the Distributive Property to find the product.

13. $\dfrac{1}{4} \times 2\dfrac{2}{7}$ 14. $\dfrac{5}{6} \times 2\dfrac{2}{5}$ 15. $\dfrac{5}{9} \times 4\dfrac{1}{2}$ 16. $\dfrac{2}{15} \times 5\dfrac{5}{8}$

Use the Distributive Property to simplify the expression.

 17. $3(x + 4)$ 18. $10(b - 6)$ 19. $6(s - 9)$ 20. $7(8 + y)$

21. $8(12 + a)$ 22. $9(2n + 1)$ 23. $12(6 - k)$ 24. $18(5 - 3w)$

25. $9(3 + c + 4)$ 26. $7(8 + x + 2)$ 27. $8(5g + 5 - 2)$ 28. $6(10 + z + 3)$

29. $4(x + y)$ 30. $25(x - y)$ 31. $7(p + q + 9)$ 32. $13(n + 4 + 7m)$

33. **ERROR ANALYSIS** Describe and correct the error in rewriting the expression.

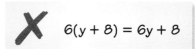

34. **ART MUSEUM** A class of 30 students visits an art museum and a special exhibit while there.

    a. Use the Distributive Property to write and simplify an expression for the cost.

    b. Estimate a reasonable value for x. Explain.

    c. Use your estimate for x to evaluate the original expression and the simplified expression in part (a). Are the values the same?

35. FITNESS Each day, you run on a treadmill for r minutes and lift weights for 15 minutes. Which expressions can you use to find how many minutes of exercise you do in 5 days? Explain your reasoning.

$5(r + 15)$ $5r + 5 \cdot 15$ $5r + 15$ $r(5 + 15)$

36. SPEED A cheetah can run 103 feet per second. A zebra can run x feet per second. Use the Distributive Property to write and simplify an expression for how much farther the cheetah can run in 10 seconds.

UNIFORMS Your baseball team has 16 players. Use the Distributive Property to write and simplify an expression for the total cost of buying the items shown for all the players.

37.

Pants: $10   Belt: $x

38.

Jersey: $12   Socks: $4   Hat: $x

5 **Simplify the expression.**

39. $6(x + 4) + 1$ **40.** $5 + 8(3 + x)$ **41.** $7(8 + 4k) + 12$

42. $x + 3 + 5x$ **43.** $7y + 6 - 1 + 12y$ **44.** $w + w + 5w$

45. $4d + 9 - d - 8$ **46.** $n + 3(n - 1)$ **47.** $2v + 8v - 5v$

48. $5(z + 4) + 5(2 - z)$ **49.** $2.7(w - 5.2)$ **50.** $\frac{2}{3}y + \frac{1}{6}y + y$

51. $\frac{3}{4}\left(z + \frac{2}{5}\right) + 2z$ **52.** $7(x + y) - 7x$ **53.** $4x + 9y + 3(x + y)$

54. ERROR ANALYSIS Describe and correct the error in simplifying the expression.

$$8x - 2x + 5x = 8x - 7x$$
$$= (8 - 7)x$$
$$= x$$

ALGEBRA Find the value of x that makes the expressions equivalent.

55. $4(x - 5); 32 - 20$ **56.** $2(x + 9); 30 + 18$ **57.** $7(8 - x); 56 - 21$

58. REASONING Simplify the expressions and compare. What do you notice? Explain.

$4(x + 6)$ $(x + 6) + (x + 6) + (x + 6) + (x + 6)$

GEOMETRY Write and simplify expressions for the area and perimeter of the rectangle.

59.  **60.** 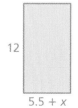 **61.**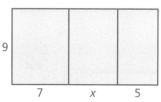

62. FUNDRAISER An art club sells 42 large candles and 56 small candles.

a. Use the Distributive Property to write and simplify an expression for the profit.

b. A large candle costs $5, and a small candle costs $3. What is the club's profit?

63. REASONING Evaluate each expression by (1) using the Distributive Property and (2) evaluating inside the parentheses first. Which method do you prefer? Is your preference the same for both expressions? Explain your reasoning.

a. $2(3.22 - 0.12)$ b. $12\left(\dfrac{1}{2} + \dfrac{2}{3}\right)$

64. REASONING Write and simplify an expression for the difference between the perimeters of the rectangle and the hexagon. Interpret your answer.

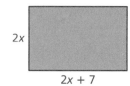 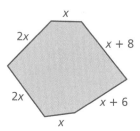

65. Puzzle Add one set of parentheses to the expression

$7 \cdot x + 3 + 8 \cdot x + 3 \cdot x + 8 - 9$ so that it is equivalent to $2(9x + 10)$.

Fair Game Review What you learned in previous grades & lessons

Evaluate the expression. *(Section 2.4, Section 2.5, and Section 2.6)*

66. $4.871 + 7.4 - 1.63$ **67.** $25.06 - 0.049 + 8.995$

68. $15.3 \cdot 9.1 - 4.017$ **69.** $29.24 \div 3.4 \cdot 0.045$

70. MULTIPLE CHOICE What is the GCF of 48, 80, and 96? *(Section 1.5)*

Ⓐ 12   Ⓑ 16

Ⓒ 24   Ⓓ 480

Extension 3.4 Factoring Expressions

Key Vocabulary
factoring an expression, *p. 140*

Key Idea

Factoring an Expression

Words Writing a numerical expression or algebraic expression as a product of factors is called **factoring the expression**. You can use the Distributive Property to factor expressions.

Numbers $3 \cdot 7 + 3 \cdot 2 = 3(7 + 2)$ **Algebra** $ab + ac = a(b + c)$

$3 \cdot 7 - 3 \cdot 2 = 3(7 - 2)$ $ab - ac = a(b - c)$

EXAMPLE 1 Factoring a Numerical Expression

Factor 20 − 12 using the GCF.

Find the GCF of 20 and 12 by listing their factors.

Factors of 20: ①, ②, ④, 5, 10, 20

Factors of 12: ①, ②, 3, ④, 6, 12

Circle the common factors.

The GCF of 20 and 12 is 4.

Study Tip

When you factor an expression, you can *factor out* any common factor.

Write each term of the expression as a product of the GCF and the remaining factor. Then use the Distributive Property to factor the expression.

$20 - 12 = 4(5) - 4(3)$ Rewrite using GCF.

$= 4(5 - 3)$ Distributive Property

EXAMPLE 2 Identifying Equivalent Expressions

Which expression is not equivalent to $16x + 24$?

Ⓐ $2(8x + 12)$ Ⓑ $4(4x + 6)$ Ⓒ $6(3x + 4)$ Ⓓ $(2x + 3)8$

Each choice is a product of two factors in which one is a whole number and the other is the sum of two terms. For an expression to be equivalent to $16x + 24$, its whole number factor must be a common factor of 16 and 24.

Equivalent Expressions

In this extension, you will
• use the Distributive Property to produce equivalent expressions.

Factors of 16: ①, ②, ④, ⑧, 16

Factors of 24: ①, ②, 3, ④, 6, ⑧, 12, 24

Circle the common factors.

The common factors of 16 and 24 are 1, 2, 4, and 8. Because 6 is not a common factor of 16 and 24, Choice C cannot be equivalent to $16x + 24$.

Check: $6(3x + 4) = 6(3x) + 6(4) = 18x + 24 \neq 16x + 24$ ✗

∴ So, the correct answer is Ⓒ.

EXAMPLE 3 Factoring an Algebraic Expression

You receive a discount on each book you buy for your electronic reader. The original price of each book is x dollars. You buy 5 books for a total of $(5x - 15)$ dollars. Factor the expression. What can you conclude about the discount?

Find the GCF of $5x$ and 15 by writing their prime factorizations.

$5x = \boxed{5} \cdot x$
$15 = \boxed{5} \cdot 3$ Circle the common prime factor.

So, the GCF of $5x$ and 15 is 5. Use the GCF to factor the expression.

$5x - 15 = 5(x) - 5(3)$ Rewrite using GCF.
$ = 5(x - 3)$ Distributive Property

The factor 5 represents the number of books purchased. The factor $(x - 3)$ represents the price of each book. This factor is a difference of two terms, showing that the price x of each book is decreased by $3.

∴ So, the factored expression shows a $3 discount for every book you buy. The original expression shows a total savings of $15.

Practice

Factor the expression using the GCF.

1. $7 + 14$
2. $44 - 11$
3. $18 - 12$
4. $70 + 95$
5. $60 - 36$
6. $100 - 80$
7. $84 + 28$
8. $48 + 80$
9. $2x + 10$
10. $15x + 6$
11. $26x - 13$
12. $50x - 60$
13. $36x + 9$
14. $14x - 98$
15. $10x - 25y$
16. $24y + 88x$

17. **REASONING** The whole numbers a and b are divisible by c. Is $a + b$ divisible by c? Is $b - a$ divisible by c? Explain your reasoning.

18. **OPEN-ENDED** Write five expressions that are equivalent to $8x + 16$.

19. **GEOMETRY** The area of the parallelogram is $(4x + 16)$ square feet. Write an expression for the base.

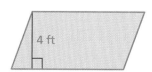

4 ft

20. **STRUCTURE** You buy 37 concert tickets for $8 each, and then sell all 37 tickets for $11 each. The work below shows two ways you can determine your profit. Describe each solution method. Which do you prefer? Explain your reasoning.

profit = 37(11) − (37)8
       = 407 − 296
       = $111

profit = 37(11) − (37)8
       = 37(11 − 8)
       = 37(3)
       = $111

Extension 3.4 Factoring Expressions

3.3–3.4 Quiz

Tell which property the statement illustrates. *(Section 3.3)*

1. $3.5 \cdot z = z \cdot 3.5$
2. $14 + (35 + w) = (14 + 35) + w$

Simplify the expression. Explain each step. *(Section 3.3)*

3. $3.2 + (b + 5.7)$
4. $6 \cdot (10 \cdot k)$

Use the Distributive Property and mental math to find the product. *(Section 3.4)*

5. 6×49
6. 7×86

Use the Distributive Property to simplify the expression. *(Section 3.4)*

7. $5(x - 8)$
8. $7(y + 3)$

Simplify the expression. *(Section 3.4)*

9. $6q + 2 + 3q + 5$
10. $4r + 3(r - 2)$

Factor the expression using the GCF. *(Section 3.4)*

11. $12 + 21$
12. $16x - 36$

13. **GEOMETRY** The expression $18 + 7 + (18 + 2x) + 7$ represents the perimeter of the trapezoid. Simplify the expression. *(Section 3.3)*

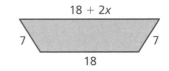

14. **MOVIES** You and four of your friends go to a movie and each buy popcorn. *(Section 3.4)*

    a. Use the Distributive Property to write an expression for the total cost to buy movie tickets and popcorn. Simplify the expression.

    b. Choose a reasonable value for x. Evaluate the expression.

15. **GEOMETRY** The length of a rectangle is 16 inches, and its area is $(32x + 48)$ square inches. Factor the expression for the area. Write an expression for the width. *(Section 3.4)*

3 Chapter Review

Review Key Vocabulary

algebraic expression, *p. 112*
terms, *p. 112*
variable, *p. 112*
coefficient, *p. 112*
constant, *p. 112*
equivalent expressions, *p. 128*
like terms, *p. 136*
factoring an expression, *p. 140*

Review Examples and Exercises

3.1 Algebraic Expressions (pp. 110–117)

a. Evaluate $a \div b$ when $a = 48$ and $b = 8$.

$a \div b = 48 \div 8$ Substitute 48 for a and 8 for b.
$ = 6$ Divide 48 by 8.

b. Evaluate $y^2 - 14$ when $y = 5$.

$y^2 - 14 = 5^2 - 14$ Substitute 5 for y.
$ = 25 - 14$ Using order of operations, evaluate 5^2.
$ = 11$ Subtract 14 from 25.

Exercises

Evaluate the expression when $x = 20$ and $y = 4$.

1. $x \div 5$
2. $y + x$
3. $8y - x$

4. **GAMING** In a video game, you score p game points and b triple bonus points. An expression for your score is $p + 3b$. What is your score when you earn 245 game points and 20 triple bonus points?

3.2 Writing Expressions (pp. 118–123)

Write the phrase as an expression.

a. a number z decreased by 18

$z - 18$ The phrase *decreased by* means *subtraction*.

b. the sum of 7 and the product of a number x and 12

$7 + 12x$ The phrase *sum of* means *addition*.
                The phrase *product of* means *multiplication*.

Exercises

Write the phrase as an expression.

5. 11 fewer than a number b

6. the product of a number d and 32

7. 18 added to a number n

8. a number t decreased by 17

9. BASKETBALL Your basketball team scored 4 fewer than twice as many points as the other team.

   a. Write an expression for the number of points your team scored.

   b. The other team scored 24 points. How many points did your team score?

3.3 Properties of Addition and Multiplication *(pp. 126–131)*

a. Simplify the expression $(x + 18) + 4$.

$(x + 18) + 4 = x + (18 + 4)$ Associative Property of Addition

$\qquad\qquad\qquad = x + 22$ Add 18 and 4.

b. Simplify the expression $(5.2 + a) + 0$.

$(5.2 + a) + 0 = 5.2 + (a + 0)$ Associative Property of Addition

$\qquad\qquad\qquad = 5.2 + a$ Addition Property of Zero

c. Simplify the expression $36 \cdot r \cdot 1$.

$36 \cdot r \cdot 1 = 36 \cdot (r \cdot 1)$ Associative Property of Multiplication

$\qquad\qquad = 36 \cdot r$ Multiplication Property of One

$\qquad\qquad = 36r$

Exercises

Simplify the expression. Explain each step.

10. $10 + (2 + y)$

11. $(21 + b) + 1$

12. $3(7x)$

13. $1(3.2w)$

14. $5.3 + (w + 1.2)$

15. $(0 + t) + 9$

16. GEOMETRY The expression $7 + 3x + 4$ represents the perimeter of the triangle. Simplify the expression.

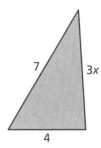

3.4 The Distributive Property (pp. 132–141)

a. Use the Distributive Property to simplify $3(n + 9)$.

$3(n + 9) = 3(n) + 3(9)$ Distributive Property

$ = 3n + 27$ Multiply.

b. Simplify $5x + 7 + 3x - 2$.

$5x + 7 + 3x - 2 = 5x + 3x + 7 - 2$ Commutative Property of Addition

$ = (5 + 3)x + 7 - 2$ Distributive Property

$ = 8x + 5$ Simplify.

c. Factor $14x - 49$ using the GCF.

Find the GCF of $14x$ and 49 by writing their prime factorizations.

$14x = 2 \cdot 7 \cdot x$ Circle the common prime factor.

$49 = 7 \cdot 7$

So, the GCF of $14x$ and 49 is 7. Use the GCF to factor the expression.

$14x - 49 = 7(2x) - 7(7)$ Rewrite using GCF.

$ = 7(2x - 7)$ Distributive Property

Exercises

Use the Distributive Property to find the product.

17. $\dfrac{3}{4} \times 2\dfrac{1}{3}$

18. $\dfrac{4}{7} \times 4\dfrac{5}{8}$

19. $\dfrac{1}{5} \times 5\dfrac{10}{11}$

Use the Distributive Property to simplify the expression.

20. $2(x + 12)$

21. $11(b - 3)$

22. $8(s - 1)$

23. $6(6 + y)$

24. $25(z - 4)$

25. $35(w - 2)$

26. **HAIRCUT** A family of four goes to a salon for haircuts. The cost of each haircut is $13. Use the Distributive Property and mental math to find the product 4×13 for the total cost.

Simplify the expression.

27. $5(n + 3) + 4n$

28. $t + 2 + 6t$

29. $3z + 4 + 5z - 9$

Factor the expression using the GCF.

30. $15 + 35$

31. $36x - 28$

32. $16x + 56y$

5. What is the value of 4.391 + 5.954?

   F. 9.12145

   G. 9.245

   H. 9.345

   I. 10.345

6. Which number pair has a greatest common factor of 6?

   A. 18, 54

   B. 30, 42

   C. 30, 60

   D. 36, 60

7. Properties of Addition and Multiplication are used to simplify an expression.

$$36 \cdot 23 + 33 \cdot 64 = 36 \cdot 23 + 64 \cdot 33$$
$$= 36 \cdot 23 + 64 \cdot (23 + 10)$$
$$= 36 \cdot 23 + 64 \cdot 23 + 64 \cdot 10$$
$$= x \cdot 23 + 64 \cdot 10$$
$$= 2300 + 640$$
$$= 2940$$

   What number belongs in place of the x?

8. Which property was used to simplify the expression?

$$(47 \times 125) \times 8 = 47 \times (125 \times 8)$$
$$= 47 \times 1000$$
$$= 47,000$$

   F. Distributive Property

   G. Multiplication Property of One

   H. Associative Property of Multiplication

   I. Commutative Property of Multiplication

9. What is the value of the expression below when $a = 5$, $b = 7$, and $c = 6$?

   $$9b - 4a + 2c$$

   A. 29

   B. 31

   C. 55

   D. 78

10. Which equation correctly demonstrates the Distributive Property?

   F. $a(b + c) = ab + c$

   G. $a(b + c) = ab + ac$

   H. $a + (b + c) = (a + b) + (a + c)$

   I. $a + (b + c) = (a + b) \cdot (a + c)$

11. Which expression is equivalent to $3\frac{3}{5} \div 6\frac{1}{2}$?

   A. $\frac{5}{18} \times \frac{13}{2}$

   B. $\frac{18}{5} \times \frac{2}{13}$

   C. $\frac{9}{5} \div \frac{6}{2}$

   D. $\frac{18}{5} \div \frac{2}{13}$

12. Which number pair does *not* have a least common multiple of 24?

   F. 2, 12

   G. 3, 8

   H. 6, 8

   I. 12, 24

13. Use the Properties of Multiplication to simplify the expression in an efficient way. Show your work and explain how you used the Properties of Multiplication.

   $(25 \times 18) \times 4$

14. You evaluated an expression using $x = 6$ and $y = 9$. You correctly got an answer of 105. Which expression did you evaluate?

   A. $3x + 6y$

   B. $5x + 10y$

   C. $6x + 9y$

   D. $10x + 5y$

15. Which number is equivalent to the expression below?

   $2 \times 12 - 8 \div 2^2$

   F. 2

   G. 4

   H. 8

   I. 22

4 Areas of Polygons

4.1 Areas of Parallelograms

4.2 Areas of Triangles

4.3 Areas of Trapezoids

4.4 Polygons in the Coordinate Plane

"Remember, Descartes, you don't have to measure area in standard units like square inches or square centimeters."

"You can also use nonstandard units ... like the length of your paw squared."

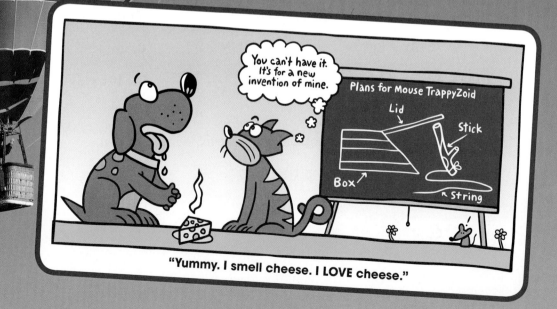

"Yummy. I smell cheese. I LOVE cheese."

What You Learned Before

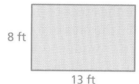

"To find the area of Fuzzy's new poster, would you use 'base times height' or 'length times width'?"

● Finding Areas of Squares and Rectangles

Example 1 Find the area of the square or rectangle.

a.
15 cm
15 cm

b.
8 ft
13 ft

$A = s^2$ Write formula. $A = \ell w$

$= 15^2$ Substitute. $= 13(8)$

$= 225$ Simplify. $= 104$

∴ The area of the square is 225 square centimeters.

∴ The area of the rectangle is 104 square feet.

Try It Yourself
Find the area of the square or rectangle.

1.
7 m
7 m

2.
9 yd
20 yd

3. 65 mm
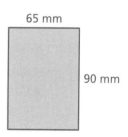
90 mm

● Plotting Ordered Pairs

Example 2 Plot (2, 3) in a coordinate plane.

Start at the origin. Move 2 units right and 3 units up. Then plot the point.

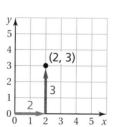

Try It Yourself
Plot the ordered pair in a coordinate plane.

4. (1, 4)   5. (3, 2)   6. (5, 1)

4.1 Areas of Parallelograms

Essential Question How can you derive a formula for the area of a parallelogram?

A **polygon** is a closed figure in a plane that is made up of three or more line segments that intersect only at their endpoints. Several examples of polygons are parallelograms, triangles, and trapezoids.

The formulas for the areas of polygons can be derived from one area formula, the area of a rectangle. Recall that the area of a rectangle is the product of its length ℓ and its width w. The process you use to derive these other formulas is called *deductive reasoning*.

Area = ℓw

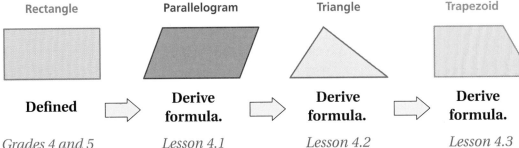

1 ACTIVITY: Deriving the Area Formula of a Parallelogram

Work with a partner.

a. Draw *any* rectangle on a piece of grid paper. An example is shown below. Label the length and width. Then find the area of your rectangle.

Geometry

In this lesson, you will
- find areas of parallelograms.
- solve real-life problems.

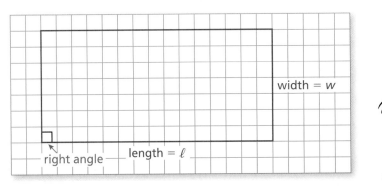

b. Cut your rectangle into two pieces to form a parallelogram. Compare the area of the rectangle with the area of the parallelogram. What do you notice? Use your results to write a formula for the area A of a parallelogram.

$A = $ _____   Formula

152   Chapter 4   Areas of Polygons

2 ACTIVITY: Finding Areas of Parallelograms

Work with a partner.

Math Practice

Use Assumptions
How are rectangles and parallelograms similar? How can you use this information to solve the problem?

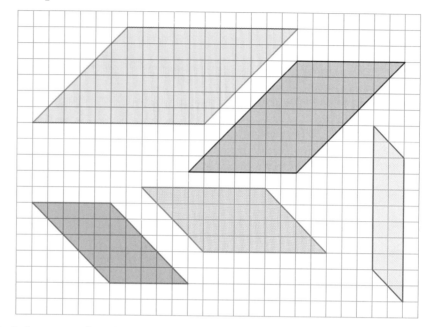

a. Find the area of each parallelogram by cutting it into two pieces to form a rectangle.
b. Use the formula you wrote in Activity 1 to find the area of each parallelogram. Compare your answers to those in part (a).
c. Count unit squares for each parallelogram to check your results.

What Is Your Answer?

3. **IN YOUR OWN WORDS** How can you derive a formula for the area of a parallelogram?

4. **REASONING** The areas of a rectangle and a parallelogram are equal. The length of a rectangle is equal to the base of the parallelogram. What can you say about the width of the rectangle and the height of the parallelogram? Draw a diagram to support your answer.

5. What is the height of the parallelogram shown? How do you know?

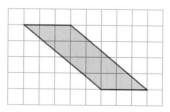

Practice

Use what you learned about the areas of parallelograms to complete Exercises 3–5 on page 156.

Section 4.1 Areas of Parallelograms 153

4.1 Lesson

Key Vocabulary
polygon, *p. 152*

The *area* of a polygon is the amount of surface it covers. You can find the area of a parallelogram in much the same way as you can find the area of a rectangle.

🔑 Key Idea

Area of a Parallelogram

Words The area A of a parallelogram is the product of its base b and its height h.

Algebra $A = bh$

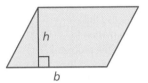

EXAMPLE 1 Finding Areas of Parallelograms

Find the area of each parallelogram.

a.
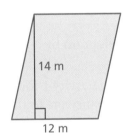

b.
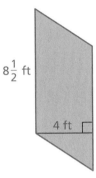

Remember
Area is measured in square units.

$A = bh$	Write formula.	$A = bh$	
$= 12(14)$	Substitute values.	$= 8\dfrac{1}{2}(4)$	
$= 168$	Multiply.	$= 34$	

∴ The area of the parallelogram is 168 square meters.

∴ The area of the parallelogram is 34 square feet.

On Your Own

Now You're Ready
Exercises 3–8

Find the area of the parallelogram.

1.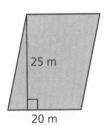

2.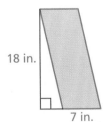

3.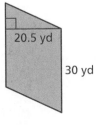

154   Chapter 4   Areas of Polygons

EXAMPLE 2 **Finding the Area of a Parallelogram on a Grid**

Find the area of the parallelogram.

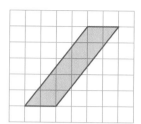

Count grid lines to find the dimensions.
The base b is 2 units, and the height h is 5 units.

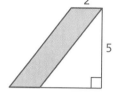

$A = bh$ Write formula.

$ = 2(5)$ Substitute values.

$ = 10$ Multiply.

The area of the parallelogram is 10 square units.

On Your Own

Exercises 11–13

4. Find the area of the parallelogram.

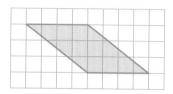

EXAMPLE 3 **Real-Life Application**

You make a photo prop for a school fair. You cut a 10-inch square out of a parallelogram-shaped piece of wood. What is the area of the photo prop?

Convert the dimensions of the piece of wood to inches.

There are 12 inches in 1 foot, so the base is 4 • 12 = 48 inches and the height is 8 • 12 = 96 inches.

Use a verbal model to solve the problem.

area of photo prop	=	area of wood	−	area of square	
	=	96(48)	−	10^2	Substitute.
	=	96(48)	−	100	Evaluate 10^2.
	=	4608	−	100	Multiply 96 and 48.
	=	4508			Subtract 100 from 4608.

The area of the photo prop is 4508 square inches.

On Your Own

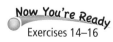
Exercises 14–16

5. Find the area of the shaded region.

6. **WHAT IF?** In Example 3, you cut a 12-inch square out of the piece of wood. What is the area of the photo prop?

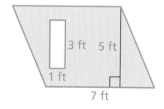

Section 4.1   Areas of Parallelograms   **155**

4.1 Exercises

Vocabulary and Concept Check

1. **WRITING** What is the area of a polygon? Explain how the perimeter and the area of the polygon are different.

2. **CHOOSE TOOLS** Construct a parallelogram that has an area of 24 square inches. Explain your method.

Practice and Problem Solving

Find the area of the parallelogram.

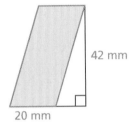

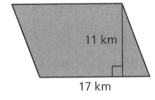

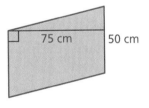

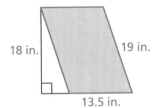

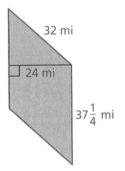

9. **ERROR ANALYSIS** Describe and correct the error in finding the area of the parallelogram.

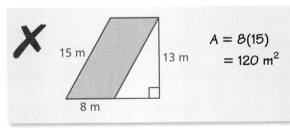

10. **CERAMIC TILE** A ceramic tile in the shape of a parallelogram has a base of 4 inches and a height of 1.5 inches. What is the area of the tile?

Find the area of the parallelogram.

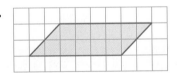

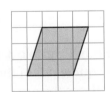

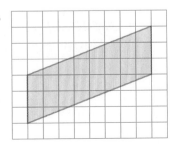

156 Chapter 4 Areas of Polygons

Find the area of the shaded region.

14.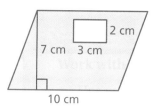

15.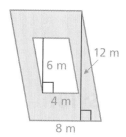

16.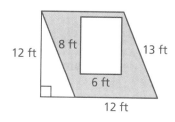

17. **DECK** Your deck has an area of 128 square feet. After adding a section, the area will be $s^2 + 128$ square feet. Draw a diagram of how this can happen.

18. **T-SHIRT DESIGN** You use the parallelogram-shaped sponge to create the T-shirt design. The area of the design is 66 square inches. How many times do you use the sponge to create the design? Draw a diagram to support your answer.

19. **STAIRCASE** The staircase has three parallelogram-shaped panels that are the same size. The horizontal distance between each panel is 4.25 inches. What is the area of one panel?

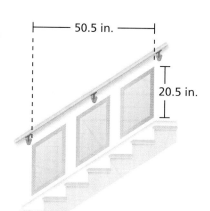

20. **REASONING** Find the missing dimensions in the table.

Parallelogram	Base	Height	Area
A	$x + 4$		$5x + 20$
B		8	$8x - 24$
C	6		$12x + 6y$

21. Each dimension of a parallelogram is multiplied by a positive number n. Write an expression for the area of the new parallelogram.

Fair Game Review What you learned in previous grades & lessons

Use mental math to multiply. *(Skills Review Handbook)*

22. $\dfrac{1}{2} \times 26$

23. 82×20

24. 16×30

25. $\dfrac{1}{2} \times 236$

26. **MULTIPLE CHOICE** Which of the following describes angle B? *(Skills Review Handbook)*

    Ⓐ acute
    Ⓑ obtuse
    Ⓒ right
    Ⓓ isosceles

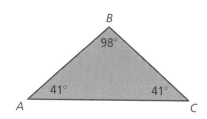

4.2 Lesson

Key Idea

Area of a Triangle

Words The area A of a triangle is one-half the product of its base b and its height h.

Algebra $A = \dfrac{1}{2}bh$

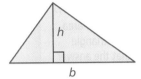

EXAMPLE 1 Finding the Area of a Triangle

Find the area of the triangle.

$A = \dfrac{1}{2}bh$ Write formula.

$= \dfrac{1}{2}(5)(8)$ Substitute 5 for b and 8 for h.

$= \dfrac{1}{2}(40)$ Multiply 5 and 8.

$= 20$ Multiply $\dfrac{1}{2}$ and 40.

Remember
In Example 1, use the Associative Property of Multiplication to multiply 5 and 8 first.

∴ The area of the triangle is 20 square inches.

Reasonable? Draw the triangle on grid paper and count unit squares. Each square in the grid represents 1 square inch.

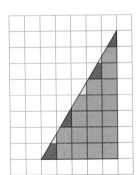

Squares full or nearly full: 18

Squares about half full: 4

The area is $18(1) + 4\left(\dfrac{1}{2}\right) = 20$ square inches.

So, the answer is reasonable. ✓

On Your Own

Now You're Ready
Exercises 3–8

Find the area of the triangle.

1. 4 ft, 11 ft

2. 10 m, 22 m

160   Chapter 4   Areas of Polygons

EXAMPLE 2 Finding the Area of a Triangle

Find the area of the triangle.

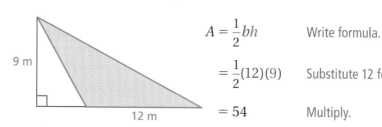

$A = \dfrac{1}{2}bh$ Write formula.

$= \dfrac{1}{2}(12)(9)$ Substitute 12 for *b* and 9 for *h*.

$= 54$ Multiply.

∴ The area of the triangle is 54 square meters.

EXAMPLE 3 Real-Life Application

The base and height of the red butterfly wing are two times greater than the base and height of the blue butterfly wing. How many times greater is the area of the red wing than the area of the blue wing?

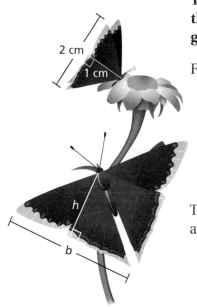

Find the area of the blue wing.

$A = \dfrac{1}{2}bh$ Write formula.

$= \dfrac{1}{2}(2)(1)$ Substitute 2 for *b* and 1 for *h*.

$= 1 \text{ cm}^2$ Multiply.

The red wing dimensions are 2 times greater, so the base is 2 × 2 = 4 cm and the height is 2 × 1 = 2 cm. Find the area of the red wing.

$A = \dfrac{1}{2}bh$ Write formula.

$= \dfrac{1}{2}(4)(2)$ Substitute 4 for *b* and 2 for *h*.

$= 4 \text{ cm}^2$ Multiply.

∴ Because $\dfrac{4 \text{ cm}^2}{1 \text{ cm}^2} = 4$, the area of the red wing is 4 times greater.

On Your Own

Exercises 12–14

3. Find the area of the triangle.

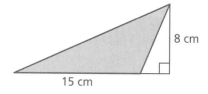

4. **WHAT IF?** In Example 3, the base and the height of the red butterfly wing are three times greater than those of the blue wing. How many times greater is the area of the red wing?

4.2 Exercises

Vocabulary and Concept Check

1. **CRITICAL THINKING** Can *any* side of a triangle be labeled as its base? Explain.

2. **DIFFERENT WORDS, SAME QUESTION** Which is different? Find "both" answers.

What is the area of the triangle?	What is the distance around the triangle?
How many unit squares fit in the triangle?	What is one-half the product of the base and the height?

Practice and Problem Solving

Find the area of the triangle.

3.

4 cm, 3 cm

4. 5 ft, 16 ft

5. 54 in., 60 in.

6. 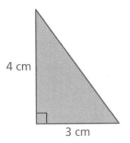 22 yd, 14 yd

7.  30 cm, 75 cm

8. 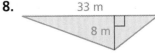 33 m, 8 m

9. **ERROR ANALYSIS** Describe and correct the error in finding the area of the triangle.

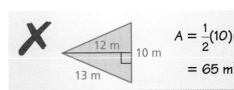

$A = \dfrac{1}{2}(10)(13)$

$= 65 \text{ m}^2$

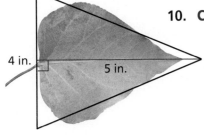

10. **COTTONWOOD LEAF** Estimate the area of the cottonwood leaf.

11. **CORNER SHELF** A shelf has the shape of a triangle. The base of the shelf is 36 centimeters, and the height is 18 centimeters. Find the area of the shelf.

162   Chapter 4   Areas of Polygons

Find the area of the triangle.

12.
13.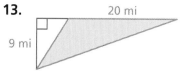
14.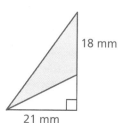

15. **OPEN-ENDED** Draw and label two triangles that each have an area of 24 square feet.

16. **HANG GLIDING** The wingspan of the triangular hang glider is 30 feet.

    a. How much fabric is needed to make the sail?
    b. **RESEARCH** Use the Internet or some other source to find how the area of the sail is related to the weight limit of the pilot.

17. **SAILBOATS** The base and the height of Sail B are x times greater than the base and the height of Sail A. How many times greater is the area of Sail B? Write your answer as a power.

18. **WRITING** You know the height and the perimeter of an equilateral triangle. Explain how to find the area of the triangle. Draw a diagram to support your reasoning.

19. **REASONING** The base and the height of Triangle A are half the base and the height of Triangle B. How many times greater is the area of Triangle B?

20. **Critical Thinking** The total area of the polygon is 176 square feet. Find the value of x.

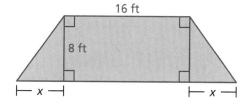

Fair Game Review *What you learned in previous grades & lessons*

Tell which property is illustrated by the statement. *(Section 3.3)*

21. $n \cdot 1 = n$
22. $4 \cdot m = m \cdot 4$
23. $(x + 2) + 5 = x + (2 + 5)$

24. **MULTIPLE CHOICE** What is the first step when using order of operations? *(Section 1.3)*

    Ⓐ Multiply or divide from left to right.
    Ⓑ Add or subtract from left to right.
    Ⓒ Perform operations in parentheses.
    Ⓓ Evaluate numbers with exponents.

4 Study Help

You can use a **four square** to organize information about a topic. Each of the four squares can be a category, such as *definition, vocabulary, example, non-example, words, algebra, table, numbers, visual, graph,* or *equation*. Here is an example of a four square for the area of a parallelogram.

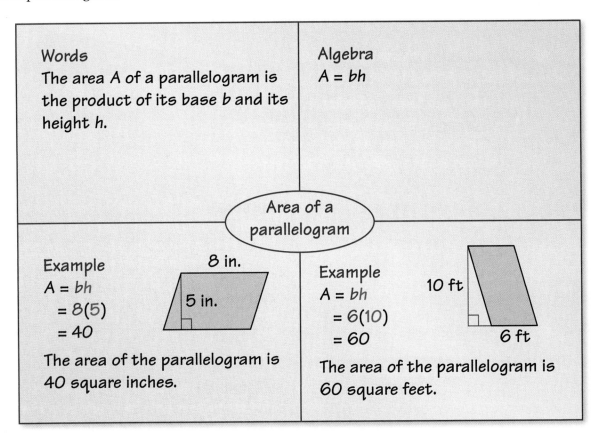

On Your Own

Make a four square to help you study the topic.

1. area of a triangle

After you complete this chapter, make four squares for the following topics.

2. area of a trapezoid
3. area of a composite figure
4. drawing a polygon in a coordinate plane
5. finding distances in the first quadrant

"Sorry, but I have limited space in my four square. I needed pet names with only three letters."

4.1–4.2 Quiz

Find the area of the parallelogram. *(Section 4.1)*

1.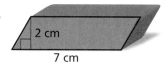

2.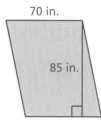

3.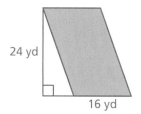

4.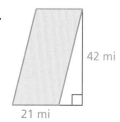

Find the area of the triangle. *(Section 4.2)*

5.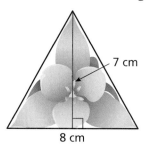

6.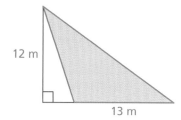

7. **LAND** A wildlife conservation group buys a plot of land. How much land does it buy? *(Section 4.2)*

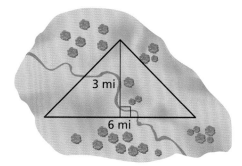

8. **FRAMING** A sheet of plywood is 4 feet wide by 8 feet long. What is the minimum number of sheets of plywood needed to cover the frame? Justify your answer. *(Section 4.2)*

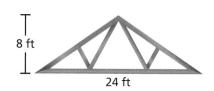

4.3 Areas of Trapezoids

Essential Question How can you derive a formula for the area of a trapezoid?

1 ACTIVITY: Deriving the Area Formula of a Trapezoid

Work with a partner. Use a piece of centimeter grid paper.

a. Draw *any* trapezoid so that its base lies on one of the horizontal lines of the paper.

b. Estimate the area of your trapezoid (in square centimeters) by counting unit squares.

   Area ≈ _____   Estimate

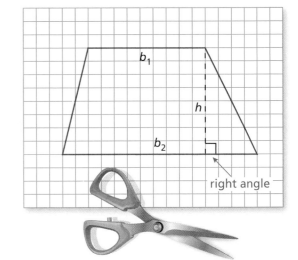

c. Label the height and the bases *inside* the trapezoid.

d. Cut out the trapezoid. Mark the midpoint of the side opposite the height. Draw a line from the midpoint to the opposite upper vertex.

e. Cut along the line. You will end up with a triangle and a quadrilateral. Arrange these two figures to form a figure whose area you know.

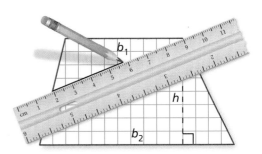

Geometry

In this lesson, you will
- find areas of trapezoids.
- solve real-life problems.

f. Use your result to write a *formula* for the area of a trapezoid.

   Area = _____   Formula

g. Use your formula to find the area of your trapezoid (in square centimeters).

   Area = _____   Exact Area

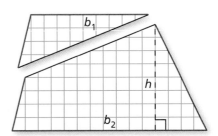

h. Compare this area with your estimate in part (b).

166   Chapter 4   Areas of Polygons

2 ACTIVITY: Writing a Math Lesson

Work with a partner. Use your results from Activity 1 to write a lesson on finding the area of a trapezoid.

Math Practice

Use Clear Definitions

Do your steps for the *Key Idea* help another person understand how to solve the problem? Do the examples follow your steps?

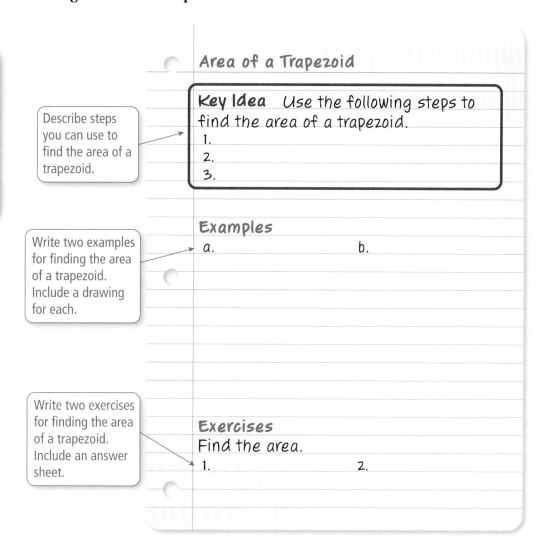

Describe steps you can use to find the area of a trapezoid.

Write two examples for finding the area of a trapezoid. Include a drawing for each.

Write two exercises for finding the area of a trapezoid. Include an answer sheet.

What Is Your Answer?

3. **IN YOUR OWN WORDS** How can you derive a formula for the area of a trapezoid?

4. In this chapter, you used deductive reasoning to derive new area formulas from area formulas you have already learned. Describe a real-life career in which deductive reasoning is important.

Practice ➤ Use what you learned about the areas of trapezoids to complete Exercises 4–6 on page 170.

4.3 Lesson

Key Idea

Area of a Trapezoid

Words The area A of a trapezoid is one-half the product of its height h and the sum of its bases b_1 and b_2.

Algebra $A = \dfrac{1}{2}h(b_1 + b_2)$

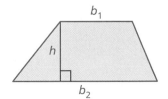

EXAMPLE 1 Finding Areas of Trapezoids

Find the area of each trapezoid.

a.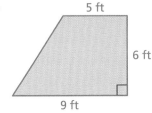

b.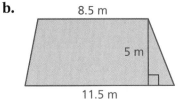

$A = \dfrac{1}{2}h(b_1 + b_2)$ Write formula. $A = \dfrac{1}{2}h(b_1 + b_2)$

$= \dfrac{1}{2}(6)(5 + 9)$ Substitute. $= \dfrac{1}{2}(5)(8.5 + 11.5)$

$= \dfrac{1}{2}(6)(14)$ Add. $= \dfrac{1}{2}(5)(20)$

$= 42$ Multiply. $= 50$

∴ The area of the trapezoid is 42 square feet.

∴ The area of the trapezoid is 50 square meters.

On Your Own

Now You're Ready
Exercises 7–9

Find the area of the trapezoid.

1.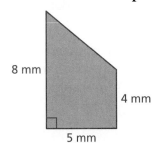

2.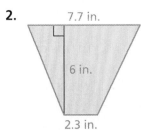

168 Chapter 4 Areas of Polygons

EXAMPLE 2 **Finding the Area of a Trapezoid on a Grid**

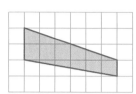

What is the area of the trapezoid?

A 6 units^2 **B** 7 units^2 **C** 9 units^2 **D** 12 units^2

Count grid lines to find the dimensions. The height h is 6 units, base b_1 is 1 unit, and base b_2 is 2 units.

$$A = \frac{1}{2}h(b_1 + b_2) \quad \text{Write formula.}$$

$$= \frac{1}{2}(6)(1+2) \quad \text{Substitute values.}$$

$$= \frac{1}{2}(6)(3) \quad \text{Add.}$$

$$= 9 \quad \text{Multiply.}$$

∴ The area of the trapezoid is 9 square units. The correct answer is **C**.

EXAMPLE 3 **Real-Life Application**

You can use a trapezoid to approximate the shape of Scott County, Virginia. The population is about 23,200. About how many people are there per square mile?

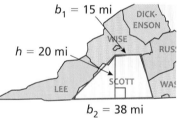

Find the area of Scott County.

$$A = \frac{1}{2}h(b_1 + b_2) \quad \text{Write formula for area of a trapezoid.}$$

$$= \frac{1}{2}(20)(15 + 38) \quad \text{Substitute 20 for } h, 15 \text{ for } b_1, \text{ and } 38 \text{ for } b_2.$$

$$= \frac{1}{2}(20)(53) = 530 \quad \text{Simplify.}$$

The area of Scott County is about 530 square miles. Divide the population by the area to find the number of people per square mile.

∴ So, there are about $\frac{23{,}200 \text{ people}}{530 \text{ mi}^2} \approx 44$ people per square mile.

On Your Own

Exercises 11–13

3. Find the area of the trapezoid.

4. **WHAT IF?** In Example 3, the population of Scott County decreases by 550. By how much does the number of people per square mile change? Explain.

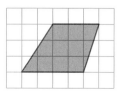

Section 4.3 Areas of Trapezoids

4.3 Exercises

Vocabulary and Concept Check

1. **VOCABULARY** Identify the bases and the height of the trapezoid.

2. **REASONING** What measures do you need to find the area of a trapezoid?

3. **WHICH ONE DOESN'T BELONG?** Which one does *not* belong with the other three? Explain your reasoning.

$\frac{1}{2}bh$ ℓw $2\ell + 2w$ $\frac{1}{2}h(b_1 + b_2)$

Practice and Problem Solving

Find the area of the trapezoid.

4. $b_1 = 4, b_2 = 8, h = 2$

5. $b_1 = 5, b_2 = 7, h = 4$

6. $b_1 = 12, b_2 = 6, h = 3$

7.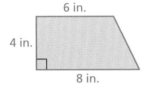

8.

9.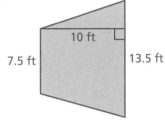

10. **ERROR ANALYSIS** Describe and correct the error in finding the area of the trapezoid.

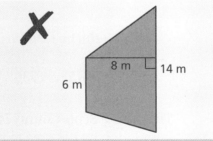

$Area = \frac{1}{2}(6 + 14)$
$= 10 \text{ m}^2$

Find the area of the trapezoid.

11.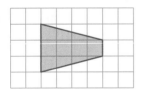

12.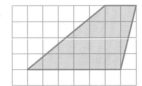

13.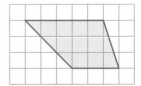

14. **LIGHT** Light shines through a window. What is the area of the trapezoid-shaped region created by the light?

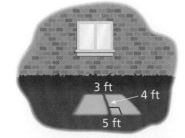

170   Chapter 4   Areas of Polygons

Find the area of a trapezoid with height h and bases b_1 and b_2.

15. $h = 6$ in.
$b_1 = 9$ in.
$b_2 = 11$ in.

16. $h = 22$ cm
$b_1 = 10.5$ cm
$b_2 = 12.5$ cm

17. $h = 12$ mi
$b_1 = 5.6$ mi
$b_2 = 7.4$ mi

18. $h = 14$ m
$b_1 = 21$ m
$b_2 = 22$ m

19. REASONING The rectangle and the trapezoid have the same area. What is the length ℓ of the rectangle?

20. OPEN-ENDED The area of the trapezoidal student election sign is 5 square feet. Find two possible values for each base length.

21. AUDIO How many times greater is the area of the floor covered by the larger speaker than by the smaller speaker?

22. Critical Thinking The triangle and the trapezoid share a 15-inch base and a height of 10 inches.

  a. The area of the trapezoid is less than twice the area of the triangle. Find the values of x. Explain your reasoning.

  b. Can the area of the *trapezoid* be exactly twice the area of the triangle? Explain your reasoning.

Fair Game Review *What you learned in previous grades & lessons*

Plot the ordered pair in a coordinate plane. *(Skills Review Handbook)*

23. $(5, 0)$ **24.** $(2, 4)$ **25.** $(0, 3)$ **26.** $(6, 1)$

27. MULTIPLE CHOICE Which expression represents "6 more than x"? *(Section 3.2)*

  Ⓐ $6 - x$ **Ⓑ** $6x$ **Ⓒ** $x + 6$ **Ⓓ** $\dfrac{6}{x}$

Section 4.3 Areas of Trapezoids

Extension 4.3 Areas of Composite Figures

Key Vocabulary
composite figure, p. 172

A **composite figure** is made up of triangles, squares, rectangles, and other two-dimensional figures. Here are two examples.

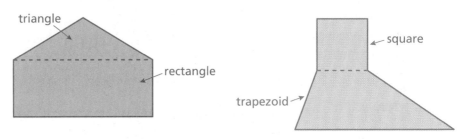

To find the area of a composite figure, separate it into figures with areas you know how to find. Then find the sum of the areas of those figures.

EXAMPLE 1 Finding the Area of a Composite Figure

Find the area of the purple figure.

You can separate the figure into a rectangle and a trapezoid. Count grid lines to find the dimensions of each figure. Then find the area of each figure.

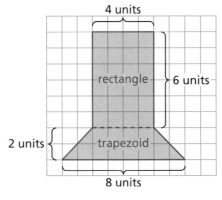

Study Tip
There is often more than one way to separate composite figures. In Example 1, you can separate the figure into one rectangle and two triangles.

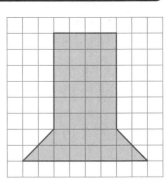

Area of Rectangle	Area of Trapezoid
$A = \ell w$	$A = \frac{1}{2}h(b_1 + b_2)$
$= 6(4)$	$= \frac{1}{2}(2)(4 + 8)$
$= 24$	$= 12$

∴ So, the area of the purple figure is $24 + 12 = 36$ square units.

Reasonable? You can check your result by counting unit squares.

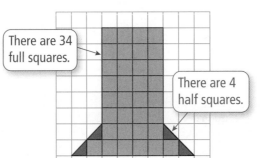

There are 34 full squares.

There are 4 half squares.

Full squares: 34

Half squares: 4

The area is

$$34(1) + 4\left(\frac{1}{2}\right) = 36 \text{ square units.}$$

So, the answer is reasonable. ✓

Geometry
In this extension, you will
- find areas of composite figures.
- solve real-life problems.

172 Chapter 4 Areas of Polygons

Multi-Language Glossary at BigIdeasMath.com

EXAMPLE 2 Real-Life Application

Find the area of the fairway between two streams on a golf course.

There are several ways to separate the fairway into figures whose areas you can find using formulas. It appears that one way is to separate it into a right triangle and a rectangle.

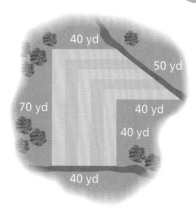

Identify each shape and find any missing dimensions.

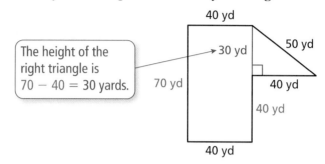

The height of the right triangle is 70 − 40 = 30 yards.

Area of Rectangle

$A = \ell w$

$= 70(40)$

$= 2800$

Area of Right Triangle

$A = \frac{1}{2}bh$

$= \frac{1}{2}(40)(30)$

$= 600$

∴ So, the area of the fairway is 2800 + 600 = 3400 square yards.

Practice

Find the area of the shaded figure.

1.

2.

3.

Find the area of the figure.

4.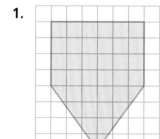

5.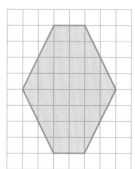

6.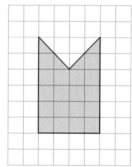

7. **ANOTHER METHOD** Find the area in Example 2 using a different method.

4.4 Polygons in the Coordinate Plane

Essential Question How can you find the lengths of line segments in a coordinate plane?

1 ACTIVITY: Finding Distances on a Map

Work with a partner. The coordinate grid shows a portion of a city. Each square on the grid represents one square mile.

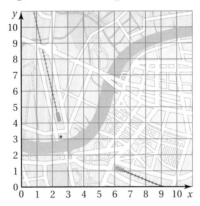

a. A public library is located at (4, 5). City Hall is located at (7, 5). Plot and label these points.

b. How far is the public library from City Hall?

c. A stadium is located 4 miles from the public library. Give the coordinates of several possible locations of the stadium. Justify your answers by graphing.

d. Connect the three locations of the public library, City Hall, and the stadium using your answers in part (c). What shapes are formed?

2 ACTIVITY: Graphing Polygons

Work with a partner. Plot and label each set of points in the coordinate plane. Then connect each set of points to form a polygon.

Rectangle: $A(2, 3)$, $B(2, 10)$, $C(6, 10)$, $D(6, 3)$

Triangle: $E(8, 3)$, $F(14, 8)$, $G(14, 3)$

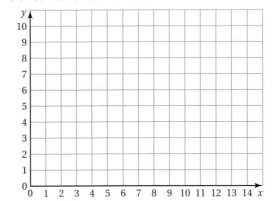

Geometry

In this lesson, you will
- draw polygons in the coordinate plane.
- find distances in the coordinate plane.
- solve real-life problems.

174 Chapter 4 Areas of Polygons

3 ACTIVITY: Finding Distances in a Coordinate Plane

Work with a partner.

a. Find the length of each horizontal line segment in Activity 2.

b. STRUCTURE What relationship do you notice between the lengths of the line segments in part (a) and the coordinates of their endpoints? Explain.

c. Find the length of each vertical line segment in Activity 2.

d. STRUCTURE What relationship do you notice between the lengths of the line segments in part (c) and the coordinates of their endpoints? Explain.

e. Plot and label the points below in the coordinate plane. Then connect each pair of points with a line segment. Use the relationships you discovered in parts (b) and (d) above to find the length of each line segment. Show your work.

$S(3, 1)$ and $T(14, 1)$ $U(9, 8)$ and $V(9, 0)$

$W(0, 7)$ and $X(0, 10)$ $Y(1, 9)$ and $Z(7, 9)$

f. Check your answers in part (e) by counting grid lines.

> **Math Practice**
>
> **Repeat Calculations**
>
> What calculations are repeated? How can you use this information to write a rule about the length of a line segment?

What Is Your Answer?

4. IN YOUR OWN WORDS How can you find the lengths of line segments in a coordinate plane? Give examples to support your explanation.

5. Do the methods you used in Activity 3 work for diagonal line segments? Explain why or why not.

6. Use the Internet or some other reference to find an example of how "finding distances in a coordinate plane" is helpful in each of the following careers.

a.   b.   c.

Archaeologist   Surveyor   Pilot

Practice Use what you learned about finding the lengths of line segments to complete Exercises 3–5 on page 178.

Section 4.4 Polygons in the Coordinate Plane 175

4.4 Lesson

You can use ordered pairs to represent vertices of polygons. To draw a polygon in a coordinate plane, plot and connect the ordered pairs.

EXAMPLE 1   Drawing a Polygon in a Coordinate Plane

The vertices of a quadrilateral are $A(2, 4)$, $B(3, 9)$, $C(7, 8)$, and $D(8, 1)$. Draw the quadrilateral in a coordinate plane.

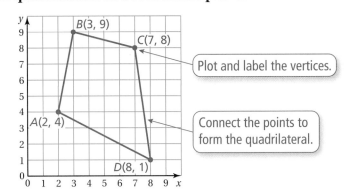

Study Tip

After you plot the vertices, connect them *in order* to draw the polygon.

On Your Own

Exercises 6–11

Draw the polygon with the given vertices in a coordinate plane.

1. $A(0, 0)$, $B(5, 7)$, $C(7, 4)$
2. $W(4, 4)$, $X(7, 4)$, $Y(7, 1)$, $Z(4, 1)$
3. $F(1, 3)$, $G(3, 6)$, $H(5, 6)$, $J(3, 3)$
4. $P(1, 4)$, $Q(3, 5)$, $R(7, 3)$, $S\left(6, \frac{1}{2}\right)$, $T\left(2, \frac{1}{2}\right)$

Key Idea

Finding Distances in the First Quadrant

You can find the length of a horizontal or vertical line segment in a coordinate plane by using the coordinates of the endpoints.

- When the *x*-coordinates are the same, the vertical distance between the points is the difference of the *y*-coordinates.
- When the *y*-coordinates are the same, the horizontal distance between the points is the difference of the *x*-coordinates.

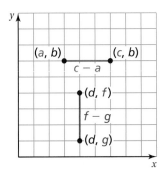

Be sure to subtract the lesser coordinate from the greater coordinate.

EXAMPLE 2 Finding a Perimeter

The vertices of a rectangle are $F(1, 6)$, $G(7, 6)$, $H(7, 2)$, and $J(1, 2)$. Draw the rectangle in a coordinate plane and find its perimeter.

Draw the rectangle and use the vertices to find its dimensions.

Study Tip
You can also find the length using vertices H and J. You can find the width using vertices F and J.

The length is the horizontal distance between $F(1, 6)$ and $G(7, 6)$, which is the difference of the *x*-coordinates.

length = 7 − 1 = 6 units

The width is the vertical distance between $G(7, 6)$ and $H(7, 2)$, which is the difference of the *y*-coordinates.

width = 6 − 2 = 4 units

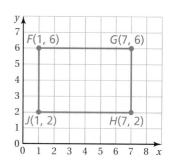

∴ So, the perimeter of the rectangle is 2(6) + 2(4) = 20 units.

EXAMPLE 3 Real-Life Application

In a grid of the exhibits at a zoo, the vertices of the giraffe exhibit are $E(0, 90)$, $F(60, 90)$, $G(100, 30)$, and $H(0, 30)$. The coordinates are measured in feet. What is the area of the giraffe exhibit?

Plot and connect the vertices using a coordinate grid to form a trapezoid. Use the coordinates to find the lengths of the bases and the height.

$b_1 = 60 − 0 = 60$

$b_2 = 100 − 0 = 100$

$h = 90 − 30 = 60$

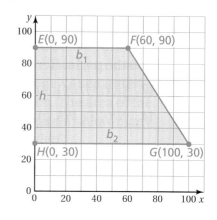

Use the formula for the area of a trapezoid.

$A = \dfrac{1}{2}(60)(60 + 100)$

$= \dfrac{1}{2}(60)(160) = 4800$

Common Error
You can count grid lines to find the dimensions, but make sure you consider the scale of the axes.

∴ The area of the giraffe exhibit is 4800 square feet.

On Your Own

Now You're Ready
Exercises 12–15

5. The vertices of a rectangle are $J(2, 7)$, $K(4, 7)$, $L(4, 1.5)$, and $M(2, 1.5)$. Find the perimeter and the area of the rectangle.

6. **WHAT IF?** In Example 3, the giraffe exhibit is enlarged by moving vertex *F* to (80, 90). How does this affect the area? Explain.

4.3–4.4 Quiz

Find the area of the trapezoid. *(Section 4.3)*

1.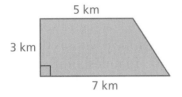

2.

3.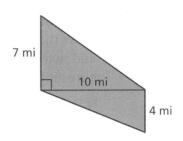

Find the area of the figure. *(Section 4.3)*

4.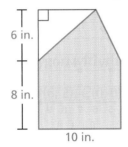

5.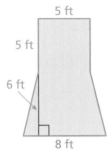

6.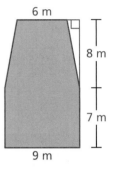

Draw the polygon with the given vertices in a coordinate plane. *(Section 4.4)*

7. $A(1, 2)$, $B(3, 5)$, $C(6, 1)$

8. $E(1, 2)$, $F(3, 6)$, $G(8, 6)$, $H(6, 2)$

Find the perimeter and the area of the polygon with the given vertices. *(Section 4.4)*

9. $J(1, 3)$, $K(1, 8)$, $L(5, 8)$, $M(5, 3)$

10. $P(1, 2)$, $Q(1, 7)$, $R(7, 7)$, $S(7, 2)$

11. **BACK POCKET** How much material do you need to make two back pockets? *(Section 4.3)*

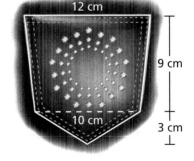

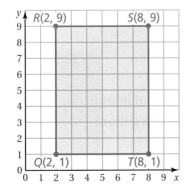

12. **PATIO** Plans for a patio are shown in the coordinate plane at the left. The coordinates are measured in feet. Find the perimeter and the area of the patio. *(Section 4.4)*

180   Chapter 4   Areas of Polygons

4 Chapter Review

Review Key Vocabulary

polygon, *p. 152*   composite figure, *p. 172*

Review Examples and Exercises

4.1 Areas of Parallelograms *(pp. 152–157)*

Find the area of the parallelogram.

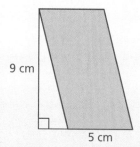

$A = bh$ Write formula.
$ = 5(9)$ Substitute 5 for *b* and 9 for *h*.
$ = 45$ Multiply.

∴ The area of the parallelogram is 45 square centimeters.

Exercises

Find the area of the parallelogram.

1.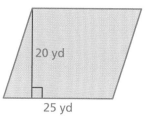

2.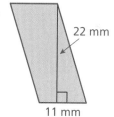

4.2 Areas of Triangles *(pp. 158–163)*

Find the area of the triangle.

$A = \frac{1}{2}bh$ Write formula.
$ = \frac{1}{2}(10)(7)$ Substitute.
$ = 35$ Multiply.

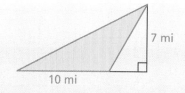

∴ The area of the triangle is 35 square miles.

Chapter Review   **181**

Exercises

Find the area of the triangle.

3.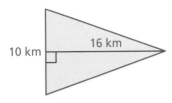

4.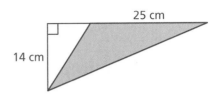

4.3 Areas of Trapezoids (pp. 166–173)

Find the area of the trapezoid.

$$A = \frac{1}{2}h(b_1 + b_2)$$ Write formula.

$$= \frac{1}{2}(10)(8 + 18)$$ Substitute.

$$= \frac{1}{2}(10)(26) = 130$$ Multiply.

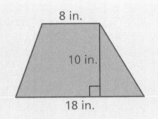

∴ The area of the trapezoid is 130 square inches.

Exercises

Find the area of the trapezoid.

5.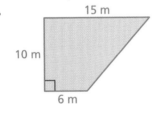

6.

7.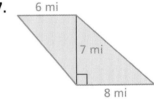

Find the area of the figure.

8.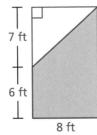

9.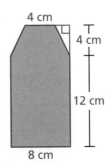

10.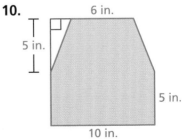

4.4 Polygons in the Coordinate Plane (pp. 174–179)

a. The vertices of a triangle are $A(1, 3)$, $B(5, 9)$, and $C(8, 2)$. Draw the triangle in a coordinate plane.

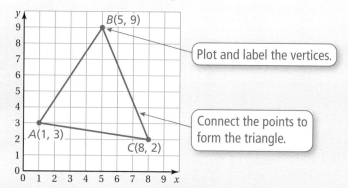

Plot and label the vertices.

Connect the points to form the triangle.

b. The vertices of a rectangle are $F(2, 6)$, $G(8, 6)$, $H(8, 1)$, and $J(2, 1)$. Draw the rectangle in a coordinate plane and find its perimeter.

Draw the rectangle and use the vertices to find its dimensions.

The length is the horizontal distance between $F(2, 6)$ and $G(8, 6)$, which is the difference of the *x*-coordinates.

length = 8 − 2 = 6 units

The width is the vertical distance between $G(8, 6)$ and $H(8, 1)$, which is the difference of the *y*-coordinates.

width = 6 − 1 = 5 units

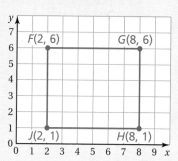

∴ So, the perimeter of the rectangle is 2(6) + 2(5) = 22 units.

Exercises

Draw the polygon with the given vertices in a coordinate plane.

11. $A(3, 2)$, $B(4, 7)$, $C(6, 0)$

12. $D(1, 1)$, $E(1, 5)$, $F(4, 5)$, $G(4, 1)$

13. $J(1, 2)$, $K(1, 7)$, $L(5, 7)$, $M(8, 2)$

14. $K\left(3, 3\frac{1}{2}\right)$, $L(5, 7)$, $M(8, 7)$, $N\left(6, 3\frac{1}{2}\right)$

Find the perimeter and the area of the polygon with the given vertices.

15. $P(4, 3)$, $Q(4, 7)$, $R(9, 7)$, $S(9, 3)$

16. $T(2, 7)$, $U(2, 9)$, $V(5, 9)$, $W(5, 7)$

17. $W(11, 2)$, $X(11, 8)$, $Y(14, 8)$, $Z(14, 2)$

18. $A(12, 2)$, $B(12, 13)$, $C(15, 13)$, $D(15, 2)$

4 Chapter Test

Find the area of the parallelogram, triangle, or trapezoid.

1.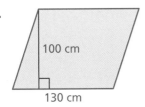

2.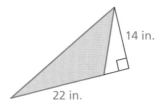

3.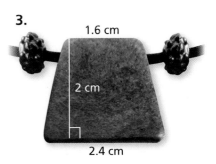

Find the area of the figure.

4.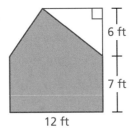

5.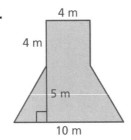

6.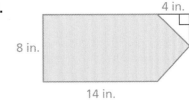

Draw the polygon with the given vertices in a coordinate plane.

7. $A(4, 2)$, $B(5, 6)$, $C(7, 4)$

8. $D(3, 4)$, $E(5, 8)$, $F(8, 8)$, $G(6, 4)$

Find the perimeter and the area of the polygon with the given vertices.

9. $Q(5, 6)$, $R(5, 10)$, $S(9, 10)$, $T(9, 6)$

10. $W(2, 8)$, $X(2, 16)$, $Y(8, 16)$, $Z(8, 8)$

11. **TABLETOP** The base lengths of a trapezoidal tabletop are 6 feet and 8 feet. The height is 5 feet. What is the area of the tabletop?

12. **PENTAGON** The Pentagon in Arlington, Virginia, is the headquarters of the U.S. Department of Defense.

    a. Find the perimeter of the Pentagon.

    b. A pentagon is made of a triangle and a trapezoid. The height of the triangle shown is about 541 feet, and the height of the trapezoid shown is about 876 feet. Estimate the land area of the Pentagon.

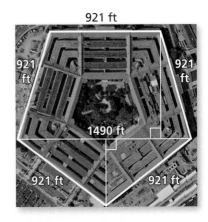

13. **CAMPING** The vertices of a campsite are (25, 15), (25, 30), (55, 30), and (55, 15). The vertices of your tent are (30, 20), (30, 25), (40, 25), and (40, 20). The coordinates are measured in feet. What is the area of the campsite not covered by your tent?

4 Cumulative Assessment

1. What is the area of the shaded figure shown below?

   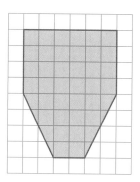

   A. 32 units²

   B. 40 units²

   C. 44 units²

   D. 56 units²

2. What is the value of the expression below?

   18^3

3. You have 36 red apples and 42 green apples. What is the greatest number of identical fruit baskets you can make with no apples left over?

   F. 6   H. 12

   G. 9   I. 18

4. What is the perimeter of the rectangle with the vertices shown below?

   $A(4, 7)$, $B(4, 15)$, $C(9, 15)$, $D(9, 7)$

   A. 8 units   C. 26 units

   B. 13 units   D. 70 units

Cumulative Assessment   185

5. What property was used to simplify the expression?

$$5 \times 78 = 5(70 + 8)$$
$$= 5(70) + 5(8)$$
$$= 350 + 40$$
$$= 390$$

F. Associative Property of Multiplication

G. Commutative Property of Addition

H. Distributive Property

I. Multiplication Property of One

6. What is the area, in square yards, of the triangle below?

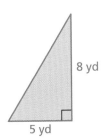

7. Which of the following is equivalent to $\dfrac{12}{35}$?

A. $\dfrac{5}{6} \div \dfrac{2}{7}$

B. $\dfrac{2}{7} \div \dfrac{6}{5}$

C. $\dfrac{2}{7} \div \dfrac{5}{6}$

D. $\dfrac{5}{6} \div \dfrac{7}{2}$

8. The description below represents the area of which polygon?

"one-half the product of its height and the sum of the lengths of its bases"

F. rectangle

G. square

H. trapezoid

I. triangle

9. Edward was evaluating the expression in the box.

$$180 \div 9 + 3^4 - 1 = 180 \div 9 + 81 - 1$$
$$= 180 \div 90 - 1$$
$$= 2 - 1$$
$$= 1$$

What should Edward do to correct the error that he made?

A. Add 9 and 81 then subtract 1 before dividing.

B. Divide 180 by 9 before adding or subtracting.

C. Divide 180 by 9 then subtract 1 before adding 3^4.

D. Subtract 1 from 90 before dividing.

10. You have 3 times as many guitar picks as your cousin. Let v be the number of guitar picks that your cousin has. Which expression represents the number of guitar picks you have?

F. $3v$

G. $v + 3$

H. $3 - v$

I. $\dfrac{v}{3}$

11. Your family hires a company to install invisible fencing around your yard.

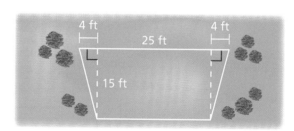

Part A Find the area of the yard using only the area formulas for rectangles and triangles. Show your work.

Part B Find the area of the yard using the area formula for trapezoids.

Part C Explain why the two methods of finding the area of the yard give the same result. Describe the advantages of each method.

Cumulative Assessment

5 Ratios and Rates

5.1 Ratios
5.2 Ratio Tables
5.3 Rates
5.4 Comparing and Graphing Ratios
5.5 Percents
5.6 Solving Percent Problems
5.7 Converting Measures

"By my records, I ate 1460 dog biscuits last year."

"So I calculated that my unit rate is 4 biscuits per day."

"It says 75% tomatoes, 15% sugar, 5% vinegar, 4% water, and 1% salt."

"See... no cats in catsup."

What You Learned Before

Identifying Patterns

Example 1 Using the numbers from the table, find and state the rule in words. Then find the missing value.

x	y
1	6
2	12
3	18
4	

Each y-value is 6 times the x-value.

∴ The x-value times 6 equals the y-value. The missing value is 6(4) = 24.

"I like seventy-five percent of them."

Try It Yourself

Using the numbers from the table, find and state the rule in words. Then find the missing value.

1.
x	y
1	2
3	6
5	10
7	

2.
x	y
2	8
4	16
6	24
8	

3.
x	y
1	5
2	10
3	15
4	

Multiplying and Dividing by Fractions

Example 2 Find $\dfrac{5}{6} \cdot \dfrac{3}{4}$.

$$\dfrac{5}{6} \cdot \dfrac{3}{4} = \dfrac{5 \cdot \overset{1}{\cancel{3}}}{\underset{2}{\cancel{6}} \cdot 4}$$
$$= \dfrac{5}{8}$$

Example 3 Find $2 \div \dfrac{9}{10}$.

$$2 \div \dfrac{9}{10} = 2 \cdot \dfrac{10}{9}$$ ← Multiply by the reciprocal of the divisor.
$$= \dfrac{2 \cdot 10}{9}$$
$$= \dfrac{20}{9}$$

Try It Yourself

Evaluate the expression. Write the answer in simplest form.

4. $\dfrac{1}{5} \cdot \dfrac{13}{20}$

5. $\dfrac{3}{4} \cdot \dfrac{13}{25}$

6. $7 \div \dfrac{9}{10}$

7. $4 \div \dfrac{16}{17}$

5.1 Ratios

Essential Question How can you represent a relationship between two quantities?

1 ACTIVITY: Comparing Quantities

Work with a partner. Use the collection of objects to complete each statement.

There are ____ graphing calculators to ____ protractors.

There are ____ protractors to ____ graphing calculators.

There are ____ compasses to ____ protractors.

There are ____ graphing calculators to ____ compasses.

There are ____ protractors to ____ total objects.

The number of graphing calculators is $\dfrac{}{}$ of the total number of objects.

2 ACTIVITY: Playing Garbage Basketball

Work with a partner.

- Take turns shooting a ball or other object into a wastebasket from a reasonable distance.
- Organize the numbers of shots you made and shots you missed in a chart.

Ratios

In this lesson, you will
- understand the concept of a ratio.
- use ratios to describe the relationship between two quantities.

a. Write a statement similar to those in Activity 1 that describes the relationship between the number of shots you made and the number of shots you missed.

b. Write a statement similar to those in Activity 1 that describes the relationship between the number of shots you made and the total number of shots.

c. What fraction of your shots did you make? What fraction did you miss?

3 ACTIVITY: Reading a Diagram

Work with a partner. You mix different amounts of paint to create new colors. Write a statement that describes the relationship between the amounts of paint shown in each diagram.

a. Blue [4 parts]
   Green [3 parts]

   There are ___ parts blue for every ___ parts green.

b. Orange [2 parts]
   Yellow [3 parts]

   There are _____ for every _____.

c. Red [4 parts]
   Blue [2 parts]

   _____.

d. White [1 part]
   Purple [5 parts]

   _____.

4 ACTIVITY: Describing Relationships

Work with a partner. Use a table or a diagram to represent the relationship between the two quantities.

a. For every 3 boys standing in a line, there are 4 girls.
b. For each vote Brian received, Sasha received 6 votes.
c. A class counts the number of vehicles that pass by its school from 1:00 to 2:00 P.M. There are 3 times as many cars as trucks.
d. A hand sanitizer contains 5 parts aloe for every 2 parts distilled water.

Math Practice

Use a Table or Diagram

What are the quantities in this problem? How does a table or diagram represent the relationship between the quantities?

What Is Your Answer?

5. **IN YOUR OWN WORDS** How can you represent a relationship between two quantities? Give examples to support your explanation.

6. **MODELING** You make 48 pints of pink paint by using 5 pints of red paint for every 3 pints of white paint. Use a diagram to find the number of pints of red paint and white paint in your mixture. Explain.

Practice — Use what you learned about comparing two quantities to complete Exercises 4 and 5 on page 194.

Section 5.1 Ratios 191

5.1 Lesson

Check It Out
Lesson Tutorials
BigIdeasMath.com

Key Vocabulary
ratio, p. 192

Key Idea

Ratio

Words A **ratio** is a comparison of two quantities. Ratios can be part-to-part, part-to-whole, or whole-to-part comparisons.

Examples
2 red crayons *to* 6 blue crayons
1 red crayon *for every* 3 blue crayons
3 blue crayons *per* 1 red crayon
3 blue crayons *for each* red crayon
3 blue crayons *out of every* 4 crayons
2 red crayons *out of* 8 crayons

Algebra The ratio of *a* to *b* can be written as *a* : *b*.

EXAMPLE 1 Writing Ratios

You have the coins shown.

Remember

Part-to-whole relationships compare a part of a whole to the whole. Fractions represent part-to-whole relationships. Part-to-part relationships compare a part of a whole to another part of the whole.

a. **Write the ratio of pennies to quarters.**

[6 pennies] → 6 to 7 ← [7 quarters]

∴ So, the ratio of pennies to quarters is 6 to 7, or 6 : 7.

b. **Write the ratio of quarters to dimes.**

[7 quarters] → 7 to 3 ← [3 dimes]

∴ So, the ratio of quarters to dimes is 7 to 3, or 7 : 3.

c. **Write the ratio of dimes to the total number of coins.**

[3 dimes] → 3 to 16 ← [16 coins]

∴ So, the ratio of dimes to the total number of coins is 3 to 16, or 3 : 16.

On Your Own

Now You're Ready
Exercises 6–13

1. In Example 1, write the ratio of dimes to pennies.

2. The circle graph shows the favorite ice-cream toppings of several students. Use ratio language to compare the number of students who favor peanuts to the total number of students.

Favorite Toppings

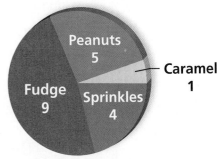

A *tape diagram* is a diagram that looks like a segment of tape. It shows the relationship between two quantities.

EXAMPLE 2 **Using a Tape Diagram**

The ratio of your monthly allowance to your friend's monthly allowance is 5 : 3. The monthly allowances total $40. How much is each allowance?

To help visualize the problem, express the ratio 5 : 3 using a tape diagram.

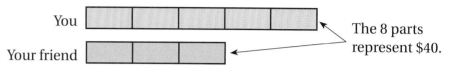

The 8 parts represent $40.

Because there are 8 parts, you know that 1 part represents $40 ÷ 8 = $5.

$$5 \text{ parts represent } \$5 \cdot 5 = \$25.$$
$$3 \text{ parts represent } \$5 \cdot 3 = \$15.$$

So, your monthly allowance is $25, and your friend's monthly allowance is $15.

EXAMPLE 3 **Using a Tape Diagram**

You separate 42 bulbs of garlic into two groups: one for planting and one for cooking. You will plant 3 bulbs for every 4 bulbs that you will use for cooking. Each bulb has about 8 cloves. About how many cloves will you plant?

To help visualize the problem, express the ratio *3 for every 4* using a tape diagram.

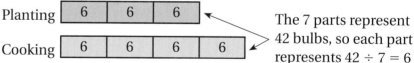

The 7 parts represent 42 bulbs, so each part represents 42 ÷ 7 = 6 bulbs.

There are 3 • 6 = 18 bulbs for planting and 4 • 6 = 24 bulbs for cooking. The group of 18 bulbs has about 18 • 8 = 144 cloves.

So, you will plant about 144 cloves.

On Your Own

Exercises 15 and 16

3. **WHAT IF?** In Example 2, the ratio is 2 to 3. How much is each allowance?

4. **WHAT IF?** In Example 3, you will plant 1 bulb for every 2 bulbs that you will use for cooking. Will you plant more or fewer cloves than originally planned? Explain your reasoning.

5.1 Exercises

Check It Out
Help with Homework
BigIdeasMath.com

Vocabulary and Concept Check

1. **VOCABULARY** The ratio of vowels to consonants in a word is 5 to 7. Are there more vowels or consonants in the word? Explain.

2. **NUMBER SENSE** You are comparing apples to oranges in a fruit bowl. Is the ratio 2 : 3 the same as the ratio 3 : 2? Explain.

3. **WHICH ONE DOESN'T BELONG?** Which ratio does *not* belong with the other three? Explain your reasoning.

   | 2 parts to 5 parts | 2 out of every 5 | 2 for each 5 | 2 for every 5 |

Practice and Problem Solving

Use a table or a diagram to represent the relationship between the two quantities.

4. For each lion, there are 7 giraffes.

5. For every 5 seats, there are 4 fans.

Write the ratio. Explain what the ratio means.

6. frogs to turtles

7. basketballs to soccer balls

8. calculators : pencils

9. shirts : pants

Use the table to write the ratio. Explain what the ratio means.

10. dramas to movies
11. comedies to movies
12. movies : action
13. movies : dramas

Movie	Number
Drama	3
Comedy	8
Action	4

Topic	Stamps
Birds	7
Celebrity	14
Horses	5
Ships	9

14. **STAMP COLLECTING** The table shows the numbers of stamps in a new stamp collection. Use ratio language to compare the number of celebrity stamps to the total number of stamps.

194 Chapter 5 Ratios and Rates

You and a friend tutor for a total of 12 hours. Use the tape diagram to find how many hours you tutor.

15.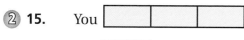

16.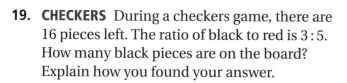

17. **REASONING** Twelve of the 28 students in a class have a dog. What is the ratio of students who have a dog to students who do not?

18. **GEOGRAPHY** In the continental United States, the ratio of states that border an ocean to states that do not border an ocean is 7 : 9. How many of the states border an ocean?

19. **CHECKERS** During a checkers game, there are 16 pieces left. The ratio of black to red is 3 : 5. How many black pieces are on the board? Explain how you found your answer.

20. **SCHOOL PLAY** There are 48 students in a school play. The ratio of boys to girls is 5 : 7. How many more girls than boys are in the play? Explain how you found your answer.

21. **GEOMETRY** Use the blue and green rectangles.

    a. Find the ratio of the length of the blue rectangle to the length of the green rectangle. Repeat this for width, perimeter, and area.

    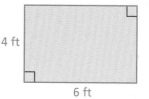

    b. Compare and contrast your ratios in part (a).

22. **PERIMETER** The ratio of the side lengths of a triangle is 2 : 3 : 4. The shortest side is 15 inches. What is the perimeter? Explain.

23. **PRECISION** You mix soda water, fruit punch concentrate, and ginger ale in the ratio of 1 : 2 : 5 to make fruit punch. How many pints of each ingredient should you use to make 4 gallons of fruit punch? Is your answer reasonable? Explain.

24. **Reasoning** There are 12 boys and 10 girls in your gym class. If 6 boys joined the class, how many girls would need to join for the ratio of boys to girls to remain the same? Justify your answer.

Fair Game Review What you learned in previous grades & lessons

Divide. *(Section 2.6)*

25. $13.8 \div 3$

26. $16.45 \div 5$

27. $53.13 \div 21$

28. $19.214 \div 13$

29. **MULTIPLE CHOICE** What is the value of the expression $x \div y$ when $x = 30$ and $y = 18$? *(Section 3.1)*

    Ⓐ $\dfrac{3}{5}$ Ⓑ $1\dfrac{2}{3}$ Ⓒ 12   Ⓓ 48

5.2 Ratio Tables

Essential Question How can you find two ratios that describe the same relationship?

1 ACTIVITY: Making a Mixture

Work with a partner. A mixture calls for 1 cup of lemonade and 3 cups of iced tea.

Lemonade   Iced Tea

a. How many total cups does the mixture contain? ▢ cups

   For every ▢ cup of lemonade, there are ▢ cups of iced tea.

b. How do you make a larger batch of this mixture? Describe your procedure and use the table below to organize your results. Add more columns to the table if needed.

Cups of Lemonade					
Cups of Iced Tea					
Total Cups					

c. Which operations did you use to complete your table? Do you think there is more than one way to complete the table? Explain.

Ratios
In this lesson, you will
- use ratio tables to find equivalent ratios.
- solve real-life problems.

d. How many total cups are in your final mixture? How many of those cups are lemonade? How many are iced tea? Compare your results with those of other groups in your class.

e. Suppose you take a sip from every group's final mixture. Do you think all the mixtures should taste the same? Do you think the color of all the mixtures should be the same? Explain your reasoning.

f. Why do you think it is useful to use a table when organizing your results in this activity? Explain.

196   Chapter 5   Ratios and Rates

5.2 Exercises

Vocabulary and Concept Check

1. **VOCABULARY** How can you tell whether two ratios are equivalent?

2. **NUMBER SENSE** Consider the ratio 3 : 5. Can you create an equivalent ratio by adding the same number to each quantity in the ratio? Explain.

3. **WHICH ONE DOESN'T BELONG?** Which ratio does *not* belong with the other three? Explain your reasoning.

   | 3 : 4 | 9 : 12 | 12 : 15 | 12 : 16 |

Practice and Problem Solving

Write several ratios that describe the collection.

4. baseballs to gloves

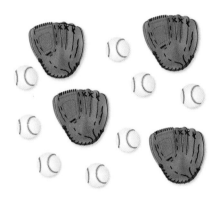

5. ladybugs to bees

Find the missing value(s) in the ratio table. Then write the equivalent ratios.

6.
Boys	1	
Girls	5	10

7.
Violins	8	24
Cellos	3	

8.
Taxis	6		36
Buses	5	15	

9.
Burgers	3		9
Hot Dogs	5	10	

10.
Towels	14	7	
Blankets	8		16

11.
Forks	16	8	
Spoons	10		30

12. **WORK** Your neighbor pays you $17 for every 2 hours you work. You work for 8 hours on Saturday. How much does your neighbor owe you?

Complete the ratio table to solve the problem.

13. For every 3 tickets you sell, your friend sells 4. You sell a total of 12 tickets. How many does your friend sell?

You	3			12
Friend	4			

14. A store sells 2 printers for every 5 computers. The store sells 40 computers. How many printers does the store sell?

Printers	2		8	
Computers	5	10		40

15. First and second place in a contest use a ratio to share a cash prize. When first place pays $100, second place pays $60. How much does first place pay when second place pays $36?

First	100		
Second	60		36

16. A grade has 81 girls and 72 boys. The grade is split into groups that have the same ratio of girls to boys as the whole grade. How many girls are in a group that has 16 boys?

Girls	81		
Boys	72		16

ERROR ANALYSIS Describe and correct the error in making the ratio table.

17.

A	3	8	13
B	7	12	17

18.

A	5	25	125
B	3	9	27

19. DONATION A sports store donates basketballs and soccer balls to the boys and girls club. The ratio of basketballs to soccer balls is 7 : 6. The store donates 24 soccer balls. How many basketballs does the store donate?

20. DOWNLOAD You are downloading songs to your MP3 player. The ratio of pop songs to rock songs is 5 : 4. You download 40 pop songs. How many rock songs do you download?

SCRAMBLED EGGS In Exercises 21–25, use the ratio table showing different batches of the same recipe for scrambled eggs.

Recipe	A	B	C	D	E	F
Servings	4	2	6	3	5	9
Eggs	8	4	12	6	10	18
Milk (cups)	$\frac{1}{2}$	$\frac{1}{4}$	$\frac{3}{4}$	$\frac{3}{8}$	$\frac{5}{8}$	$1\frac{1}{8}$

21. How can you use Recipes B and D to create Recipe E?

22. How can you use Recipes C and D to create Recipe F?

23. How can you use Recipes B and C to create Recipe A?

24. How can you use Recipes C and F to create Recipe D?

25. Describe one way to use the recipes to create a batch with 11 servings.

Two whole numbers A and B satisfy the following conditions. Find A and B.

26. $A + B = 30$
$A : B$ is equivalent to $2 : 3$.

27. $A + B = 44$
$A : B$ is equivalent to $4 : 7$.

28. $A - B = 18$
$A : B$ is equivalent to $11 : 5$.

29. $A - B = 25$
$A : B$ is equivalent to $13 : 8$.

30. CASHEWS The nutrition facts label on a container of dry roasted cashews indicates there are 161 calories in 28 grams. You eat 9 cashews totaling 12 grams.

  a. How many calories do you consume?

  b. How many cashews are in one serving?

31. REASONING The ratio of three numbers is $4 : 3 : 1$. The sum of the numbers is 64. What is the greatest number?

32. SURVEY Seven out of every 8 students surveyed owns a bike. The difference between the number of students who own a bike and those who do not is 72. How many students were surveyed?

33. BUG COLLECTION You and a classmate have a bug collection for science class. You find 5 out of every 9 bugs in the collection. You find 4 more bugs than your classmate. How many bugs are in the collection?

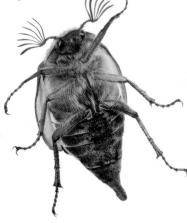

34. You and a friend each have a collection of tokens. Initially, for every 8 tokens you had, your friend had 3. After you give half of your tokens to your friend, your friend now has 18 more tokens than you. Initially, how many more tokens did you have than your friend?

Fair Game Review *What you learned in previous grades & lessons*

Factor the expression using the GCF. *(Section 3.4)*

35. $54 + 27$

36. $60x - 84$

37. $42x + 28y$

38. MULTIPLE CHOICE Which expression does *not* give the area of the shaded figure? *(Section 4.3)*

  Ⓐ $2(6) + 2\left(\dfrac{1}{2}(6)(2)\right)$

  Ⓑ $2\left(\dfrac{1}{2}(3)(2+6)\right)$

  Ⓒ $6(6) - 4\left(\dfrac{1}{2}(3)(2)\right)$

  Ⓓ $6(6) - \dfrac{1}{2}(6)(2)$

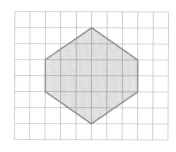

5.3 Rates

Essential Question How can you use rates to describe changes in real-life problems?

1 ACTIVITY: Stories Without Words

Work with a partner. Each diagram shows a story problem.

- Describe the story problem in your own words.
- Write the rate indicated by the diagram. What are the units?

a.

b.

c.
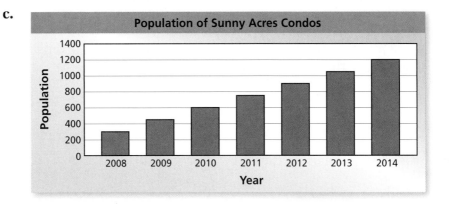

d.
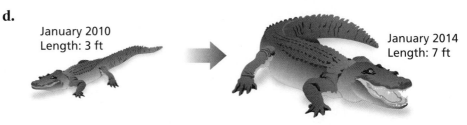

January 2010
Length: 3 ft

January 2014
Length: 7 ft

Rates

In this lesson, you will
- understand the concepts of rates and unit rates.
- write unit rates.
- solve real-life problems.

2 ACTIVITY: Finding Equivalent Rates

Math Practice

Specify Units
How do the given units help you find the units for your answer?

Work with a partner. Use the diagrams in Activity 1. Explain how you found each answer.

a. How many miles does the car travel in 1 hour?
b. How much money does the person earn every hour?
c. How much does the population of Sunny Acres Condos increase each year?
d. How many feet does the alligator grow per year?

3 ACTIVITY: Using a Double Number Line

Work with a partner. Count the number of times you can clap your hands in 12 seconds. Have your partner keep track of the time and record your results.

a. Use the results to complete the double number line.

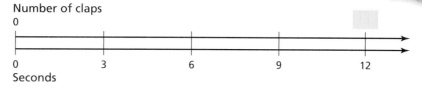

b. Explain how to use the double number line to find the number of times you clap your hands in 6 seconds and in 4 seconds.
c. Find the number of times you can clap your hands in 1 minute. Explain how you found your answer.
d. How can you find the number of times you can clap your hands in 2 minutes? 3 minutes? Explain.

What Is Your Answer?

4. **IN YOUR OWN WORDS** How can you use rates to describe changes in real-life problems? Give examples to support your explanation.

5. **MODELING** Use a double number line to model each story in Activity 1. Show how to use the double number line to answer each question in Activity 2. Why is a double number line a good problem-solving tool for these types of problems?

Practice

Use what you learned about rates to complete Exercises 3 and 4 on page 208.

5.3 Exercises

Vocabulary and Concept Check

1. **WRITING** Describe a unit rate that you use in real life.

2. **DIFFERENT WORDS, SAME QUESTION** Which is different? Find "both" answers.

   What is the cost per bagel?   What is the cost per dozen bagels?

   What is the unit cost of a bagel?   How much does each bagel cost?

Practice and Problem Solving

Write a rate that represents the situation.

3.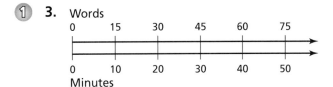

4.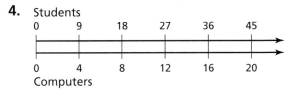

5.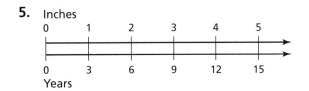

6.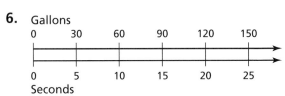

Write a unit rate for the situation.

7. $28 saved in 4 weeks

8. 18 necklaces made in 3 hours

9. 270 miles in 6 hours

10. 228 students in 12 classes

11. 2520 kilobytes in 18 seconds

12. 880 calories in 8 servings

13. 1080 miles on 15 gallons

14. $12.50 for 5 ounces

15. **LIGHTNING** Lightning strikes Earth 1000 times in 10 seconds. How many times does lightning strike per second?

16. **HEART RATE** Your heart beats 240 times in 4 minutes. How many times does your heart beat each minute?

17. **CAR WASH** You earn $35 for washing 7 cars. How much do you earn for washing 4 cars?

18. **5K RACE** You jog 2 kilometers in 12 minutes. At this rate, how long will it take you to complete a 5-kilometer race?

208   Chapter 5   Ratios and Rates

Decide whether the rates are equivalent.

19. 24 laps in 6 minutes
72 laps in 18 minutes

20. 126 points every 3 games
210 points every 5 games

21. 15 breaths every 36 seconds
90 breaths every 3 minutes

22. $16 for 4 pounds
$1 for 4 ounces

23. PRINTER A printer prints 28 photos in 8 minutes.

   a. How many minutes does it take to print 21 more photos?

   b. Construct a double number line diagram that represents the situation. How many minutes does it take to print 35 more photos?

24. SUN VISOR An athletic director pays $90 for 12 sun visors for the softball team.

   a. How much will the athletic director pay to buy 15 more sun visors?

   b. Construct a double number line diagram that represents the situation. What is the cost of 16 sun visors?

25. FOOD DRIVE The table shows the amounts of food collected by two homerooms. Homeroom A collects 21 additional items of food. How many more items does Homeroom B need to collect to have more items per student?

	Homeroom A	Homeroom B
Students	24	16
Canned Food	30	22
Dry Food	42	24

26. MARATHON A runner completed a 26.2-mile marathon in 210 minutes.

   a. Estimate the unit rate, in miles per minute.

   b. Estimate the unit rate, in minutes per mile.

   c. Another runner says, "I averaged 10-minute miles in the marathon." Is this runner talking about the kind of rate described in part (a) or in part (b)? Explain your reasoning.

27. **Logic** You can do one-half of a job in an hour. Your friend can do one-third of the same job in an hour. How long will it take to do the job if you work together?

Fair Game Review *What you learned in previous grades & lessons*

Write two fractions that are equivalent to the given fraction. *(Skills Review Handbook)*

28. $\frac{1}{3}$ **29.** $\frac{5}{6}$ **30.** $\frac{2}{5}$ **31.** $\frac{4}{9}$

32. MULTIPLE CHOICE Which expression is equivalent to $6(x) - 6(2)$? *(Section 3.4)*

  Ⓐ $2(x-6)$ **Ⓑ** $6(x-2)$ **Ⓒ** $12(x-1)$ **Ⓓ** $36(x-2)$

5.4 Comparing and Graphing Ratios

Essential Question How can you compare two ratios?

1 ACTIVITY: Comparing Ratio Tables

Work with a partner.
- You make colored frosting by adding 3 drops of red food coloring for every 1 drop of blue food coloring.
- Your teacher makes colored frosting by adding 5 drops of red food coloring for every 3 drops of blue food coloring.

a. Copy and complete the ratio table for each frosting mixture.

Your Frosting	
Drops of Blue	Drops of Red
1	
2	
3	
4	
5	

Your Teacher's Frosting	
Drops of Blue	Drops of Red
3	
6	
9	
12	
15	

b. Whose frosting is bluer? Whose frosting is redder? Justify your answers.

c. **STRUCTURE** Insert and complete a new column for each ratio table above that shows the total number of drops. How can you use this column to answer part (b)?

2 ACTIVITY: Graphing from a Ratio Table

Ratios and Rates
In this lesson, you will
- compare ratios.
- compare unit rates.
- graph ordered pairs to compare ratios and rates.

Work with a partner.

a. Explain how you can use the values from the ratio table for your frosting to create a graph in the coordinate plane.

b. Use the values in the table to plot the points. Then connect the points and describe the graph. What do you notice?

c. What does the line represent?

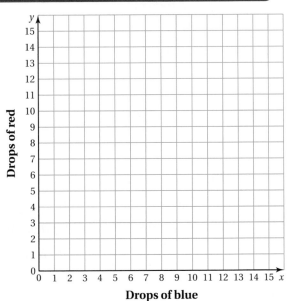

210 Chapter 5 Ratios and Rates

3 ACTIVITY: Comparing Graphs from Ratio Tables

Math Practice

Look for Patterns

What patterns do you notice in the graph? What does this tell you about the problem?

Work with a partner. The graph shows the values from the ratio table for your teacher's frosting.

a. Complete the table and the graph.

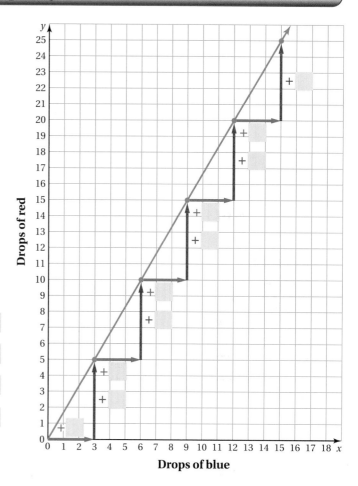

Your Teacher's Frosting	
Drops of Blue	Drops of Red
3	
6	
9	
12	
15	

b. Explain the relationship between the entries in the ratio table and the points on the graph.

c. How is this graph similar to the graph in Activity 2? How is it different?

d. How can you use the graphs to determine whose frosting has more red or blue in it? Explain.

What Is Your Answer?

4. **IN YOUR OWN WORDS** How can you compare two ratios?

5. **PRECISION** Your teacher's frosting mixture has 7 drops of blue in it. How can you use the graph to find how many drops of red are needed to make the frosting? Is your answer exact? Explain.

Practice — Use what you learned about comparing ratios to complete Exercises 3 and 4 on page 214.

5.4 Lesson

One way to compare ratios is by using ratio tables.

EXAMPLE 1 Comparing Ratios

You mix 8 tablespoons of hot sauce and 3 cups of salsa in a green bowl. You mix 12 tablespoons of hot sauce and 4 cups of salsa in an orange bowl. Which mixture is hotter?

Use ratio tables to compare the mixtures. Find a larger batch of each mixture in which the amount of hot sauce or salsa is the same.

Green Bowl

×4

Hot Sauce (tablespoons)	8	32
Salsa (cups)	3	12

×4

Orange Bowl

×3

Hot Sauce (tablespoons)	12	36
Salsa (cups)	4	12

×3

The tables show that for a larger batch of each mixture using 12 cups of salsa, the orange bowl would have 36 − 32 = 4 more tablespoons of hot sauce.

∴ So, the mixture in the orange bowl is hotter.

EXAMPLE 2 Comparing Unit Rates

Which bag of dog food is the better buy?

Use ratio tables to find and compare the unit costs.

20-Pound Bag

÷20

Cost (dollars)	17.20	0.86
Food (pounds)	20	1

÷20

30-Pound Bag

÷30

Cost (dollars)	25.20	0.84
Food (pounds)	30	1

÷30

The 20-pound bag costs $0.86 per pound, and the 30-pound bag costs $0.84 per pound.

∴ Because $0.84 is less than $0.86, the 30-pound bag is the better buy.

● **On Your Own**

Now You're Ready
Exercises 3–10

1. In Example 1, you mix 10 tablespoons of hot sauce and 3 cups of salsa in a red bowl. Which mixture is the mildest? Explain.

2. A 30-pack of paper towels costs $48.30. A 32-pack costs $49.60. Which is the better buy? Explain.

212   Chapter 5   Ratios and Rates

EXAMPLE 3 Graphing Values from Ratio Tables

A hot-air balloon rises 9 meters every 3 seconds. A blimp rises 7 meters every 2 seconds.

a. Complete the ratio table for each aircraft. Which rises faster?

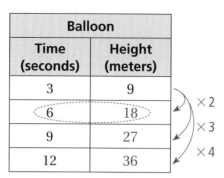

Rises 9 meters every 3 seconds.

Balloon	
Time (seconds)	Height (meters)
3	9
6	18
9	27
12	36

Blimp	
Time (seconds)	Height (meters)
2	7
4	14
6	21
8	28

Every 6 seconds, the balloon rises 18 meters and the blimp rises 21 meters.

∴ So, the blimp rises faster.

b. Graph the ordered pairs (time, height) from the tables in part (a). What can you conclude?

Write the ordered pairs.

   Balloon: (3, 9), (6, 18), (9, 27), (12, 36)

   Blimp: (2, 7), (4, 14), (6, 21), (8, 28)

Study Tip
When graphing speed, you often place time on the horizontal axis and distance on the vertical axis.

Plot and label each set of ordered pairs. Then draw a line through each set of points.

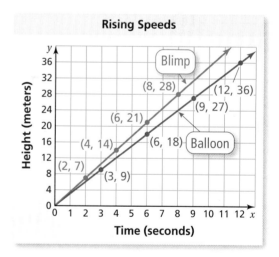

∴ Both graphs begin at (0, 0). The graph for the blimp is steeper, so the blimp rises faster than the hot-air balloon.

On Your Own

Now You're Ready
Exercises 12 and 13

3. WHAT IF? The blimp rises 6 meters every 2 seconds. How does this affect your conclusion?

Section 5.4 Comparing and Graphing Ratios 213

5.4 Exercises

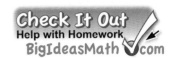

Vocabulary and Concept Check

1. **WRITING** Explain how to use tables to compare ratios.
2. **NUMBER SENSE** Just by looking at the graph, determine who earns a greater hourly wage. Explain.

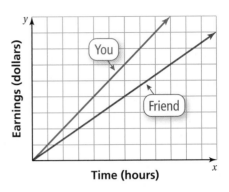

Practice and Problem Solving

Determine which car gets the better gas mileage.

3.
Car	A	B
Distance (miles)	125	120
Gallons Used	5	6

4.
Car	A	B
Distance (miles)	300	320
Gallons Used	8	10

5.
Car	A	B
Distance (miles)	450	405
Gallons Used	15	12

6.
Car	A	B
Distance (miles)	360	270
Gallons Used	20	18

Determine which is the better buy.

7.
Air Freshener	A	B
Cost (dollars)	6	12
Refills	2	3

8.
Kitten Food	A	B
Cost (dollars)	15	9
Cans	18	12

9.
Ham	A	B
Cost (dollars)	5.70	8.75
Pounds	3	5

10.
Cheese	A	B
Cost (dollars)	3.59	5.12
Slices	10	16

11. **SALT WATER GARGLE** Salt water gargle can temporarily relieve a sore throat. One recipe calls for $\frac{3}{4}$ teaspoon of salt in 1 cup of water. A second recipe calls for 1 teaspoon of salt in 2 cups of water. Which recipe will taste saltier?

Complete the ratio tables and graph the ordered pairs from the tables. What can you conclude?

12.

Water Tank	
Time (min)	Liters Leaked
2	4
4	
6	
8	

Swimming Pool	
Time (min)	Liters Leaked
3	2
6	
9	
12	

13.

Zoo	
People	Cost (dollars)
4	60
8	
12	
16	

Museum	
People	Cost (dollars)
5	95
10	
15	
20	

14. **MILK** In whole milk, 13 parts out of 400 are milk fat. In 2% milk, 1 part out of 50 is milk fat. Which type of milk has more milk fat per cup?

15. **HEART RATE** A horse's heart beats 440 times in 10 minutes. A cow's heart beats 390 times in 6 minutes. Which animal has a greater heart rate?

16. **CHOOSE TOOLS** A chemist prepares two acid solutions.

   a. Use a ratio table to determine which solution is more acidic.

   b. Use a graph to determine which solution is more acidic.

   c. Which method do you prefer? Explain.

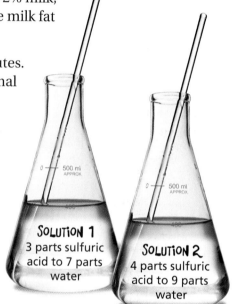

SOLUTION 1 3 parts sulfuric acid to 7 parts water

SOLUTION 2 4 parts sulfuric acid to 9 parts water

17. **NUT MIXTURE** A company offers a nut mixture with 7 peanuts for every 4 almonds. The company changes the mixture to have 8 peanuts for every 5 almonds, but the number of nuts per container does not change.

   a. Create a ratio table for each mixture. How many nuts are in the smallest possible container?

   b. Graph the ordered pairs from the tables. What can you conclude?

   c. Almonds cost more than peanuts. Should the company charge more or less for the new mixture? Explain your reasoning.

18. **Structure** The point (p, q) is on the graph of values from a ratio table. What is another point on the graph?

Fair Game Review What you learned in previous grades & lessons

Divide. *(Section 1.1)*

19. $544 \div 34$ 20. $1520 \div 83$ 21. $8439 \div 245$

22. **MULTIPLE CHOICE** Which of the following numbers is equal to 9.32 when you increase it by 4.65? *(Section 2.4)*

   Ⓐ 4.33   Ⓑ 4.67   Ⓒ 5.67   Ⓓ 13.97

5 Study Help

You can use a **definition and example chart** to organize information about a concept. Here is an example of a definition and example chart for ratio.

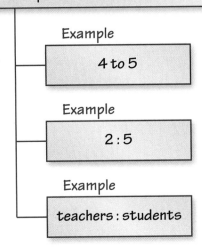

On Your Own

Make definition and example charts to help you study these topics.

1. equivalent ratios
2. ratio table
3. rate
4. unit rate
5. equivalent rates

After you complete this chapter, make definition and example charts for the following topics.

6. percent
7. U.S. customary system
8. metric system
9. conversion factor
10. unit analysis

"My math teacher taught us how to make a definition and example chart."

5.1–5.4 Quiz

Write the ratio. Explain what the ratio means. *(Section 5.1)*

1. tulips to lilies

2. crayons to markers

Find the missing values in the ratio table. Then write the equivalent ratios. *(Section 5.2)*

3.
Shoes	7		49
Boots	2	8	

4.
Trains	3	12	
Airplanes	8		48

Write a rate that represents the situation. *(Section 5.3)*

5.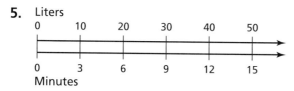

6.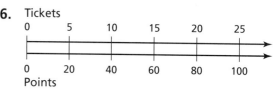

Write a unit rate for the situation. *(Section 5.3)*

7. 12 touchdowns in 6 games

8. 15 text messages in 5 minutes

9. 80 entries in 4 contests

10. 75 questions in 25 minutes

11. **DOWNLOADS** Three album downloads cost $36. How much do 5 album downloads cost? *(Section 5.3)*

12. **SHAMPOO** You can buy 20 fluid ounces of shampoo for $4.40 or 24 fluid ounces for $4.80. Which is the better buy? Explain. *(Section 5.4)*

13. **NBA CHAMPIONSHIPS** Write each ratio. Explain what the ratio means. *(Section 5.1)*

    a. Celtics championships to Lakers championships

    b. Pistons championships to Spurs championships

    c. Bulls championships to Lakers championships

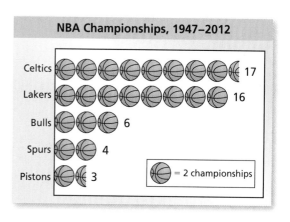

5.5 Percents

Essential Question What is the connection between ratios, fractions, and percents?

1 ACTIVITY: Writing Ratios

Work with a partner.

- Write the fraction of the squares that are shaded.
- Write the ratio of the number of shaded squares to the total number of squares.
- How are the ratios and the fractions related?
- When can you write ratios as fractions?

a. 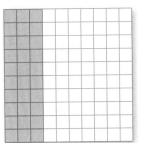 b. 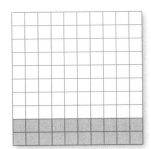 c.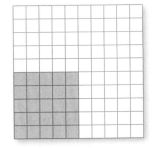

The Meaning of a Word • Percent

A century is 100 years.

A cent is one hundredth of a dollar.

In Mexico, a centavo is one hundredth of a peso.

Percents

In this lesson, you will
- write percents as fractions with denominators of 100.
- write fractions as percents.

Cent means *one hundred*, so **percent** means *per one hundred*. The symbol for percent is %.

218   Chapter 5   Ratios and Rates

2 ACTIVITY: Writing Percents as Fractions

Work with a partner.

- What percent of each diagram in Activity 1 is shaded?
- What percent of each diagram below is shaded? Write each percent as a fraction in simplest form.

a. 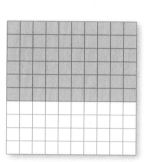 b. 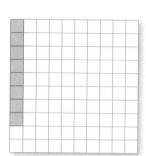 c.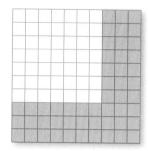

3 ACTIVITY: Writing Fractions as Percents

Work with a partner. Draw a model to represent the fraction. How can you write the fraction as a percent?

a. $\dfrac{2}{5} = \dfrac{\boxed{}}{100} = \boxed{}\%$

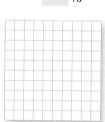

Math Practice

Consider Similar Problems

How is this problem similar to ones you have seen before? How does this help you find the solution?

b. $\dfrac{7}{10}$ c. $\dfrac{3}{5}$

d. $\dfrac{3}{4}$ e. $\dfrac{3}{25}$

What Is Your Answer?

4. **IN YOUR OWN WORDS** What is the connection between ratios, fractions, and percents? Give an example with your answer.

5. **REASONING** Your score on a test is 110%. What does this mean?

Practice Use what you learned about percents to complete Exercises 5–7 on page 222.

Section 5.5   Percents   219

5.5 Lesson

Key Vocabulary
percent, p. 220

Key Idea

Writing Percents as Fractions

Words A **percent** is a part-to-whole ratio where the whole is 100. So, you can write a percent as a fraction with a denominator of 100.

Numbers $60\% = 60 \text{ out of } 100 = \dfrac{60}{100}$ ← part / per / one hundred (whole)

Algebra $n\% = \dfrac{n}{100}$

EXAMPLE 1 — Writing Percents as Fractions

Study Tip
Equivalent fractions and percents represent the same number using different notations.

a. Write 35% as a fraction in simplest form.

$35\% = \dfrac{35}{100}$ Write as a fraction with a denominator of 100.

$= \dfrac{7}{20}$ Simplify.

∴ So, $35\% = \dfrac{7}{20}$.

b. Write 100% as a fraction in simplest form.

$100\% = \dfrac{100}{100}$ Write as a fraction with a denominator of 100.

$= 1$ Simplify.

∴ So, $100\% = 1$.

c. Write 174% as a mixed number in simplest form.

$174\% = \dfrac{174}{100}$ Write as a fraction with a denominator of 100.

$= \dfrac{87}{50}$, or $1\dfrac{37}{50}$ Simplify.

∴ So, $174\% = 1\dfrac{37}{50}$.

On Your Own

Now You're Ready
Exercises 8–19

Write the percent as a fraction or mixed number in simplest form.

1. 5%   2. 168%   3. 36%   4. 83%

220   Chapter 5   Ratios and Rates   Multi-Language Glossary at BigIdeasMath.com

Key Idea

Writing Fractions as Percents

Words Write an equivalent fraction with a denominator of 100. Then write the numerator with the percent symbol.

Numbers $\dfrac{1}{4} \xrightarrow{\times 25}_{\times 25} \dfrac{25}{100} = 25\%$

EXAMPLE 2 **Writing a Fraction as a Percent**

Write $\dfrac{3}{50}$ as a percent.

$\dfrac{3}{50} \xrightarrow{\times 2}_{\times 2} \dfrac{6}{100} = 6\%$

Because 50 × 2 = 100, multiply the numerator and denominator by 2. Write the numerator with a percent symbol.

EXAMPLE 3 **Real-Life Application**

A drought affects 9 out of 12 midwestern states. What percent of the midwestern states are affected by the drought?

Midwestern United States

$\dfrac{9}{12} = \dfrac{3}{4}$ Simplify.

$= \dfrac{75}{100}$ ← $\boxed{\dfrac{3 \times 25}{4 \times 25} = \dfrac{75}{100}}$

$= 75\%$ Write the numerator with a percent symbol.

∴ So, 75% of the midwestern states are affected by the drought.

On Your Own

Now You're Ready
Exercises 21–28

Write the fraction or mixed number as a percent.

5. $\dfrac{31}{50}$ **6.** $\dfrac{7}{25}$ **7.** $\dfrac{19}{20}$ **8.** $1\dfrac{1}{2}$

9. WHAT IF? In Example 3, it rains in all the midwestern states. In what percent of the states affected by drought does it rain?

5.5 Exercises

Vocabulary and Concept Check

1. **WRITING** Explain how you can use a 10-by-10 grid to model 42%.

2. **WHICH ONE DOESN'T BELONG?** Which one does *not* have the same value as the other three? Explain your reasoning.

   $\frac{10}{100}$ 10% $\frac{1}{10}$ 0.01

3. **OPEN-ENDED** Write three different fractions that are less than 40%.

4. **NUMBER SENSE** Can $1\frac{1}{4}$ be written as a percent? Explain.

Practice and Problem Solving

Use a 10-by-10 grid to model the percent.

5. 10%
6. 55%
7. 35%

Write the percent as a fraction or mixed number in simplest form.

8. 45%
9. 90%
10. 15%
11. 7%
12. 34%
13. 79%
14. 77.5%
15. 188%
16. 8%
17. 224%
18. 0.25%
19. 0.4%

20. **ERROR ANALYSIS** Describe and correct the error in writing 225% as a fraction.

    ✗ $225\% = \frac{225}{1000} = \frac{9}{40}$

Write the fraction or mixed number as a percent.

21. $\frac{1}{10}$
22. $\frac{1}{5}$
23. $\frac{11}{20}$
24. $\frac{2}{25}$
25. $\frac{27}{50}$
26. $\frac{18}{25}$
27. $1\frac{17}{20}$
28. $2\frac{41}{50}$

29. **ERROR ANALYSIS** Describe and correct the error in writing $\frac{14}{25}$ as a percent.

    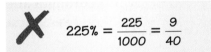
    ✗ $\frac{14}{25} = \frac{14 \times 4}{25 \times 4} = \frac{56}{100} = 0.56\%$

30. **LEFT-HANDED** Of the students in your class, 12% are left-handed. What *fraction* of the students are left-handed? Are there more right-handed or left-handed students? Explain.

31. **ARCADE** You have 125% of the tickets required for a souvenir. What *fraction* of the required tickets do you have? Do you need more tickets for the souvenir? Explain.

222 Chapter 5 Ratios and Rates

Find the percent.

32. 3 is what percent of 8?

33. 13 is what percent of 16?

34. 9 is what percent of 16?

35. 33 is what percent of 40?

36. SOCIAL NETWORKING A survey asked students to choose their favorite social networking website. The results are shown in the table.

Social Networking Website	Number of Students
Website A	35
Website B	13
Website C	22
Website D	10

  a. What fraction of the students chose Website A?

  b. What percent of the students chose Website C?

37. GEOGRAPHY The percent of the total area of the United States that is in each of four states is shown.

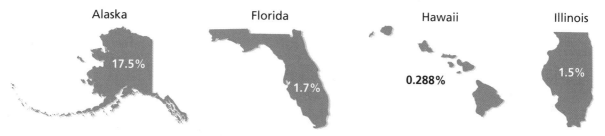

Alaska 17.5% Florida 1.7% Hawaii 0.288% Illinois 1.5%

  a. Write the percents as fractions in simplest form.

  b. How many times larger is Illinois than Hawaii?

  c. Compared to the map of Florida, is the map of Alaska the correct size? Explain your reasoning.

  d. **RESEARCH** Which of the 50 states are larger than Illinois?

38. CRITICAL THINKING A school fundraiser raised 120% of its goal last year and 125% of its goal this year. Did the fundraiser raise more money this year? Explain your reasoning.

39. CRITICAL THINKING How can you use a 10-by-10 grid to model $\frac{1}{2}$%?

40. **Reasoning** Write $\frac{1}{12}$ as a percent. Explain how you found your answer.

Fair Game Review What you learned in previous grades & lessons

Divide. Write the answer in simplest form. *(Section 2.2)*

41. $\frac{1}{6} \div \frac{1}{3}$

42. $9 \div \frac{3}{4}$

43. $10 \div \frac{5}{8}$

44. $\frac{1}{6} \div 2$

45. MULTIPLE CHOICE Which of the following is *not* equal to 15? *(Section 2.1)*

  Ⓐ $\frac{3}{4} \cdot 20$ Ⓑ $\frac{5}{9} \cdot 27$ Ⓒ $35 \cdot \frac{3}{7}$ Ⓓ $28 \cdot \frac{5}{7}$

5.6 Solving Percent Problems

Essential Question How can you use mental math to find the percent of a number?

"I have a secret way for finding 21% of 80."

"10% is 8, and 1% is 0.8."

"So, 21% is 8 + 8 + 0.8 = 16.8."

1 ACTIVITY: Finding 10% of a Number

Work with a partner.

a. How did Newton know that 10% of 80 is 8?

Write 10% as a fraction. $10\% = \dfrac{\boxed{}}{\boxed{}} = \dfrac{1}{\boxed{}}$

(10 ← per, ← cent)

Method 1: Use a model.

0% 10% 20% 30% 40% 50% 60% 70% 80% 90% 100%

0   80

Method 2: Use multiplication.

$$10\% \text{ of } 80 = \dfrac{\boxed{}}{10} \text{ of } 80 = \dfrac{\boxed{}}{10} \times \boxed{} = \dfrac{\boxed{}}{10} = \boxed{}$$

Percents

In this lesson, you will
- find percents of numbers.
- find the whole given the part and the percent.

b. How do you move the decimal point to find 10% of a number?

Move the decimal point one place to the $\boxed{}$.   10% of 80. = $\boxed{}$

2 ACTIVITY: Finding 1% of a Number

Work with a partner.

a. How did Newton know that 1% of 80 is 0.8?

b. How do you move the decimal point to find 1% of a number?

224   Chapter 5   Ratios and Rates

3 ACTIVITY: Using Mental Math

Math Practice

Evaluate Results
Does your answer seem reasonable? How can you check your answer?

Work with a partner. Use mental math to find each percent of a number.

a. 12% of 40

Think: 12% = 10% + 1% + 1%

b. 19% of 50

Think: 19% = 10% + 10% − 1%

10% of 40 = ☐   1% of 40 = ☐   10% of 50 = ☐   1% of 50 = ☐

☐ + ☐ + ☐ = ☐   ☐ + ☐ − ☐ = ☐

4 ACTIVITY: Using Mental Math

Work with a partner. Use mental math to find each percent of a number.

a. 20% tip for a $30 meal

b. 18% tip for a $30 meal

c. 6% sales tax on a $20 shirt

d. 9% sales tax on a $20 shirt

e. 6% service charge for a $200 boxing ticket

f. 2% delivery fee for a $200 boxing ticket

g. 21% bonus on a total of 40,000 points

h. 38% bonus on a total of 80,000 points

What Is Your Answer?

5. **IN YOUR OWN WORDS** How can you use mental math to find the percent of a number?

6. Describe two real-life examples of finding a percent of a number.

7. How can you use 10% of a number to find 20% of the number? 30%? Explain your reasoning.

Use what you learned about finding the percent of a number to complete Exercises 3–10 on page 229.

Section 5.6   Solving Percent Problems   225

5.6 Lesson

Key Idea

Finding the Percent of a Number

Words Write the percent as a fraction. Then multiply by the whole. The percent times the whole equals the part.

Numbers 20% of 60 is 12.

$$\frac{1}{5} \times 60 = 12$$

Model

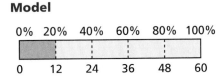

EXAMPLE 1 — Finding the Percent of a Number

25% of 40 is what number?

$$25\% \text{ of } 40 = \frac{1}{4} \cdot 40 \qquad \text{Write the percent as a fraction and multiply.}$$

$$= \frac{1 \cdot \overset{10}{\cancel{40}}}{1 \; \cancel{4}} \qquad \text{Divide out the common factor.}$$

$$= 10 \qquad \text{Simplify.}$$

So, 25% of 40 is 10.

Study Tip

You can use mental math to check your answer in Example 1.
10% of 40 = 4
5% of 40 = 2
So, 25% of 40 is
4 + 4 + 2 = 10.

You can also use a ratio table to find the percent of a number.

EXAMPLE 2 — Finding the Percent of a Number Using a Ratio Table

60% of 150 is what number?

Use a ratio table to find the part. Let one row be the *part*, and let the other be the *whole*. Find an equivalent ratio with 150 as the whole.

The first column represents the percent.

$$\frac{\text{part}}{\text{whole}} = \frac{60}{100} = 60\%$$

	÷2	×3
Part	60	30	90
Whole	100	50	150

So, 60% of 150 is 90.

On Your Own

Now You're Ready
Exercises 3–22

Find the percent of the number. Explain your method.

1. 90% of 20
2. 75% of 32
3. 10% of 110
4. 30% of 75

You can use a related division equation to find the whole given the part and the percent.

Key Idea

Finding the Whole

Write the percent as a fraction. Then divide the part by the fraction.

Words The part divided by the percent equals the whole.

Numbers 20% of 60 is 12.

$$\frac{1}{5} \times 60 = 12 \longrightarrow 12 \div \frac{1}{5} = 60$$

Multiplication equation   Related division equation

EXAMPLE 3 Finding the Whole

75% of what number is 48?

$48 \div 75\% = 48 \div \frac{3}{4}$ Write the percent as a fraction and divide.

$ = 48 \cdot \frac{4}{3}$ Multiply by the reciprocal.

$ = 64$ Simplify.

∴ So, 75% of 64 is 48.

0%	25%	50%	75%	100%
0	16	32	48	64

EXAMPLE 4 Finding the Whole Using a Ratio Table

120% of what number is 72?

Use a ratio table to find the whole. Find an equivalent ratio with 72 as the part.

The first column represents the percent.

$\frac{\text{part}}{\text{whole}} = \frac{120}{100} = 120\%$

÷ 20   × 12

Part	120	6	72
Whole	100	5	60

÷ 20   × 12

∴ So, 120% of 60 is 72.

On Your Own

Now You're Ready
Exercises 27–36

Find the whole. Explain your method.

5. 5% of what number is 10?   **6.** 62% of what number is 31?

EXAMPLE 5 Real-Life Application

The width of a rectangular room is 80% of its length. What is the area of the room?

Find 80% of 15 feet.

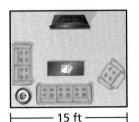

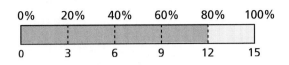

$$80\% \text{ of } 15 = \frac{4}{5} \times 15$$

$$= \frac{4 \times \overset{3}{\cancel{15}}}{\underset{1}{\cancel{5}}}$$

$$= 12 \qquad \text{The width is 12 feet.}$$

Use the formula for the area A of a rectangle.

$$A = 15 \times 12 = 180$$

∴ So, the area of the room is 180 square feet.

On Your Own

7. The width of a rectangular stage is 55% of its length. The stage is 120 feet long. What is the area?

EXAMPLE 6 Real-Life Application

You win an online auction for concert tickets. Your winning bid is 60% of your maximum bid. How much more were you willing to pay for the tickets than you actually paid?

- **Ⓐ** $72
- **Ⓑ** $80
- **Ⓒ** $120
- **Ⓓ** $200

Your maximum bid is the *whole*, and your winning bid is the *part*. Find your maximum bid by dividing the part by the percent.

$$120 \div 60\% = 120 \div \frac{3}{5} \qquad \text{Divide the part by the percent.}$$

$$= 120 \cdot \frac{5}{3} \qquad \text{Multiply by the reciprocal.}$$

$$= 200 \qquad \text{Simplify.}$$

Your maximum bid is $200, and your winning bid is $120. So, you were willing to pay $200 - 120 = \$80$ more for the tickets.

∴ The correct answer is **Ⓑ**.

On Your Own

8. **WHAT IF?** Your winning bid is 96% of your maximum bid. How much more were you willing to pay for the tickets than you actually paid?

5.6 Exercises

Vocabulary and Concept Check

1. **DIFFERENT WORDS, SAME QUESTION** Which is different? Find "both" answers.

What is twenty percent of 30?	What is one-fifth of 30?
Twenty percent of what number is 30?	What is two-tenths of 30?

2. **NUMBER SENSE** If 52 is 130% of a number, is the number greater or less than 52? Explain.

Practice and Problem Solving

Find the percent of the number. Explain your method.

3. 20% of 60
4. 10% of 40
5. 50% of 70
6. 30% of 30
7. 10% of 90
8. 15% of 20
9. 25% of 50
10. 5% of 60
11. 30% of 70
12. 75% of 48
13. 45% of 45
14. 92% of 19
15. 40% of 60
16. 38% of 22
17. 70% of 20
18. 87% of 55
19. 140% of 60
20. 120% of 33
21. 175% of 54
22. 250% of 146

23. **ERROR ANALYSIS** Describe and correct the error in finding 40% of 75.

    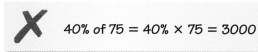
    ✗ 40% of 75 = 40% × 75 = 3000

24. **PINE TREES** A town had about 2120 acres of pine trees 40 years ago. Only about 13% of the pine trees remain. How many acres of pine trees remain?

25. **SPIDER MONKEY** The tail of the spider monkey is 64% of the length shown. What is the length of its tail?

    55 in.

26. **CABLE** A family pays $45 each month for cable television. The cost increases 7%.

    a. How many dollars is the increase?
    b. What is the new monthly cost?

Find the whole. Explain your method.

27. 10% of what number is 14? **28.** 20% of what number is 18?

29. 25% of what number is 21? **30.** 75% of what number is 27?

31. 15% of what number is 12? **32.** 85% of what number is 17?

33. 140% of what number is 35? **34.** 160% of what number is 32?

35. 125% of what number is 25? **36.** 175% of what number is 42?

37. ERROR ANALYSIS Describe and correct the error in finding the whole.

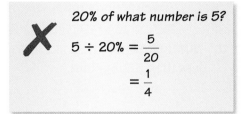

38. COUPON You have a coupon for a restaurant. You save $3 on a meal. What was the original cost of the meal?

39. SURVEY The results of a survey are shown at the right. In the survey, 12 students said that they would like to learn French.

  a. How many students were surveyed?

  b. How many of the students surveyed would like to learn Spanish?

Which language would you like to learn?
Spanish 36%
French 24%
German 18%
Other 22%

40. WEIGHT A sixth grader weighs 90 pounds, which is 120% of what he weighed in fourth grade. How much did he weigh in fourth grade?

41. PARKING LOT In a parking lot, 16% of the cars are blue. There are 4 blue cars in the parking lot. How many cars in the parking lot are *not* blue?

42. LOTION A bottle contains 20 fluid ounces of lotion and sells for $5.80. The 20-fluid-ounce bottle contains 125% of the lotion in the next smallest size, which sells for $5.12. Which is the better buy? Explain.

Copy and complete the statement using <, >, or =.

43. 80% of 60 ⬚ 60% of 80

44. 20% of 30 ⬚ 30% of 40

45. 120% of 5 ⬚ 0.8% of 250

46. 85% of 40 ⬚ 25% of 136

47. TIME How many minutes is 40% of 2 hours?

48. LENGTH How many inches is 78% of 3 feet?

49. GEOMETRY The width of the rectangle is 75% of its length.
  a. What is the area of the rectangle?
  b. The length of the rectangle is doubled. What percent of the length is the width now? Explain your reasoning.

24 in.

50. BASKETBALL To pass inspection, a new basketball should bounce back to between 68% and 75% of the starting height. A new ball is dropped from 6 feet and bounces back 4 feet 1 inch. Does the ball pass inspection? Explain.

51. REASONING You know that 15% of a number n is 12. How can you use this to find 30% of n? 45% of n? Explain.

52. SURFBOARD You have a coupon for 10% off the sale price of a surfboard. Which is the better buy? Explain your reasoning.
  • 40% off the regular price
  • 30% off the regular price and then 10% off the sale price

53. On three 150-point geography tests, you earned grades of 88%, 94%, and 90%. The final test is worth 250 points. What *percent* do you need on the final to earn 93% of the total points on all tests?

Fair Game Review What you learned in previous grades & lessons

Multiply. *(Section 2.5)*

54. 0.6×8 **55.** 3.3×5 **56.** 0.74×9 **57.** 2.19×12

58. MULTIPLE CHOICE What is the quotient of 75 and 2.4? *(Section 2.6)*

  Ⓐ 0.032   Ⓑ 0.3125   Ⓒ 3.2   Ⓓ 31.25

5.7 Converting Measures

Essential Question How can you compare lengths between the customary and metric systems?

1 ACTIVITY: Customary Measure History

Work with a partner.

a. Match the measure of length with its historical beginning.

Length	Historical Beginning
Inch	The length of a human foot
Foot	The width of a human thumb
Yard	The distance a human can walk in 1000 paces (1 pace = 2 steps)
Mile	The distance from a human nose to the end of an outstretched human arm

b. Use a ruler to measure your thumb, arm, and foot. How do your measurements compare to your answers from part (a)? Are they close to the historical measures?

You know how to convert measures within the customary and metric systems.

Converting Measures

In this lesson, you will
- use conversion factors (rates) to convert units of measurement.

Equivalent Customary Lengths

1 ft = 12 in.   1 yd = 3 ft   1 mi = 5280 ft

Equivalent Metric Lengths

1 m = 1000 mm   1 m = 100 cm   1 km = 1000 m

You will learn how to convert between the two systems.

Converting Between Systems

1 in. = 2.54 cm

1 mi ≈ 1.61 km

2 ACTIVITY: Comparing Measures

Math Practice

Analyze Givens
What is the relationship between the given quantities? What are you trying to find?

Work with a partner. Answer each question. Explain your answer. Use a diagram in your explanation.

		Metric	*Customary*
a.	**Car Speed:** Which is faster?	80 km/h	60 mi/h
b.	**Trip Distance:** Which is farther?	200 km	200 mi
c.	**Human Height:** Who is taller?	180 cm	5 ft 8 in.
d.	**Wrench Width:** Which is wider?	8 mm	5/16 in.
e.	**Swimming Pool Depth:** Which is deeper?	1.4 m	4 ft

3 ACTIVITY: Changing Units in a Rate

Work with a partner. Change the units of the rate by multiplying by a "Magic One." Write your answer as a unit rate. Show your work.

Original Rate   *Magic One*   *New Units*   *Unit Rate*

a. Sample:

$$\frac{\$120}{\cancel{h}} \times \frac{1\cancel{h}}{60\text{ min}} = \frac{\$120}{60\text{ min}} = \frac{\$2}{1\text{ min}}$$

b. $\dfrac{\$3}{\text{min}} \times 1 = \$\dfrac{\boxed{}}{1\text{ h}}$

c. $\dfrac{12\text{ in.}}{\text{ft}} \times 1 = \dfrac{\boxed{}\text{ in.}}{1\text{ yd}}$

d. $\dfrac{2\text{ ft}}{\text{week}} \times 1 = \dfrac{\boxed{}\text{ ft}}{1\text{ yr}}$

What Is Your Answer?

4. One problem-solving strategy is called *Working Backwards*. What does this mean? How can you use this strategy to find the rates in Activity 3?

5. **IN YOUR OWN WORDS** How can you compare lengths between the customary and the metric systems? Give examples with your description.

Practice

Use what you learned about converting measures between systems to complete Exercises 4 and 5 on page 236.

5.7 Lesson

Key Vocabulary
U.S. customary system, *p. 234*
metric system, *p. 234*
conversion factor, *p. 234*
unit analysis, *p. 234*

The **U.S. customary system** is a system of measurement that contains units for length, capacity, and weight. The **metric system** is a decimal system of measurement, based on powers of 10, that contains units for length, capacity, and mass.

To convert from one unit of measure to another, multiply by one or more *conversion factors*. A conversion factor can be written using fraction notation.

Conversion Factor

A **conversion factor** is a rate that equals 1.

	Relationship	*Conversion Factors*
Example	1 m ≈ 3.28 ft	$\dfrac{1 \text{ m}}{3.28 \text{ ft}}$ and $\dfrac{3.28 \text{ ft}}{1 \text{ m}}$

You can use **unit analysis** to decide which conversion factor will produce the appropriate units.

EXAMPLE 1 Converting Units

a. Convert 36 quarts to gallons.

Use a conversion factor. *(1 gal = 4 qt)*

$$36 \text{ qt} \cdot \frac{1 \text{ gal}}{4 \text{ qt}} = \frac{36 \cdot 1 \text{ gal}}{4}$$

$$= 9 \text{ gal}$$

∴ So, 36 quarts is 9 gallons.

b. Convert 20 centimeters to inches.

Use a conversion factor. *(1 in. = 2.54 cm)*

$$20 \text{ cm} \cdot \frac{1 \text{ in.}}{2.54 \text{ cm}} \approx 7.87 \text{ in.}$$

∴ So, 20 centimeters is about 7.87 inches.

On Your Own

Now You're Ready
Exercises 6–17

Copy and complete the statement. Round to the nearest hundredth if necessary.

1. 48 ft = ___ yd
2. 7 lb = ___ oz
3. 5 g = ___ mg
4. 7 mi ≈ ___ km
5. 12 qt ≈ ___ L
6. 25 kg ≈ ___ lb

EXAMPLE 2 Comparing Units

Copy and complete the statement using < or >: 25 oz ☐ 2 kg.

Convert 25 ounces to kilograms.

$$25 \text{ oz} \approx 25 \text{ oz} \times \frac{1 \text{ lb}}{16 \text{ oz}} \times \frac{0.45 \text{ kg}}{1 \text{ lb}} = \frac{25 \cdot 1 \cdot 0.45 \text{ kg}}{16 \cdot 1} \approx 0.70 \text{ kg}$$

- 1 lb = 16 oz
- 1 lb ≈ 0.45 kg

∴ Because 0.70 kilogram is less than 2 kilograms, 25 oz < 2 kg.

EXAMPLE 3 Converting a Rate: Changing One Unit

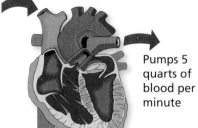

Pumps 5 quarts of blood per minute

How many liters does the human heart pump per minute?

$$\frac{5 \text{ qt}}{1 \text{ min}} \approx \frac{5 \text{ qt}}{1 \text{ min}} \cdot \frac{0.95 \text{ L}}{1 \text{ qt}} = \frac{4.75 \text{ L}}{1 \text{ min}}$$

1 qt ≈ 0.95 L

∴ The rate of 5 quarts per minute is about 4.75 liters per minute.

EXAMPLE 4 Converting a Speed: Changing Both Units

You are riding on a zip line. Your speed is 15 miles per hour. What is your speed in feet per second?

$$\frac{15 \text{ mi}}{1 \text{ h}} \left(\frac{5280 \text{ ft}}{1 \text{ mi}} \right) \left(\frac{1 \text{ h}}{3600 \text{ sec}} \right) = \frac{15 \cdot 5280 \text{ ft}}{3600 \text{ sec}}$$

$$= \frac{79{,}200 \text{ ft}}{3600 \text{ sec}}$$

$$= \frac{22 \text{ ft}}{1 \text{ sec}}$$

- 1 mi = 5280 ft
- 1 h = 3600 sec

∴ Your speed is 22 feet per second.

On Your Own

Now You're Ready
Exercises 20–31

Copy and complete the statement using < or >.

7. 7 cm ☐ 3 in.   8. 8 c ☐ 2 L   9. 3 oz ☐ 70 g

10. An oil tanker is leaking oil at a rate of 300 gallons per minute. What is this rate in gallons per second?

11. A tennis ball travels at a speed of 120 miles per hour. What is this rate in feet per second?

Section 5.7 Converting Measures

5.7 Exercises

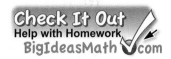

Vocabulary and Concept Check

1. **VOCABULARY** Is $\frac{10 \text{ mm}}{1 \text{ cm}}$ a conversion factor? Explain.

2. **WRITING** Describe how to convert 2 liters per hour to milliliters per second.

3. **DIFFERENT WORDS, SAME QUESTION** Which is different? Find "both" answers.

   Convert 5 inches to centimeters.   Find the number of inches in 5 centimeters.

   How many centimeters are in 5 inches?   Five inches equals how many centimeters?

Practice and Problem Solving

Answer the question. Explain your answer.

4. Which juice container is larger: 2 L or 1 gal?

5. Which person is heavier: 75 kg or 110 lb?

Copy and complete the statement. Round to the nearest hundredth if necessary.

6. 3 pt = ___ c
7. 1500 mL = ___ L
8. 40 oz = ___ lb
9. 12 L ≈ ___ qt
10. 14 m ≈ ___ ft
11. 4 ft ≈ ___ m
12. 64 lb ≈ ___ kg
13. 0.3 km ≈ ___ mi
14. 75.2 in. ≈ ___ cm
15. 17 kg ≈ ___ lb
16. 15 cm ≈ ___ in.
17. 9 mi ≈ ___ km

18. **ERROR ANALYSIS** Describe and correct the error in converting the units.

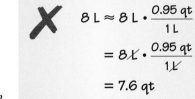

19. **BRIDGE** The Mackinac Bridge in Michigan is the third-longest suspension bridge in the United States.

    a. How high above the water is the roadway in meters?

    b. The bridge has a length of 26,372 feet. What is the length in kilometers?

Copy and complete the statement using < or >.

20. 8 kg ▢ 30 oz
21. 6 ft ▢ 300 cm
22. 3 gal ▢ 6 L
23. 10 in. ▢ 200 mm
24. 1200 g ▢ 5 lb
25. 1500 m ▢ 3000 ft

Copy and complete the statement.

26. $\dfrac{13 \text{ km}}{\text{h}} \approx \dfrac{\text{▢ mi}}{\text{h}}$

27. $\dfrac{22 \text{ L}}{\text{min}} = \dfrac{\text{▢ L}}{\text{h}}$

28. $\dfrac{63 \text{ mi}}{\text{h}} = \dfrac{\text{▢ mi}}{\text{sec}}$

29. $\dfrac{3 \text{ km}}{\text{min}} \approx \dfrac{\text{▢ mi}}{\text{h}}$

30. $\dfrac{17 \text{ gal}}{\text{h}} \approx \dfrac{\text{▢ qt}}{\text{min}}$

31. $\dfrac{6 \text{ cm}}{\text{min}} = \dfrac{\text{▢ m}}{\text{sec}}$

32. **BOTTLE** Can you pour the water from a full 2-liter bottle into a 2-quart pitcher without spilling any? Explain.

33. **AUTOBAHN** Germany suggests a speed limit of 130 kilometers per hour on highways.

   a. Is the speed shown greater than the suggested limit?

   b. Suppose the speed shown drops 30 miles per hour. Is the new speed below the suggested limit?

34. **BIRDS** The table shows the flying speeds of several birds.

   a. Which bird is the fastest? Which is the slowest?

   b. The peregrine falcon has a dive speed of 322 kilometers per hour. Is the dive speed of the peregrine falcon faster than the flying speed of any of the birds? Explain.

35. **SPEED OF LIGHT** The speed of light is about 300,000 kilometers per second. Convert the speed to miles per hour.

36. **Critical Thinking** One liter of paint covers 100 square feet. How many gallons of paint does it take to cover a room whose walls have an area of 800 square meters?

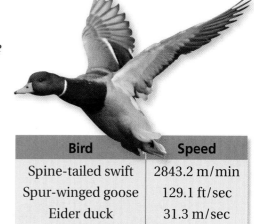

Bird	Speed
Spine-tailed swift	2843.2 m/min
Spur-winged goose	129.1 ft/sec
Eider duck	31.3 m/sec
Mallard	65 mi/h

Fair Game Review   What you learned in previous grades & lessons

Find the percent of the number. *(Section 5.6)*

37. 25% of 120
38. 65% of 47
39. 120% of 15
40. 3.2% of 80

41. **MULTIPLE CHOICE** What is the area of a parallelogram with a base of 15 centimeters and a height of 12 centimeters? *(Section 4.1)*

   Ⓐ 90 cm²   Ⓑ 175 cm²   Ⓒ 180 cm²   Ⓓ 205 cm²

5.5–5.7 Quiz

Write the percent as a fraction or mixed number in simplest form. *(Section 5.5)*

1. 14%
2. 124%

Write the fraction or mixed number as a percent. *(Section 5.5)*

3. $\frac{13}{20}$
4. $1\frac{1}{4}$

Find the percent of the number. Explain your method. *(Section 5.6)*

5. 25% of 64
6. 120% of 50

Find the whole. Explain your method. *(Section 5.6)*

7. 60% of what number is 24?
8. 160% of what number is 80?

Copy and complete the statement. Round to the nearest hundredth if necessary. *(Section 5.7)*

9. 6.4 in. ≈ ▢ cm
10. 4 qt ≈ ▢ L
11. 10 kg ≈ ▢ lb

12. **ANATOMY** About 62% of the human body is composed of water. Write this percent as a fraction in simplest form. *(Section 5.5)*

13. **SAVES** A goalie's saves (•) and goals scored against (×) are shown. What percent of shots did the goalie save? Explain. *(Section 5.5)*

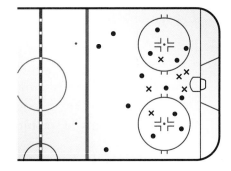

14. **SHOPPING** You went to the mall with $80. You spent 25% of your money on a pair of shorts and 65% of the remainder on sandals. How much did you spend on the sandals? Explain how you found your answer. *(Section 5.6)*

15. **WINDSURFING** Determine which windsurfer is traveling faster. Explain your reasoning. *(Section 5.7)*

Speed:
5 meters per second

Speed:
720 feet per minute

5 Chapter Review

Review Key Vocabulary

ratio, *p. 192*
equivalent ratios, *p. 198*
ratio table, *p. 198*
rate, *p. 206*
unit rate, *p. 206*
equivalent rates, *p. 206*
percent, *p. 220*
U.S. customary system, *p. 234*
metric system, *p. 234*
conversion factor, *p. 234*
unit analysis, *p. 234*

Review Examples and Exercises

5.1 Ratios (pp. 190–195)

Write the ratio of apples to oranges. Explain what the ratio means.

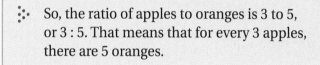

So, the ratio of apples to oranges is 3 to 5, or 3 : 5. That means that for every 3 apples, there are 5 oranges.

Exercises

Write the ratio. Explain what the ratio means.

1. butterflies : caterpillars
2. saxophones : trumpets

5.2 Ratio Tables (pp. 196–203)

Find the missing values in the ratio table. Then write the equivalent ratios.

Trees	2	6	
Birds	5		30

You can use multiplication to find the missing values.

The equivalent ratios are 2 : 5, 6 : 15, and 12 : 30.

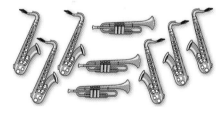

Exercises

Find the missing values in the ratio table. Then write the equivalent ratios.

3.
Levers	6		18
Pulleys	3	6	

4.
Cars	3	6	
Trucks	4		24

5.3 Rates (pp. 204–209)

A horse can run 165 feet in 3 seconds. At this rate, how far can the horse run in 5 seconds?

Using a ratio table, divide to find the unit rate. Then multiply to find the distance that the horse can run in 5 seconds.

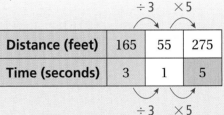

∴ So, the horse can run 275 feet in 5 seconds.

Exercises

Write a unit rate for the situation.

5. 12 stunts in 4 movies

6. 3600 stitches in 3 minutes

7. **MUSIC** A song has 28 beats in 4 seconds. At this rate, how many beats are there in 30 seconds?

5.4 Comparing and Graphing Ratios (pp. 210–215)

There are 24 grams of sugar in 6 fluid ounces of Soft Drink A, and there are 15 grams of sugar in 4 fluid ounces of Soft Drink B. Which soft drink contains more sugar in a 12-ounce can?

Use ratio tables to compare the soft drinks.

Soft Drink A   ×2

Sugar (grams)	24	48
Volume (fluid ounces)	6	12

×2

Soft Drink B   ×3

Sugar (grams)	15	45
Volume (fluid ounces)	4	12

×3

The tables show that a 12-ounce can of Soft Drink A has 48 − 45 = 3 more grams of sugar than Soft Drink B.

∴ So, a 12-ounce can of Soft Drink A has more sugar.

Exercises

8. **TUNA** A 5-ounce can of tuna costs $0.90. A 12-ounce can of tuna costs $2.40. Which is the better buy?

5.5 Percents (pp. 218–223)

Write $\dfrac{3}{20}$ as a percent.

$\dfrac{3}{20} = \dfrac{15}{100} = 15\%$ (×5 numerator and denominator)

Because 20 × 5 = 100, multiply the numerator and denominator by 5. Write the numerator with a percent symbol.

Exercises

Write the percent as a fraction or mixed number in simplest form.

9. 12%
10. 88%
11. 0.8%

Write the fraction or mixed number as a percent.

12. $\dfrac{3}{5}$
13. $\dfrac{43}{25}$
14. $1\dfrac{21}{50}$

5.6 Solving Percent Problems (pp. 224–231)

a. 75% of 80 is what number?

$75\% \text{ of } 80 = \dfrac{3}{4} \times 80 = \dfrac{3 \times \overset{20}{\cancel{80}}}{\underset{1}{\cancel{4}}} = 60$

So, 75% of 80 is 60.

b. 30% of what number is 27?

$27 \div 30\% = 27 \div \dfrac{3}{10} = \overset{9}{\cancel{27}} \cdot \dfrac{10}{\underset{1}{\cancel{3}}} = 90$

So, 30% of 90 is 27.

Exercises

Find the percent of the number. Explain your method.

15. 60% of 80
16. 80% of 55
17. 150% of 48

Find the whole. Explain your method.

18. 70% of what number is 35?
19. 140% of what number is 56?

5.7 Converting Measures (pp. 232–237)

Convert 8 kilometers to miles.

$8 \text{ km} \times \dfrac{1 \text{ mi}}{1.6 \text{ km}} \approx 5 \text{ mi}$

Because 1 mi ≈ 1.6 km, use the ratio $\dfrac{1 \text{ mi}}{1.6 \text{ km}}$.

Exercises

Copy and complete the statement. Round to the nearest hundredth if necessary.

20. 3 L ≈ ___ qt
21. 9.2 in. ≈ ___ cm
22. 15 lb ≈ ___ kg

5 Chapter Test

Write the ratio. Explain what the ratio means.

1. scooters : bikes

2. starfish : seashells

Find the missing values in the ratio table. Then write the equivalent ratios.

3.
Lemons	4		36
Limes	2	6	

4.
Rabbits	2	4	
Hamsters		9	54

Write a unit rate for the situation.

5. $54.00 for 3 tickets

6. 210 miles in 3 hours

Write the fraction or mixed number as a percent.

7. $\frac{21}{25}$

8. $\frac{17}{20}$

9. $1\frac{2}{5}$

Find the percent of the number. Explain your method.

10. 80% of 90

11. 30% of 50

12. 120% of 75

Find the whole. Explain your method.

13. 34 is 40% of what number?

14. 52 is 130% of what number?

Copy and complete the statement. Round to the nearest hundredth if necessary.

15. 5 L ≈ ___ qt

16. 56 lb ≈ ___ kg

17. **SOUP** There are 600 milligrams of sodium in 4 ounces of Soup A, and there are 720 milligrams of sodium in 6 ounces of Soup B. You prepare an 18-ounce bowl of each soup. Which bowl of soup contains more sodium?

18. **ORANGE JUICE** A 48-fluid-ounce container of orange juice costs $2.40. A 60-fluid-ounce container of orange juice costs $3.60. Which is the better buy?

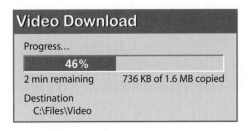

19. **DOWNLOAD** Your computer displays the progress of a downloading video. What fraction of the video is downloaded? Write your answer in simplest form.

20. **GLASSES** In a class of 20 students, 40% are boys. Twenty-five percent of the boys and 50% of the girls wear glasses. How many students wear glasses?

5 Cumulative Assessment

1. What is the value of the expression below?

$$8\frac{4}{9} \div 4\frac{2}{3}$$

   A. $1\frac{17}{21}$

   B. $2\frac{2}{3}$

   C. $32\frac{8}{27}$

   D. $39\frac{11}{27}$

2. Which fraction is *not* equivalent to 25%?

   F. $\frac{1}{4}$

   G. $\frac{2}{5}$

   H. $\frac{5}{20}$

   I. $\frac{25}{100}$

3. The school store sells 4 pencils for $0.50. At that rate, what would be the cost of 10 pencils?

   A. $1.10

   B. $1.25

   C. $2.00

   D. $5.00

4. Which expression is equivalent to the expression below?

$$2(m + n)$$

   F. $2m \times 2n$

   G. $2m + 2n$

   H. $(2 + m) \times (2 + n)$

   I. $(2 + m) + (2 + n)$

5. A service club wants to buy tickets to a baseball game. Tickets are available for the grandstand and for the bleachers.

Grandstand Ticket $25	Bleachers Ticket $15

   Which expression represents the total cost, in dollars, for g grandstand tickets and b bleachers tickets?

   A. $375(g + b)$

   B. $40(g \times b)$

   C. $25g + 15b$

   D. $25g \times 15b$

6. What property was used to simplify the expression?

$$12 \times 47 = 12 \times (40 + 7)$$
$$= 12 \times 40 + 12 \times 7$$
$$= 480 + 84$$
$$= 564$$

   F. Distributive Property

   G. Identity Property of Addition

   H. Commutative Property of Addition

   I. Associative Property of Multiplication

7. What is 15% of 36?

8. If 5 dogs share equally a bag of dog treats, each dog gets 24 treats. Suppose 8 dogs share equally the bag of treats. How many treats does each dog get?

   A. 3

   B. 15

   C. 21

   D. 38

9. The figure below consists of a rectangle and a right triangle.

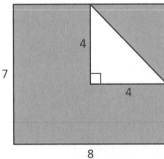

   What is the area of the shaded region?

   F. 23 units^2

   G. 40 units^2

   H. 48 units^2

   I. 60 units^2

10. What is the area, in square inches, of the trapezoid-shaped award?

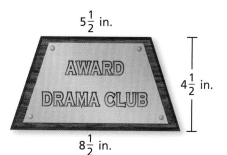

11. Your friend evaluated an expression using $k = 0.5$ and $p = 1.6$ and got an answer of 12. Which expression did your friend evaluate?

　A. $5p + 8k$　　　**C.** $0.5k + 1.6p$

　B. $8p + 5k$　　　**D.** $0.8k + 0.5p$

12. For a party, you made a gelatin dessert in a rectangular pan and cut the dessert into equal-sized pieces as shown below.

The dessert consisted of 5 layers of equal height. Each layer was a different flavor, as shown below by a side view of the pan.

Your guests ate $\frac{3}{5}$ of the pieces of the dessert.

　Part A　Write the amount of cherry gelatin eaten by your guests as a fraction of the total dessert. Justify your answer.

　Part B　Write the amount of cherry gelatin eaten by your guests as a percent of the total dessert. Justify your answer.

6 Integers and the Coordinate Plane

- **6.1** Integers
- **6.2** Comparing and Ordering Integers
- **6.3** Fractions and Decimals on the Number Line
- **6.4** Absolute Value
- **6.5** The Coordinate Plane

"Don't worry. At negative 20 miles per hour, we're still under the speed limit."

"Dear Sir: You asked me to 'find' the opposite of −1."

"I didn't know it was missing."

What You Learned Before

Ordering Decimals

Example 1 Use a number line to order 0.25, 1.15, 0.2, and 0.34 from least to greatest.

Try It Yourself
Use a number line to order the numbers from least to greatest.

1. 0.01, 0.42, 0.2, 0.5
2. 1.05, 0.95, 0.75, 1.01

Comparing Numbers

Complete the number sentence with <, >, or =.

Example 2   10 ▇ 15

On a number line, 10 is closer to zero than 15.

∴ So, 10 < 15.

Example 3 0.875 ▇ $\dfrac{7}{8}$

$$0.875 = \dfrac{875}{1000} = \dfrac{875 \div 125}{1000 \div 125} = \dfrac{7}{8}$$

∴ So, $0.875 = \dfrac{7}{8}$.

Example 4 Find three numbers that make the number sentence $1\dfrac{2}{5} \leq$ ▇ true.

∴ *Sample answer:* $1\dfrac{3}{5}, \dfrac{5}{2}, 2$

Try It Yourself
Complete the number sentence with <, >, or =.

3. 2.01 ▇ 2.001
4. 4.5 ▇ $\dfrac{9}{2}$
5. 3.18 ▇ 3.2

Find three numbers that make the number sentence true.

6. $\dfrac{17}{2} \leq$ ▇
7. $1\dfrac{1}{2} >$ ▇
8. $0.75 \geq$ ▇

6.1 Integers

Essential Question How can you represent numbers that are less than 0?

1 ACTIVITY: Reading Thermometers

Work with a partner. The thermometers show the temperatures in four cities.

Honolulu, Hawaii *Anchorage, Alaska*
Death Valley, California *Seattle, Washington*

Write each temperature. Then match each temperature with its most appropriate location.

a.    b.    c. 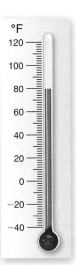   d.

e. How would you describe all the temperatures in relation to 0°F?

2 ACTIVITY: Describing a Temperature

Integers
In this lesson, you will
- understand positive and negative integers and use them to describe real-life situations.
- graph integers on a number line.

Work with a partner. The thermometer shows the coldest temperature ever recorded in Seattle, Washington.

a. What is the temperature?
b. How do you write temperatures that are colder than this?
c. Suppose the record for the coldest temperature in Seattle is broken by 10 degrees. What is the new coldest temperature? Draw a thermometer that shows the new coldest temperature.
d. How is the new coldest temperature different from the temperatures in Activity 1?

248 Chapter 6 Integers and the Coordinate Plane

3 ACTIVITY: Extending the System of Whole Numbers

Math Practice

Maintain Oversight

How does this activity help you represent numbers less than 0?

Work with a partner.

a. Copy and complete the number line using whole numbers only.

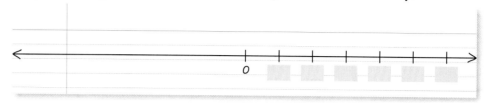

b. Fold the paper with your number line around 0 so that the lines overlap. Make tick marks on the other side of the number line to match the tick marks for the whole numbers.

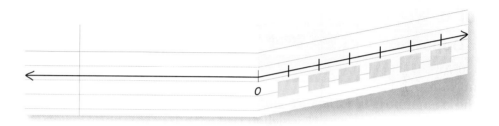

c. **STRUCTURE** Compare this number line to the thermometers from Activities 1 and 2. What do you think the new tick marks represent? How would you label them?

What Is Your Answer?

4. **IN YOUR OWN WORDS** How can you represent numbers that are less than 0?

5. Describe another real-life example that uses numbers that are less than 0.

6. **REASONING** How are the temperatures shown by the thermometers at the right similar? How are they different?

7. **WRITING** The temperature in a town on Thursday evening is 25°F. On Sunday morning, the temperature drops below 0°F. Write a story to describe what may have happened in the town. Be sure to include the temperatures for each day.

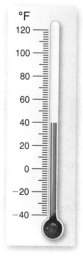
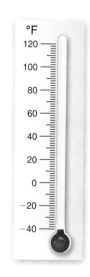

Practice

Use what you learned about positive and negative numbers to complete Exercises 4–7 on page 252.

6.1 Exercises

Vocabulary and Concept Check

1. **VOCABULARY** Which of the following numbers are integers?

   $8, -4.1, -9, \dfrac{1}{6}, 1.75, 22$

2. **OPEN-ENDED** Describe a real-life example that you can represent by -1200.

3. **VOCABULARY** List three words or phrases used in real life that indicate negative numbers.

Practice and Problem Solving

Graph the number that represents the situation on a number line.

4. A football team loses 3 yards.
5. The temperature is 6 degrees below zero.
6. A person climbs 600 feet up a mountain.
7. You earn $15 raking leaves.

Write a positive or negative integer that represents the situation.

8. You withdraw $42 from an account.
9. An airplane climbs to 37,500 feet.
10. The temperature rises 17 degrees.
11. You lose 56 points in a video game.
12. A ball falls 350 centimeters.
13. You receive 5 bonus points in class.

14. **STOCK MARKET** A stock market gains 83 points. The next day, the stock market loses 47 points. Write each amount as an integer.

15. **SCUBA DIVING** The world record for scuba diving is 318 meters below sea level. Write this as an integer.

Graph the integer and its opposite.

16. -5
17. -8
18. 14
19. 9
20. 30
21. -150
22. -32
23. 400

24. **ERROR ANALYSIS** Describe and correct the error in describing positive integers.

   ✗ The positive integers are 0, 1, 2, 3, ...

25. **TEMPERATURE** The highest temperature in February is 25°F. The lowest temperature in February is the opposite of the highest temperature. Graph both temperatures.

252 Chapter 6 Integers and the Coordinate Plane

Identify the integer represented by the point on the number line.

26. A **27.** B **28.** C **29.** D

30. TIDES Use the information below.

- Low tide is 1 foot below the average water level.
- High tide is 5 feet higher than low tide.

Write an integer that represents the average water level relative to high tide.

31. REPEATED REASONING Choose any positive integer.

   a. Find the opposite of the integer. **b.** Find the opposite of the integer in part (a).

   c. What can you conclude about the opposite of the opposite of the integer? Is this true for all integers? Use a number line to justify your answer.

   d. Describe the meaning of $-(-(-6))$. Find its value.

32. **Number Sense** In a game of tug-of-war, a team wins by pulling the flag over its goal line. The flag begins at 0. During a game, the flag moves 8 feet to the right, 12 feet to the left, and 13 feet back to the right. Did a team win? Explain.

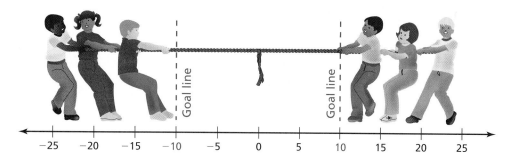

 Fair Game Review What you learned in previous grades & lessons

Order the numbers from least to greatest. *(Skills Review Handbook)*

33. $\frac{7}{8}, \frac{1}{2}, \frac{3}{8}, \frac{3}{4}$

34. 4.5, 4.316, 4.32, 4.312

35. MULTIPLE CHOICE The height of a statue is 276 inches. What is the height of the statue in meters? Round your answer to the nearest hundredth. *(Section 5.7)*

  Ⓐ 1.09 m   **Ⓑ** 7.01 m   **Ⓒ** 108.66 m   **Ⓓ** 701.04 m

6.2 Comparing and Ordering Integers

Essential Question How can you use a number line to order real-life events?

1 ACTIVITY: Seconds to Takeoff

Work with a partner. You are listening to a command center before the liftoff of a rocket.

You hear the following:

"T minus 10 seconds ... go for main engine start ... T minus 9 ... 8 ... 7 ... 6 ... 5 ... 4 ... 3 ... 2 ... 1 ... we have liftoff."

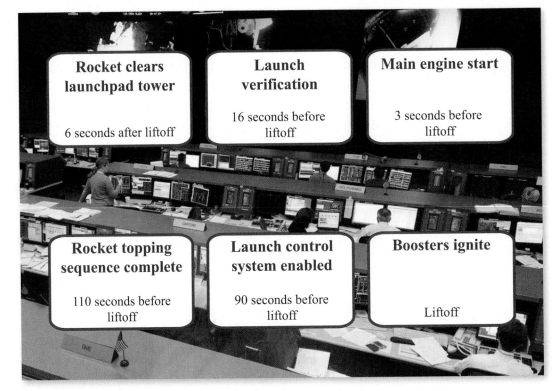

Rocket clears launchpad tower — 6 seconds after liftoff

Launch verification — 16 seconds before liftoff

Main engine start — 3 seconds before liftoff

Rocket topping sequence complete — 110 seconds before liftoff

Launch control system enabled — 90 seconds before liftoff

Boosters ignite — Liftoff

Integers
In this lesson, you will
- use a number line to compare positive and negative integers.
- use a number line to order positive and negative integers for real-life situations.

a. Draw a number line. Then locate the events shown above at appropriate points on the number line.

b. Which event occurs at zero on your number line? Explain.

c. Which of the events occurs first? Which of the events occurs last? How do you know?

d. List the events in the order they occurred.

2 ACTIVITY: Being Careful with Terminology

Work with a partner.

a. Use a number line to show that the phrase "3 seconds away from liftoff" can have two meanings.

b. Reword the phrase "3 seconds away from liftoff" in two ways so that each meaning is absolutely clear.

c. Explain why you must be very careful with terminology if you are working in the command center for a rocket launch.

3 ACTIVITY: A Day in the Life of an Astronaut

Math Practice

Recognize the Usefulness of Tools

Which sources would give you the most accurate information? How do you know you can trust the information you find?

Make a time line that shows a day in the life of an astronaut. Use the Internet or another reference source to gather information.

- Use a number line with units representing hours. Start at 12 hours before liftoff and end at 12 hours after liftoff. Locate the liftoff at 0. Assume liftoff occurs at noon.

- Include at least five events before liftoff, such as when the astronauts suit up.

- Include at least five events after liftoff, such as when the rocket enters Earth's orbit.

- How do you determine where each event occurs on the number line?

What Is Your Answer?

4. **IN YOUR OWN WORDS** How can you use a number line to order real-life events?

5. Describe how you can use a number line to create a time line.

Practice

Use what you learned about number lines to complete Exercises 4–7 on page 258.

Section 6.2 Comparing and Ordering Integers

6.2 Lesson

On a horizontal number line, numbers to the left are less than numbers to the right. Numbers to the right are greater than numbers to the left.

EXAMPLE 1 Comparing Integers on a Horizontal Number Line

Compare 2 and −6.

∴ 2 is to the right of −6. So, 2 > −6.

On a vertical number line, numbers below are less than numbers above. Numbers above are greater than numbers below.

EXAMPLE 2 Comparing Integers on a Vertical Number Line

Compare −5 and −3.

∴ −5 is below −3. So, −5 < −3.

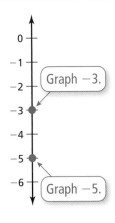

On Your Own

Now You're Ready
Exercises 4–11

Copy and complete the statement using < or >.

1. 0 ▢ −4
2. −5 ▢ 5
3. −8 ▢ −7

EXAMPLE 3 Ordering Integers

Order −4, 3, 0, −1, −2 from least to greatest.

Graph each integer on a number line.

Write the integers as they appear on the number line from left to right.

∴ So, the order from least to greatest is −4, −2, −1, 0, 3.

EXAMPLE 4 — Reasoning with Integers

A number is greater than −8 and less than 0. What is the greatest possible integer value of this number?

Ⓐ −10   Ⓑ −7   Ⓒ −1   Ⓓ 2

Study Tip
In Example 4, you can eliminate Choices A and D because −10 is to the left of −8 and 2 is to the right of 0.

The number is greater than −8 and less than 0. So, the number must be to the right of −8 and to the left of 0 on a horizontal number line.

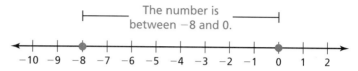

The greatest possible integer value between −8 and 0 is the integer farthest to the right on the number line between these values, which is −1.

∴ So, the correct answer is Ⓒ.

EXAMPLE 5 — Real-Life Application

The diagram shows the coldest recorded temperatures for several cities in Virginia.

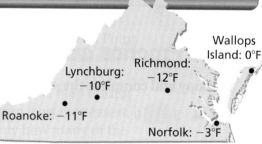

a. Which city has the coldest recorded temperature?

Graph each integer on a vertical number line.

∴ −12 is the lowest on the number line. So, Richmond has the coldest recorded temperature.

b. Has a negative Fahrenheit temperature ever been recorded on Wallops Island? Explain.

∴ The coldest recorded temperature on Wallops Island is 0°F, which is greater than every negative temperature. So, a negative temperature has never been recorded on Wallops Island.

On Your Own

Now You're Ready Exercises 14–19 and 22

Order the integers from least to greatest.

4. −2, −3, 3, 1, −1
5. 4, −7, −8, 6, 1

6. In Example 4, what is the least possible integer value of the number?

7. In Example 5, Norfolk recorded a new record low last night. The new record low is greater than the record low in Lynchburg. What integers can represent the new record low in Norfolk?

6.3 Fractions and Decimals on the Number Line

Essential Question How can you use a number line to compare positive and negative fractions and decimals?

1 ACTIVITY: Locating Fractions on a Number Line

On your time line for "A Day in the Life of an Astronaut" from Activity 3 in Section 6.2, include the following events. Represent each using a fraction or a mixed number.

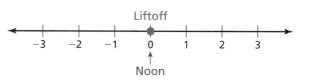

a. Radio Transmission: 10:30 A.M.

b. Space Walk: 7:30 P.M.

c. Physical Exam: 4:45 A.M.

d. Photograph Taken: 3:15 A.M.

Fractions and Decimals
In this lesson, you will
- understand positive and negative numbers and use them to describe real-life situations.
- graph numbers on a number line.

e. Float in the Cabin: 6:20 P.M.

f. Eat Dinner: 8:40 P.M.

2 ACTIVITY: Fractions and Decimals on a Number Line

Math Practice

Make a Plan

How can you find a number between two given numbers?

Work with a partner. Find a number that is between the two numbers. The number must be greater than the green number *and* less than the blue number.

a.

b.

c.

3 ACTIVITY: Decimals on a Number Line

Work with a partner.

Snorkeling: -5 meters

Scuba diving: -50 meters

Deep-sea diving: -700 meters

a. Write the position of each diver in kilometers.

b. **CHOOSE TOOLS** Would a horizontal or a vertical number line be more appropriate for representing these data? Why?

c. Use a number line to order the positions from deepest to shallowest.

What Is Your Answer?

4. **IN YOUR OWN WORDS** How can you use a number line to compare positive and negative fractions and decimals?

5. Draw a number line. Graph and label three values between -2 and -1.

Practice — Use what you learned about fractions and decimals on a number line to complete Exercises 4 and 5 on page 264.

Section 6.3 Fractions and Decimals on the Number Line

6.3 Lesson

In Section 6.1, you learned that integers can be negative. Fractions and decimals can also be negative.

EXAMPLE 1 Graphing Negative Fractions and Decimals

Graph each number and its opposite.

a. $\frac{3}{4}$

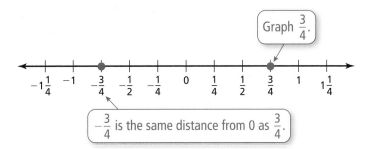

b. -1.6

On Your Own

Now You're Ready
Exercises 6–9

Graph the number and its opposite.

1. $2\frac{1}{2}$
2. $-\frac{4}{5}$
3. -3.5
4. 5.25

EXAMPLE 2 Comparing Fractions and Mixed Numbers

a. Compare $-\frac{1}{2}$ and $-\frac{3}{4}$. b. Compare $-4\frac{5}{6}$ and $-4\frac{1}{6}$.

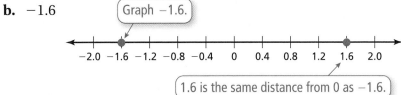

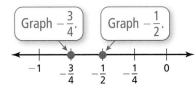

$-\frac{1}{2}$ is to the right of $-\frac{3}{4}$.

∴ So, $-\frac{1}{2} > -\frac{3}{4}$.

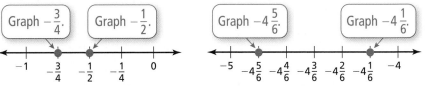

$-4\frac{5}{6}$ is to the left of $-4\frac{1}{6}$.

∴ So, $-4\frac{5}{6} < -4\frac{1}{6}$.

EXAMPLE 3 **Comparing Decimals**

Compare −3.08 and −3.8.

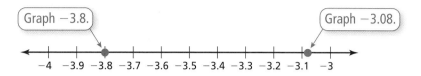

−3.08 is to the right of −3.8.

∴ So, −3.08 > −3.8.

EXAMPLE 4 **Real-Life Application**

A *Chinook wind* is a warm mountain wind that can cause rapid temperature changes. The table shows three of the greatest temperature drops ever recorded after a Chinook wind occurred. On which date did the temperature drop the fastest? Explain.

Date	Temperature Change
January 10, 1911	$-3\frac{1}{10}$ °F per minute
November 10, 1911	$-\frac{5}{8}$ °F per minute
January 22, 1943	$-2\frac{1}{5}$ °F per minute

Graph the numbers on a number line.

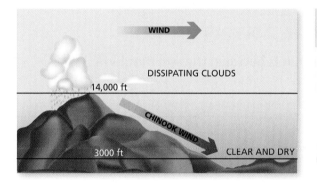

$-3\frac{1}{10}$ is farthest to the left.

∴ So, the temperature dropped the fastest on January 10, 1911.

On Your Own

Now You're Ready
Exercises 10–18
and 20–23

Copy and complete the statement using < or >.

5. $-\frac{4}{7}$ ⬚ $-\frac{1}{7}$

6. $-1\frac{2}{3}$ ⬚ $-1\frac{5}{6}$

7. -0.5 ⬚ 0.3

8. **WHAT IF?** In Example 4, a temperature change of $-3\frac{2}{5}$°F per minute is recorded. How does this temperature change compare with the other temperature changes? Explain.

6.3 Exercises

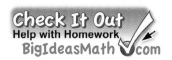

Vocabulary and Concept Check

1. **NUMBER SENSE** Which statement is *not* true?
   a. On a number line, $-2\frac{1}{6}$ is to the left of $-2\frac{2}{3}$.
   b. $-2\frac{2}{3}$ is less than $-2\frac{1}{6}$.
   c. $-2\frac{1}{6}$ is greater than $-2\frac{2}{3}$.
   d. On a number line, $-2\frac{2}{3}$ is to the left of $-2\frac{1}{6}$.

2. **NUMBER SENSE** Is a negative decimal *always*, *sometimes*, or *never* equal to a positive decimal? Explain.

3. **NUMBER SENSE** On a number line, is -2.06 or -2.6 farther to the left?

Practice and Problem Solving

Find a fraction or mixed number that is between the two numbers.

4.

5.

Graph the number and its opposite.

6. $\frac{2}{3}$
7. $-2\frac{1}{4}$
8. -3.8
9. 2.15

Copy and complete the statement using < or >.

10. $-3\frac{1}{3}$ ▢ $-3\frac{2}{3}$
11. $-\frac{1}{2}$ ▢ $-\frac{1}{6}$
12. $-\frac{3}{4}$ ▢ $\frac{5}{8}$

13. $-2\frac{2}{3}$ ▢ $-2\frac{1}{2}$
14. $-1\frac{5}{6}$ ▢ $-1\frac{3}{4}$
15. -4.6 ▢ -4.8

16. -0.12 ▢ -0.05
17. 2.41 ▢ -3.16
18. -3.524 ▢ -3.542

19. **SAND DOLLARS** In rough water, a small sand dollar burrows $-\frac{1}{2}$ centimeter into the sand. A larger sand dollar burrows $-1\frac{1}{4}$ centimeters into the sand. Which sand dollar burrowed farther?

264 Chapter 6 Integers and the Coordinate Plane

Order the numbers from least to greatest.

20. $-2\frac{3}{10}, -2\frac{2}{5}, -2, -2\frac{1}{2}, -3$

21. $-\frac{1}{20}, -\frac{5}{8}, 0, -1, -\frac{3}{4}$

22. $1.3, -2, -1.8, 0, -1.75$

23. $-4, -4.35, -4.9, -5, -4.3$

24. STARS The *apparent magnitude* of a star measures how bright the star appears as seen from Earth. The brighter the star, the lesser the number. Which star is the brightest?

Star	Alpha Centauri	Antares	Canopus	Deneb	Sirius
Apparent Magnitude	-0.27	0.96	-0.72	1.25	-1.46

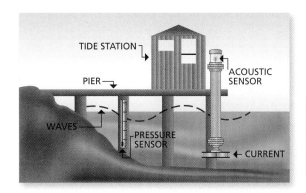

25. LOW TIDE The daily water level is recorded for seven straight days at a tide station on the Big Marco River in Florida. On which days is the water level higher than on the previous day? On which days is it lower?

Day	Sun.	Mon.	Tues.	Wed.	Thurs.	Fri.	Sat.
Water Level of the Day (feet)	$-\frac{3}{25}$	$-\frac{7}{20}$	$-\frac{27}{50}$	$-\frac{13}{20}$	$-\frac{16}{25}$	$-\frac{53}{100}$	$-\frac{1}{3}$

26. PROBLEM SOLVING A guitar tuner allows you to tune a guitar string to its correct pitch. The units on a tuner are measured in *cents*. The units tell you how far the string tone is above or below the correct pitch.

Guitar String	6	5	4	3	2	1
Number of Cents Away from the Correct Pitch	-0.3	1.6	-2.3	2.8	2.4	-3.6

  a. What number on the tuner represents a correctly tuned guitar string?
  b. Which strings have a pitch below the correct pitch?
  c. Which string has a pitch closest to its correct pitch?
  d. Which string has a pitch farthest from its correct pitch?
  e. The tuner is rated to be accurate to within 0.5 cent of the true pitch. Which string could possibly be correct?

27. Number Sense What integer values of x make the statement $-\frac{3}{x} < -\frac{x}{3}$ true?

Fair Game Review *What you learned in previous grades & lessons*

Graph the integer and its opposite. *(Section 6.1)*

28. -7 **29.** 40 **30.** 100 **31.** -15

32. MULTIPLE CHOICE You pay $48 for 8 pounds of chicken. Which is an equivalent rate? *(Section 5.3)*

  Ⓐ $44 for 4 pounds   Ⓑ $28 for 4 pounds

  Ⓒ $15 for 3 pounds   Ⓓ $30 for 5 pounds

6 Study Help

You can use a **summary triangle** to explain a concept. Here is an example of a summary triangle for integers.

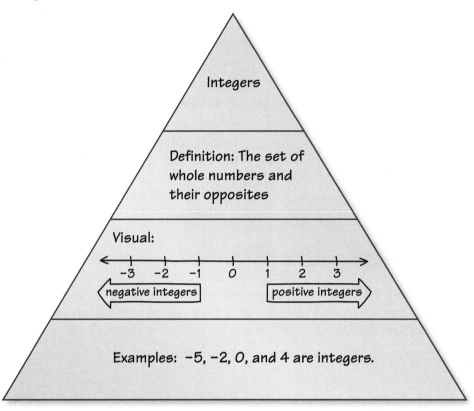

On Your Own

Make summary triangles to help you study these topics.

1. positive integers
2. negative integers
3. opposites

After you complete this chapter, make summary triangles for the following topics.

4. absolute value
5. coordinate plane
6. origin
7. quadrants

"I'm posting my new summary triangle on my daily blog. Do you think it will get me more hits?"

6.1–6.3 Quiz

Write a positive or negative integer that represents the situation. *(Section 6.1)*

1. The price of a stock goes up $2.
2. You descend 15 feet.

Graph the integer and its opposite. *(Section 6.1)*

3. 8
4. −3

Copy and complete the statement using < or >. *(Section 6.2)*

5. −5 ⬚ 0
6. −7 ⬚ −9

Order the integers from least to greatest. *(Section 6.2)*

7. 3, −2, 0, −1, −5
8. −6, 5, 3, −8, 7

Graph the number and its opposite. *(Section 6.3)*

9. $\frac{3}{5}$
10. −1.4

Copy and complete the statement using < or >. *(Section 6.3)*

11. $-2\frac{2}{5}$ ⬚ $-2\frac{1}{4}$
12. −3.28 ⬚ −3.72

13. **ROLLER COASTER** At the top of a roller coaster hill, you are 210 feet above ground. At the bottom of the hill, you are 15 feet above ground. Write an integer that represents the change in height from the top to the bottom. *(Section 6.1)*

14. **PLANETS** The table shows the average surface temperatures of four planets. Which planet is the coldest? Explain. *(Section 6.2)*

Planet	Jupiter	Neptune	Saturn	Uranus
Temperature (°C)	−150	−220	−180	−214

15. **STOCK** The table shows the changes in the value of a stock over several days. Order the numbers from least to greatest. *(Section 6.3)*

Day	Change (dollars)
1	−0.42
2	0.26
3	−0.45
4	0.37

6.4 Absolute Value

Essential Question How can you describe how far an object is from sea level?

1 ACTIVITY: Sea Level

Work with a partner. Write an integer that represents the elevation of each object. How far is each object from sea level? Explain your reasoning.

a. Boeing 747
b. Seaplane
c. Bald eagle
d. Leatherback turtle
e. U.S.S. Dolphin
f. Whale
g. Jason Jr.
h. Alvin
i. Kaiko

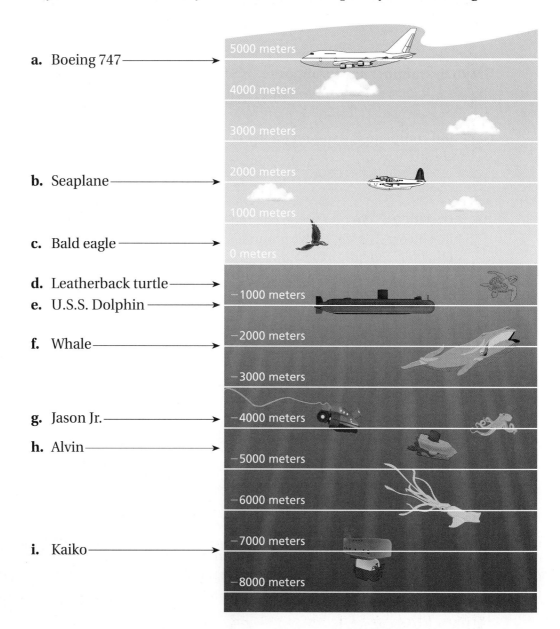

Absolute Value

In this lesson, you will
- find the absolute value of numbers.
- use absolute value to compare numbers in real-life situations.

268   Chapter 6   Integers and the Coordinate Plane

2 ACTIVITY: Finding a Distance

Work with a partner. Use the diagram in Activity 1.

a. What integer represents sea level?

b. The vessel *Kaiko* ascends to the same depth as the U.S.S. *Dolphin*. About how many meters did *Kaiko* travel? Explain how you found your answer.

c. The vessel *Jason Jr.* descends to the same depth as the *Alvin*. About how many meters did *Jason Jr.* travel? Explain how you found your answer.

d. **REASONING** Which pairs of objects are the same distance from sea level? How do you know?

e. **REASONING** An airplane is the same distance from sea level as the *Kaiko*. How far is the airplane from sea level?

3 ACTIVITY: Oceanography Project

Math Practice

Use Technology to Explore

How can you find more information on oceanography? What information is useful to your report?

Work with a partner. Use the Internet or some other resource to write a report that describes two ways in which mathematics is used in oceanography.

Here are two possible ideas. You can use one or both of these, or you can use other ideas.

Diving Bell

Mine Neutralization Vehicle

What Is Your Answer?

4. **IN YOUR OWN WORDS** How can you describe how far an object is from sea level?

5. **PRECISION** In Activity 1, an object has an elevation of -7500 meters. Is -7500 greater than or less than -7000? Does this object have a depth greater than or less than 7000 meters? Explain your reasoning.

Practice — Use what you learned about elevation and sea level to complete Exercises 4–6 on page 272.

6.4 Lesson

Key Vocabulary
absolute value, p. 270

Key Idea

Absolute Value

Words The **absolute value** of a number is the distance between the number and 0 on a number line. The absolute value of a number a is written as $|a|$.

Numbers $|-2| = 2$ $\quad$ $|2| = 2$

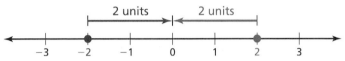

EXAMPLE 1 — Finding Absolute Value

a. Find the absolute value of 3.

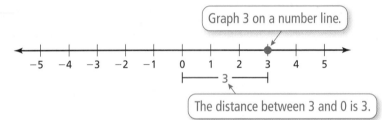

∴ So, $|3| = 3$.

b. Find the absolute value of $-2\frac{1}{2}$.

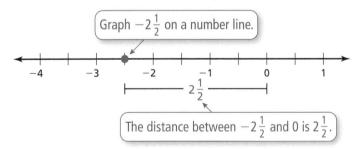

∴ So, $\left|-2\frac{1}{2}\right| = 2\frac{1}{2}$.

On Your Own

Now You're Ready
Exercises 7–14

Find the absolute value.

1. $|8|$
2. $|-6|$
3. $|0|$
4. $\left|\frac{1}{4}\right|$
5. $\left|-7\frac{1}{3}\right|$
6. $|-12.9|$

270 Chapter 6 Integers and the Coordinate Plane

EXAMPLE 2 Comparing Values

Compare 2 and $|-5|$.

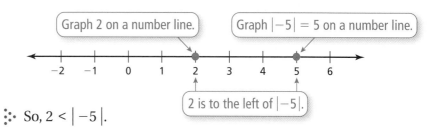

So, $2 < |-5|$.

On Your Own

Now You're Ready
Exercises 17–22

Copy and complete the statement using <, >, or =.

7. $|-4|$ ☐ -2

8. -5 ☐ $|5|$

9. $|9|$ ☐ 10

10. 3.9 ☐ $|-3.9|$

EXAMPLE 3 Real-Life Application

Animal	Elevation (ft)
Shark	-4
Sea lion	5
Seagull	56
Shrimp	-65
Turtle	-22

The table shows the elevations of several animals.

a. **Which animal is the deepest? Explain.**

 Graph each elevation.

 The lowest elevation represents the animal that is the deepest. The integer that is lowest on the number line is -65.

 So, the shrimp is the deepest.

b. **Is the shark or the sea lion closer to sea level?**

 Because sea level is at 0 feet, use absolute values.

 Shark: $|-4| = 4$ **Sea lion:** $|5| = 5$

 Because 4 is less than 5, the shark is closer to sea level than the sea lion.

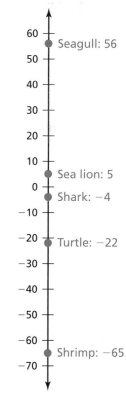

On Your Own

11. Is the seagull or the shrimp closer to sea level? Explain your reasoning.

6.4 Exercises

Vocabulary and Concept Check

1. **VOCABULARY** Explain how to find the absolute value of an integer.

2. **REASONING** Which integer is greater, −50 or 25? Which has the greater absolute value? Explain.

3. **DIFFERENT WORDS, SAME QUESTION** Which is different? Find "both" answers.

   How far is −3 from 0?

   What integer is 3 units to the left of 0?

   What is the absolute value of −3?

   What is the distance between −3 and 0?

Practice and Problem Solving

Use a vertical number line to graph the location of each object. Then tell which object is farther from sea level.

4. Scuba diver: −15 m
   Dolphin: −22 m

5. Seagull: 12 m
   School of fish: −4 m

6. Shark: −40 m
   Flag on a ship: 32 m

Find the absolute value.

 7. $|-2|$

8. $|23|$

9. $|-8.35|$

10. $\left|\dfrac{1}{6}\right|$

11. $\left|-3\dfrac{2}{5}\right|$

12. $|11|$

13. $|14.06|$

14. $|-68|$

15. **REASONING** Write two integers that have an absolute value of 10.

16. **ERROR ANALYSIS** Describe and correct the error in finding the absolute value.

    ✗ $|14| = -14$

Copy and complete the statement using <, >, or =.

 17. $6 \; \_\_ \; |-8|$

18. $|-3| \; \_\_ \; 3$

19. $|-5.5| \; \_\_ \; |-3.1|$

20. $\dfrac{3}{4} \; \_\_ \; \left|-\dfrac{2}{5}\right|$

21. $|-6.8| \; \_\_ \; |8.25|$

22. $-12 \; \_\_ \; |12|$

23. **CAVES** Three scientists explore a cave. Which scientist is farthest underground?

   Scientist A: −48 ft   Scientist B: −62 ft   Scientist C: −53 ft

MATCHING Match the account balance with the debt that it represents. Explain your reasoning.

24. account balance = −$25

25. account balance < −$25

26. account balance > −$25

   A. debt > $25

   B. debt = $25

   C. debt < $25

Order the values from least to greatest.

27. 5, 0, $|-1|$, $|4|$, -2

28. $|-3|$, $|5|$, -3, -4, $|-4|$

29. 10, $|-6|$, 9, $|3|$, -11, 0

30. -18, $|30|$, -19, $|-22|$, -20, $|-18|$

Simplify the expression.

31. $|0|$

32. $-|6|$

33. $-|-1|$

Absolute Zero
Thermometers compare Fahrenheit, Celsius, and Kelvin scales.

	Fahrenheit	Celsius	Kelvin
Water Boils	212°F	100°C	373 K
Water Freezes	32°F	0°C	273 K
Absolute Zero	−459°F	−273°C	0 K

34. ABSOLUTE ZERO The coldest possible temperature is called *absolute zero*. It is represented by 0 K on the Kelvin temperature scale.

   a. Which temperature is closer to 0 K: 32°F or −50°C?

   b. What do absolute values and temperatures on the Kelvin scale have in common?

Tell whether the statement is *always*, *sometimes*, or *never* true. Explain.

35. The absolute value of a number is greater than the number.

36. The absolute value of a negative number is positive.

37. The absolute value of a positive number is its opposite.

38. PALINDROME A *palindrome* is a word or sentence that reads the same forward as it does backward.

   a. Graph and label the following points on a number line: $A = -2$, $C = -1$, $E = 0$, $R = -3$. Then graph and label the absolute value of each point on the *same* number line.

   b. What word do the letters spell? Is this a palindrome?

   c. Make up your own palindrome.

39. **Critical Thinking** Find values of x and y so that $|x| < |y|$ and $x > y$.

Fair Game Review *What you learned in previous grades & lessons*

Draw the polygon with the given vertices in a coordinate plane. *(Section 4.4)*

40. $A(1, 1)$, $B(3, 5)$, $C(5, 0)$

41. $D(0, 6)$, $E(2, 1)$, $F(6, 3)$

42. $P(2, 1)$, $Q(4, 4)$, $R(8, 4)$, $S(6, 1)$

43. $W(1, 6)$, $X(9, 6)$, $Y(9, 1)$, $Z(4, 1)$

44. MULTIPLE CHOICE Which expression represents "6 less than the product of 4 and a number x"? *(Section 3.2)*

   Ⓐ $(6 - 4)x$ **Ⓑ** $6 - 4x$ **Ⓒ** $\dfrac{6}{4x}$ **Ⓓ** $4x - 6$

6.4–6.5 Quiz

Find the absolute value. *(Section 6.4)*

1. $|-12|$
2. $|4|$

Copy and complete the statement using <, >, or =.
(Section 6.4)

3. $5 \;\square\; |-9|$
4. $|-11| \;\square\; |-10|$

Write an ordered pair corresponding to the point.
(Section 6.5)

5. Point A
6. Point B
7. Point C
8. Point D

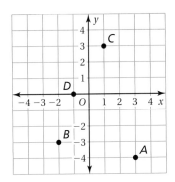

Plot the ordered pair in a coordinate plane. Describe the location of the point.
(Section 6.5)

9. $Q(2, 5)$
10. $R(1, -4)$
11. $S(-2.5, 3.5)$
12. $T\left(0, -1\tfrac{1}{2}\right)$

Reflect the point in (a) the *x*-axis and (b) the *y*-axis. *(Section 6.5)*

13. $(1, 3)$
14. $(-2, 6)$

Reflect the point in the *x*-axis followed by the *y*-axis. *(Section 6.5)*

15. $(3, -2)$
16. $(-4, -5)$

17. **HIKING** The table shows the elevations of several checkpoints along a hiking trail. *(Section 6.4)*

    a. Which checkpoint is farthest from sea level?
    b. Which checkpoint is closest to sea level?
    c. Is Checkpoint 2 or Checkpoint 3 closer to sea level? Explain.

Checkpoint	Elevation (feet)
1	110
2	38
3	−24
4	12
5	−142

18. **GEOMETRY** The points $A(-4, 2)$, $B(1, 2)$, $C(1, -1)$, and $D(-4, -1)$ are the vertices of a figure. *(Section 6.5)*

    a. Draw the figure in a coordinate plane.
    b. Find the perimeter of the figure.
    c. Find the area of the figure.

6 Chapter Review

Review Key Vocabulary

positive numbers, *p. 250*
negative numbers, *p. 250*
opposites, *p. 250*
integers, *p. 250*
absolute value, *p. 270*
coordinate plane, *p. 276*
origin, *p. 276*
quadrants, *p. 276*

Review Examples and Exercises

6.1 Integers (pp. 248–253)

Write a positive or negative integer to represent losing 150 points in a pinball game.

"Lose" indicates a number less than 0. So, use a negative integer.

∴ −150

Exercises

Write a positive or negative integer that represents the situation.

1. An elevator goes down 8 floors.
2. You earn $12.

Graph the integer and its opposite.

3. −7
4. 13
5. 4
6. −100

6.2 Comparing and Ordering Integers (pp. 254–259)

Order −3, −4, 2, 0, −1 from least to greatest.

Graph each integer on a number line.

Write the integers as they appear on the number line from left to right.

∴ So, the order from least to greatest is −4, −3, −1, 0, 2.

Exercises

Order the integers from least to greatest.

7. −5, 4, 2, −3, −1
8. 5, −20, −10, 10, 15

9. Order the temperatures −3°C, 8°C, −12°C, −7°C, and 0°C from coldest to warmest.

Chapter Review **285**

6.3 Fractions and Decimals on the Number Line (pp. 260–265)

Compare $-3\frac{7}{8}$ and $-3\frac{3}{8}$.

$-3\frac{7}{8}$ is to the left of $-3\frac{3}{8}$.

So, $-3\frac{7}{8} < -3\frac{3}{8}$.

Exercises

Graph the number and its opposite.

10. $-\frac{2}{5}$
11. $1\frac{3}{4}$
12. -1.6
13. 2.75

Copy and complete the statement using < or >.

14. $-2\frac{1}{6}$ ⬚ $-2\frac{5}{6}$
15. $-\frac{1}{3}$ ⬚ $-\frac{1}{8}$
16. -3.27 ⬚ -2.68

6.4 Absolute Value (pp. 268–273)

Find the absolute value of -3.

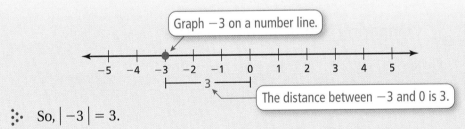

So, $|-3| = 3$.

Exercises

Find the absolute value.

17. $|-8|$
18. $|13|$
19. $\left|3\frac{6}{7}\right|$
20. $|-1.34|$

Copy and complete the statement using <, >, or =.

21. $|-2|$ ⬚ 2
22. $|4.4|$ ⬚ $|-2.8|$
23. $\left|\frac{1}{6}\right|$ ⬚ $\left|-\frac{2}{9}\right|$

6.5 The Coordinate Plane (pp. 274–283)

a. Plot (−3, 0) and (4, −4) in a coordinate plane. Describe the location of each point.

To plot (−3, 0), start at the origin. Move 3 units left. Then plot the point.

To plot (4, −4), start at the origin. Move 4 units right and 4 units down. Then plot the point.

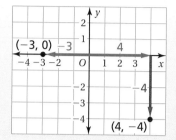

∴ The point (−3, 0) is on the x-axis.
  The point (4, −4) is in Quadrant IV.

b. Reflect (2, −3) in the x-axis.

Plot (2, −3).

To reflect (2, −3) in the x-axis, use the same x-coordinate, 2, and take the opposite of the y-coordinate. The opposite of −3 is 3.

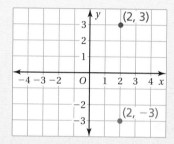

∴ So, the reflection of (2, −3) in the x-axis is (2, 3).

c. Reflect (2, −3) in the y-axis.

Plot (2, −3).

To reflect (2, −3) in the y-axis, use the same y-coordinate, −3, and take the opposite of the x-coordinate. The opposite of 2 is −2.

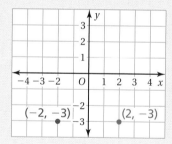

∴ So, the reflection of (2, −3) in the y-axis is (−2, −3).

Exercises

Plot the ordered pair in a coordinate plane. Describe the location of the point.

24. $A(1, 3)$
25. $B(0, -3)$
26. $C(-4, -2)$
27. $D(-1, 2)$

Reflect the point in (a) the x-axis and (b) the y-axis.

28. $(4, 1)$
29. $(-2, 3)$
30. $(2, -5)$
31. $(-3.5, -2.5)$

Reflect the point in the x-axis followed by the y-axis.

32. $(1, 2)$
33. $(-4, 6)$
34. $(3, -4)$
35. $(-3, -3)$

6 Chapter Test

Order the integers from least to greatest.

1. $0, -2, 3, 1, -4$
2. $-8, -3, 5, 4, -5$

Graph the number and its opposite.

3. 14
4. -40
5. $-1\frac{1}{3}$
6. 1.75

Find the absolute value.

7. $|-7|$
8. $|-11|$

Copy and complete the statement using <, >, or =.

9. $-\frac{2}{3}$ ⬚ $-\frac{3}{5}$
10. 1.55 ⬚ -2.46
11. $|-6|$ ⬚ -3
12. -2.5 ⬚ $|2.5|$

Plot the ordered pair in a coordinate plane. Describe the location of the point.

13. $J(4, 0)$
14. $K(-3, 5)$
15. $L(1.5, -3.5)$
16. $M(-2, -3)$

Reflect the point in the x-axis followed by the y-axis.

17. $(2, 4)$
18. $(-5, 1)$

19. **POOL** A diver is on a springboard that is 3 meters above the surface of a pool. Another diver is 2 meters below the surface of the pool.

    a. Write an integer for the position of each diver relative to the surface of the pool.
    b. Find the absolute value of each integer.
    c. Who is farther from the surface of the pool?

20. **OPEN-ENDED** Two vertices of a triangle are $F(1, -4)$ and $G(6, -4)$. List two possible coordinates of the third vertex so that the triangle has an area of 20 square units.

21. **MELTING POINT** The table shows the melting points (in degrees Celsius) of several elements. Compare the melting point of mercury to the melting point of each of the other elements.

Element	Mercury	Radon	Bromine	Cesium	Francium
Melting Point (°C)	−38.83	−71	−7.2	28.5	27

6 Cumulative Assessment

1. What is the value of the expression below when $a = 6$, $b = 5$, and $c = 4$?

   $$8a - 3c + 5b$$

   A. 11   **C.** 61

   B. 53   **D.** 107

2. Point P is plotted in the coordinate plane below.

   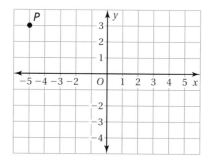

   What are the coordinates of point P?

   F. $(-5, -3)$ **H.** $(-3, -5)$

   G. $(-5, 3)$ **I.** $(3, -5)$

3. What is the value of the expression below?

   $$4\frac{1}{8} \div 5\frac{1}{2}$$

4. Which list of numbers is in order from least to greatest?

   A. $2, |-3|, |4|, -6$ **C.** $-6, |-3|, 2, |4|$

   B. $-6, |4|, 2, |-3|$ **D.** $-6, 2, |-3|, |4|$

5. Which percent is equivalent to $\frac{4}{5}$?

   F. 20%   **H.** 80%

   G. 45%   **I.** 125%

6. Which property is illustrated by the statement below?

$$4 + (6 + n) = (4 + 6) + n$$

  A. Associative Property of Addition

  B. Commutative Property of Addition

  C. Associative Property of Multiplication

  D. Distributive Property

7. You bought 0.875 kilogram of mixed nuts. What was the total cost, in dollars, of the mixed nuts that you bought?

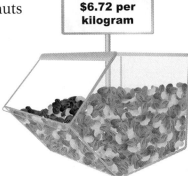

8. On Saturday, you earned $35 mowing lawns. This was x dollars more than you earned on Thursday. Which expression represents the amount, in dollars, you earned mowing lawns on Thursday?

  F. $35x$

  G. $x + 35$

  H. $x - 35$

  I. $35 - x$

9. Helene was finding the percent of a number in the box below.

> 75% of 24 is what number?
>
> 75% of 24 = $24 \div \frac{3}{4}$
>
> = 32

What should Helene do to correct the error that she made?

  A. Divide 24 by 75.

  B. Divide $\frac{3}{4}$ by 24.

  C. Multiply 24 by 75.

  D. Multiply 24 by $\frac{3}{4}$.

Chapter 6 Integers and the Coordinate Plane

10. In the mural below, the squares that are painted red are marked with the letter R.

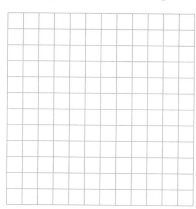

What percent of the mural is painted red?

F. 24%   H. 48%

G. 25%   I. 50%

11. Use grid paper to complete the following.

Part A Draw an *x*-axis and a *y*-axis in the coordinate plane. Then plot and label the point (2, −3).

Part B Plot and label *four* points that are 3 units away from (2, −3).

12. Which expression is equivalent to the expression below?

$$k \div 3\frac{1}{3}$$

A. $k \cdot \dfrac{3}{10}$

B. $k \cdot \dfrac{10}{3}$

C. $k \div \dfrac{3}{10}$

D. $k \div \dfrac{7}{3}$

7 Equations and Inequalities

- **7.1** Writing Equations in One Variable
- **7.2** Solving Equations Using Addition or Subtraction
- **7.3** Solving Equations Using Multiplication or Division
- **7.4** Writing Equations in Two Variables
- **7.5** Writing and Graphing Inequalities
- **7.6** Solving Inequalities Using Addition or Subtraction
- **7.7** Solving Inequalities Using Multiplication or Division

"What is 7Q plus 3Q?"

"You're welcome."

"With the help of your twin brother, I think I have figured it out."

"You weigh 36 dog biscuits."

What You Learned Before

Evaluating Expressions

Example 1 Evaluate $7x + 3y$ when $x = 2$ and $y = 4$.

$7x + 3y = 7 \cdot 2 + 3 \cdot 4$ Substitute 2 for x and 4 for y.
$= 14 + 12$ Using order of operations, multiply from left to right.
$= 26$ Add 14 and 12.

Example 2 Evaluate $5x^2 - 2(y + 1) + 9$ when $x = 2$ and $y = 1$.

$5x^2 - 2(y + 1) + 9 = 5(2)^2 - 2(1 + 1) + 9$ Substitute 2 for x and 1 for y.
$= 5(2)^2 - 2 \cdot 2 + 9$ Using order of operations, evaluate within the parentheses.
$= 5 \cdot 4 - 2 \cdot 2 + 9$ Using order of operations, evaluate the exponent.
$= 20 - 4 + 9$ Using order of operations, multiply from left to right.
$= 25$ Subtract 4 from 20. Add the result to 9.

Try It Yourself

Evaluate the expression when $a = \dfrac{1}{2}$ and $b = 7$.

1. $6ab$
2. $16a - b$
3. $3b - 2a - 9$
4. $b^2 - 16a + 5$

Writing Expressions

Example 3 Write the phrase as an expression.

a. the sum of twice a number n and five

$2n + 5$

b. twelve less than four times a number y

$4y - 12$

Try It Yourself

Write the phrase as an expression.

5. six more than three times a number w
6. the quotient of seven and a number p
7. two less than a number t
8. the product of a number x and five
9. five more than six divided by a number r
10. four less than three times a number b

7.2 Solving Equations Using Addition or Subtraction

Essential Question How can you use addition or subtraction to solve an equation?

When two sides of a scale weigh the same, the scale will balance.

When you add or subtract the same amount on each side of the scale, it will still balance.

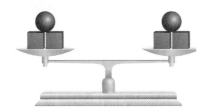

1 ACTIVITY: Solving an Equation

Work with a partner.

a. Use a model to solve $n + 3 = 7$.
  - Explain how the model represents the equation $n + 3 = 7$.

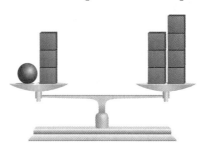

  - How much does one ● weigh? How do you know?
  - The solution is $n = $ ____ .

Solving Equations
In this lesson, you will
- use addition or subtraction to solve equations.
- use substitution to check answers.
- solve real-life problems.

b. Describe how you could check your answer in part (a).

c. Which model below represents the solution of $n + 1 = 9$? How do you know?

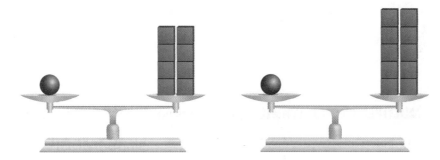

Chapter 7 Equations and Inequalities

2 ACTIVITY: Solving Equations

Work with a partner. Solve the equation using the method in Activity 1.

Math Practice

Understand Quantities
What does the variable represent in the equation?

a.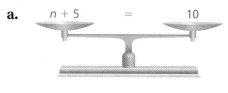

b.

c.

d.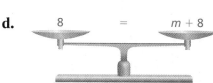

3 ACTIVITY: Solving Equations Using Mental Math

Work with a partner. Write a question that represents the equation. Use mental math to answer the question. Then check your solution.

Equation	Question	Solution	Check
a. $x + 1 = 5$			
b. $4 + m = 11$			
c. $8 = a + 3$			
d. $x - 9 = 21$			
e. $13 = p - 4$			

What Is Your Answer?

4. **REPEATED REASONING** In Activity 3, how are parts (d) and (e) different from parts (a)–(c)? Did your process to find the solution change? Explain.

5. Decide whether the statement is *true* or *false*. If false, explain your reasoning.
   a. In an equation, you can use any letter as a variable.
   b. The goal in solving an equation is to get the variable by itself.
   c. In the solution, the variable must always be on the left side of the equal sign.
   d. If you add a number to one side, you should subtract it from the other side.

6. **IN YOUR OWN WORDS** How can you use addition or subtraction to solve an equation? Give two examples to show how your procedure works.

7. Are the following equations equivalent? Explain your reasoning.
$$x - 5 = 12 \quad \text{and} \quad 12 = x - 5$$

Practice

Use what you learned about solving equations to complete Exercises 12–17 on page 305.

Section 7.2 Solving Equations Using Addition or Subtraction

7.2 Lesson

Key Vocabulary
solution, p. 302
inverse operations, p. 303

Equations may be true for some values and false for others. A **solution** of an equation is a value that makes the equation true.

Value of x	x + 3 = 7	Are both sides equal?
3	$3 + 3 \stackrel{?}{=} 7$ $6 \neq 7$ ✗	no
4	$4 + 3 \stackrel{?}{=} 7$ $7 = 7$ ✓	yes
5	$5 + 3 \stackrel{?}{=} 7$ $8 \neq 7$ ✗	no

Reading
The symbol ≠ means *is not equal to*.

So, the value $x = 4$ is a solution of the equation $x + 3 = 7$.

EXAMPLE 1   Checking Solutions

Tell whether the given value is a solution of the equation.

a. $p + 10 = 38$; $p = 18$

$18 + 10 \stackrel{?}{=} 38$ Substitute 18 for *p*.

$28 \neq 38$ ✗   Sides are *not* equal.

∴ So, $p = 18$ is *not* a solution.

b. $4y = 56$; $y = 14$

$4(14) \stackrel{?}{=} 56$ Substitute 14 for *y*.

$56 = 56$ ✓   Sides are equal.

∴ So, $y = 14$ is a solution.

On Your Own

Now You're Ready
Exercises 6–11

Tell whether the given value is a solution of the equation.

1. $a + 6 = 17$; $a = 9$
2. $9 - g = 5$; $g = 3$
3. $35 = 7n$; $n = 5$
4. $\dfrac{q}{2} = 28$; $q = 14$

You can use *inverse operations* to solve equations. **Inverse operations** "undo" each other. Addition and subtraction are inverse operations.

Key Ideas

Addition Property of Equality

Words When you add the same number to each side of an equation, the two sides remain equal.

Numbers
$$8 = 8$$
$$+5 \quad +5$$
$$13 = 13$$

Algebra
$$x - 4 = 5$$
$$+4 \quad +4$$
$$x = 9$$

Subtraction Property of Equality

Words When you subtract the same number from each side of an equation, the two sides remain equal.

Numbers
$$8 = 8$$
$$-5 \quad -5$$
$$3 = 3$$

Algebra
$$x + 4 = 5$$
$$-4 \quad -4$$
$$x = 1$$

EXAMPLE 2 Solving Equations Using Addition

a. Solve $x - 2 = 6$.

	$x - 2 =$	6	Write the equation.
Undo the subtraction. →	$+ 2$	$+ 2$	Addition Property of Equality
	$x =$	8	Simplify.

∴ The solution is $x = 8$.

Check
$$x - 2 = 6$$
$$8 - 2 \stackrel{?}{=} 6$$
$$6 = 6 \checkmark$$

Study Tip
You can check your solution by substituting it for the variable in the original equation.

b. Solve $18 = x - 7$.

$$18 = x - 7 \quad \text{Write the equation.}$$
$$+ 7 \quad + 7 \quad \text{Addition Property of Equality}$$
$$25 = x \quad \text{Simplify.}$$

∴ The solution is $x = 25$.

Check
$$18 = x - 7$$
$$18 \stackrel{?}{=} 25 - 7$$
$$18 = 18 \checkmark$$

On Your Own

Now You're Ready
Exercises 18–20

Solve the equation. Check your solution.

5. $k - 3 = 1$ **6.** $n - 10 = 4$ **7.** $15 = r - 6$

EXAMPLE 3 Solving Equations Using Subtraction

a. Solve $x + 2 = 9$.

$$x + 2 = 9 \quad \text{Write the equation.}$$

Undo the addition. → $-2 \quad -2 \quad$ Subtraction Property of Equality

$$x = 7 \quad \text{Simplify.}$$

∴ The solution is $x = 7$.

Check
$x + 2 = 9$
$7 + 2 \stackrel{?}{=} 9$
$9 = 9$ ✓

b. Solve $26 = 11 + x$.

$$26 = 11 + x \quad \text{Write the equation.}$$
$$-11 \quad -11 \quad \text{Subtraction Property of Equality}$$
$$15 = x \quad \text{Simplify.}$$

∴ The solution is $x = 15$.

Check
$26 = 11 + x$
$26 \stackrel{?}{=} 11 + 15$
$26 = 26$ ✓

EXAMPLE 4 Real-Life Application

Your parents give you $20 to help buy the new pair of shoes shown. After you buy the shoes, you have $5.50 left. Write and solve an equation to find how much money you had before your parents gave you $20.

Words The starting amount plus the amount your parents gave you minus the cost of the shoes is the amount left.

Variable Let s be the starting amount.

Equation $s + 20 - 59.95 = 5.50$

$$s + 20 - 59.95 = 5.50 \quad \text{Write the equation.}$$
$$s + 20 - 59.95 + 59.95 = 5.50 + 59.95 \quad \text{Addition Property of Equality}$$
$$s + 20 = 65.45 \quad \text{Simplify.}$$
$$s + 20 - 20 = 65.45 - 20 \quad \text{Subtraction Property of Equality}$$
$$s = 45.45 \quad \text{Simplify.}$$

∴ You had $45.45 before your parents gave you money.

Study Tip
In Example 4, you can solve the problem arithmetically by working backwards from $5.50.
$5.50 + 59.95 - 20 = 45.45$
So, your answer is reasonable.

On Your Own

Now You're Ready Exercises 21–23

Solve the equation. Check your solution.

8. $s + 8 = 17$
9. $9 = y + 6$
10. $13 + m = 20$

11. You eat 8 blueberries and your friend eats 11 blueberries from a package. There are 23 blueberries left. Write and solve an equation to find the number of blueberries in a full package.

7.2 Exercises

Vocabulary and Concept Check

1. **WRITING** How can you check the solution of an equation?

Name the inverse operation you can use to solve the equation.

2. $x - 8 = 12$
3. $n + 3 = 13$
4. $b + 14 = 33$

5. **WRITING** When solving $x + 5 = 16$, why do you subtract 5 from the left side of the equation? Why do you subtract 5 from the right side of the equation?

Practice and Problem Solving

Tell whether the given value is a solution of the equation.

6. $x + 42 = 85;\ x = 43$
7. $8b = 48;\ b = 6$
8. $19 - g = 7;\ g = 15$
9. $\dfrac{m}{4} = 16;\ m = 4$
10. $w + 23 = 41;\ w = 28$
11. $s - 68 = 11;\ s = 79$

Use a scale to model and solve the equation.

12. $n + 7 = 9$
13. $t + 4 = 5$
14. $c + 2 = 8$

Write a question that represents the equation. Use mental math to answer the question. Then check your solution.

15. $a + 5 = 12$
16. $v + 9 = 18$
17. $20 = d - 6$

Solve the equation. Check your solution.

18. $y - 7 = 3$
19. $z - 3 = 13$
20. $8 = r - 14$
21. $p + 5 = 8$
22. $k + 6 = 18$
23. $64 = h + 30$
24. $f - 27 = 19$
25. $25 = q + 14$
26. $\dfrac{3}{4} = j - \dfrac{1}{2}$
27. $x + \dfrac{2}{3} = \dfrac{9}{10}$
28. $1.2 = m - 2.5$
29. $a + 5.5 = 17.3$

ERROR ANALYSIS Describe and correct the error in solving the equation.

30.
$$\begin{array}{r} x + 7 = 13 \\ +7 \quad +7 \\ \hline x \quad\ \ = 20 \end{array}$$

31.
$$\begin{array}{r} 34 = y - 12 \\ -12 \quad +12 \\ \hline 22 = y \end{array}$$

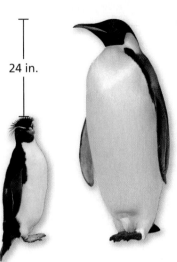

32. PENGUINS An emperor penguin is 45 inches tall. It is 24 inches taller than a rockhopper penguin. Write and solve an equation to find the height of a rockhopper penguin. Is your answer reasonable? Explain.

33. ELEVATOR You get in an elevator and go down 8 floors. You exit on the 16th floor. Write and solve an equation to find what floor you got on the elevator.

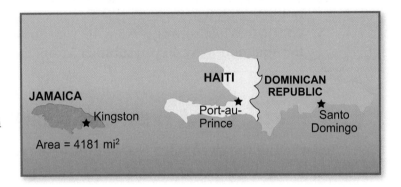

34. AREA The area of Jamaica is 6460 square miles less than the area of Haiti. Write and solve an equation to find the area of Haiti.

35. REASONING The solution of the equation $x + 3 = 12$ is shown. Explain each step. Use a property, if possible.

$$x + 3 = 12$$
$$x + 3 - 3 = 12 - 3$$
$$x + 0 = 9$$
$$x = 9$$

Write the equation.

Write the word sentence as an equation. Then solve the equation.

36. 13 subtracted from a number w is 15.   **37.** A number k increased by 7 is 34.

38. 9 is the difference of a number n and 7.   **39.** 93 is the sum of a number g and 58.

Solve the equation. Check your solution.

40. $b + 7 + 12 = 30$ **41.** $y + 4 - 1 = 18$ **42.** $m + 18 + 23 = 71$

43. $v - 7 = 9 + 12$ **44.** $5 + 44 = 2 + r$ **45.** $22 + 15 = d - 17$

GEOMETRY Write and solve an addition equation to find x.

46. Perimeter = 48 ft   **47.** Perimeter = 132 in.   **48.** Perimeter = 93 ft

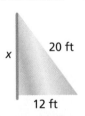
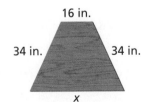
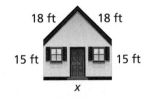

49. REASONING Explain why the equations $x + 4 = 13$ and $4 + x = 13$ have the same solution.

50. REASONING Explain why the equations $x - 13 = 4$ and $13 - x = 4$ do *not* have the same solution.

51. SIMPLIFYING AND SOLVING Compare and contrast the two problems.

Simplify the expression $2(x + 3) - 4$.
$$2(x + 3) - 4 = 2x + 6 - 4$$
$$= 2x + 2$$

Solve the equation $x + 3 = 4$.
$$x + 3 = 4$$
$$\underline{-3 \quad -3}$$
$$x = 1$$

52. PUZZLE In a *magic square*, the sum of the numbers in each row, column, and diagonal is the same. Write and solve equations to find the values of *a*, *b*, and *c*.

a	37	16
19	25	b
34	c	28

53. FUNDRAISER You participate in a dance-a-thon fundraiser. After your parents pledge $15.50 and your neighbor pledges $8.75, you have $66.55. Write and solve an equation to find how much money you had before your parents and neighbor pledged.

54. MONEY On Saturday, you spend $33, give $15 to a friend, and receive $20 for mowing your neighbor's lawn. You have $21 left. Use two methods to find how much money you started with that day.

Bumper Cars: $1.75
Roller Coaster: $1.25 more than Ferris Wheel
Giant Slide: $0.50 less than Bumper Cars
Ferris Wheel: $1.50 more than Giant Slide

55. AMUSEMENT PARK You have $15.

 a. How much money do you have left if you ride each ride once?

 b. Do you have enough money to ride each ride twice? Explain.

56. **Critical Thinking** Consider the equation $x + y = 15$. The value of *x* increases by 3. What has to happen to the value of *y* so that $x + y = 15$ remains true?

Fair Game Review *What you learned in previous grades & lessons*

Find the value of the expression. Use estimation to check your answer. *(Section 1.1)*

57. 12×8 **58.** 13×16 **59.** $75 \div 15$ **60.** $72 \div 3$

61. MULTIPLE CHOICE What is the area of the parallelogram? *(Section 4.1)*

 Ⓐ 25 in.²   Ⓑ 30 in.²
 Ⓒ 50 in.²   Ⓓ 100 in.²

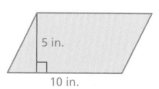

7.3 Solving Equations Using Multiplication or Division

Essential Question How can you use multiplication or division to solve an equation?

1 ACTIVITY: Finding Missing Dimensions

Work with a partner. Describe how you would find the value of x. Then find the value and check your result.

a. rectangle

Area = 24 square units

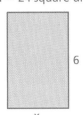

b. parallelogram

Area = 20 square units

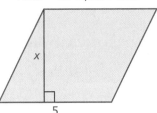

c. triangle

Area = 28 square units

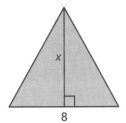

2 ACTIVITY: Using an Equation to Model a Story

Work with a partner.

a. Use a model to solve the problem.

> Three people go out to lunch. They decide to share the $12 bill evenly. How much does each person pay?

- What equation does the model represent? Explain how this represents the problem.

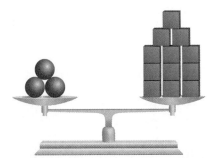

- How much does one ● weigh? How do you know?

∴ Each person pays ___ .

b. Describe how you can check your answer in part (a).

Solving Equations

In this lesson, you will
- use multiplication or division to solve equations.
- use substitution to check answers.
- solve real-life problems.

308   Chapter 7   Equations and Inequalities

3 ACTIVITY: Using Equations to Model a Story

Work with a partner.
- What is the unknown?
- Write an equation that represents each problem.
- What does the variable in your equation represent?
- Explain how you can solve the equation.
- Answer the question.

Problem	Equation
a. Three robots go out to lunch. They decide to share the $11.91 bill evenly. How much does each robot pay?	
b. On Earth, objects weigh 6 times what they weigh on the Moon. A robot weighs 96 pounds on Earth. What does it weigh on the Moon?	
c. At maximum speed, a robot runs 6 feet in 1 second. How many feet does the robot run in 1 minute?	
d. Four identical robots lie on the ground head-to-toe and measure 14 feet. How tall is each robot?	

Math Practice

Interpret Results
What does the solution represent? Does the answer make sense?

What Is Your Answer?

4. Complete each sentence by matching.
   - The inverse operation of addition
   - The inverse operation of subtraction
   - The inverse operation of multiplication
   - The inverse operation of division

   - is multiplication.
   - is subtraction.
   - is addition.
   - is division.

5. **IN YOUR OWN WORDS** How can you use multiplication or division to solve an equation? Give two examples to show how your procedure works.

Practice Use what you learned about solving equations to complete Exercises 15–18 on page 312.

Section 7.3 Solving Equations Using Multiplication or Division

7.3 Lesson

Check It Out
Lesson Tutorials
BigIdeasMath.com

Key Ideas

Remember
Inverse operations "undo" each other. Multiplication and division are inverse operations.

Multiplication Property of Equality

Words When you multiply each side of an equation by the same nonzero number, the two sides remain equal.

Numbers $\frac{8}{4} = 2$ **Algebra** $\frac{x}{4} = 2$

$\frac{8}{4} \cdot 4 = 2 \cdot 4$ $\frac{x}{4} \cdot 4 = 2 \cdot 4$

$8 = 8$ $x = 8$

Multiplicative Inverse Property

Words The product of a nonzero number n and its reciprocal, $\frac{1}{n}$, is 1.

Numbers $5 \cdot \frac{1}{5} = 1$ **Algebra** $n \cdot \frac{1}{n} = \frac{1}{n} \cdot n = 1, n \neq 0$

EXAMPLE 1 Solving Equations Using Multiplication

a. Solve $\frac{w}{4} = 12$.

$\frac{w}{4} = 12$ Write the equation.

Undo the division. $\frac{w}{4} \cdot 4 = 12 \cdot 4$ Multiplication Property of Equality

$w = 48$ Simplify.

∴ The solution is $w = 48$.

Check
$\frac{w}{4} = 12$

$\frac{48}{4} \stackrel{?}{=} 12$

$12 = 12$ ✓

b. Solve $\frac{2}{7}x = 6$.

$\frac{2}{7}x = 6$ Write the equation.

Use the Multiplicative Inverse Property. $\frac{7}{2} \cdot \left(\frac{2}{7}x\right) = \frac{7}{2} \cdot 6$ Multiplication Property of Equality

$x = 21$ Simplify.

∴ The solution is $x = 21$.

On Your Own

Now You're Ready
Exercises 7–10

Solve the equation. Check your solution.

1. $\frac{a}{8} = 6$

2. $14 = \frac{2y}{5}$

3. $3z \div 2 = 9$

Key Idea

Division Property of Equality

Words When you divide each side of an equation by the same nonzero number, the two sides remain equal.

Numbers $8 \cdot 4 = 32$ **Algebra** $4x = 32$

$8 \cdot 4 \div 4 = 32 \div 4$ $\dfrac{4x}{4} = \dfrac{32}{4}$

$8 = 8$ $x = 8$

EXAMPLE 2 Solving an Equation Using Division

Solve $5b = 65$.

Undo the multiplication.

$5b = 65$ Write the equation.

$\dfrac{5b}{5} = \dfrac{65}{5}$ Division Property of Equality

$b = 13$ Simplify.

Check
$5b = 65$
$5(13) \stackrel{?}{=} 65$
$65 = 65$ ✓

∴ The solution is $b = 13$.

EXAMPLE 3 Real-Life Application

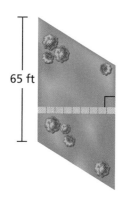

65 ft

The area of the parallelogram-shaped courtyard is 2730 square feet. What is the length of the sidewalk?

The height of the parallelogram represents the length of the sidewalk.

$A = bh$ Use the formula for area of a parallelogram.

$2730 = 65h$ Substitute 2730 for A and 65 for b.

$\dfrac{2730}{65} = \dfrac{65h}{65}$ Division Property of Equality

$42 = h$ Simplify.

∴ So, the sidewalk is 42 feet long.

On Your Own

Now You're Ready
Exercises 11–14

Solve the equation. Check your solution.

4. $p \cdot 3 = 18$ **5.** $12q = 60$ **6.** $81 = 9r$

7. You and four friends buy tickets to a baseball game. The total cost is $70. Write and solve an equation to find the cost of each ticket.

7.3 Exercises

Vocabulary and Concept Check

1. **NUMBER SENSE** What number divided by 12 equals 1?

2. **WRITING** What property of equality would you use to solve $\frac{x}{6} = 7$? Explain how you would use the property.

Copy and complete the first step in the solution.

3. $4x = 24$

   $\dfrac{4x}{\square} = \dfrac{24}{\square}$

4. $\dfrac{x}{3} = 11$

   $\dfrac{x}{3} \cdot \square = 11 \cdot \square$

5. $8 = n \div 3$

   $8 \cdot \square = (n \div 3) \cdot \square$

6. **OPEN-ENDED** Write an equation that can be solved using the Division Property of Equality.

Practice and Problem Solving

Solve the equation. Check your solution.

① 7. $\dfrac{s}{10} = 7$ 8. $6 = \dfrac{t}{5}$ 9. $5x \div 6 = 20$ 10. $24 = \dfrac{3r}{4}$

② 11. $3a = 12$ 12. $5 \cdot z = 35$ 13. $40 = 4y$ 14. $42 = 7k$

15. $7x = 105$ 16. $75 = 6 \cdot w$ 17. $13 = d \div 6$ 18. $9 = v \div 5$

19. $\dfrac{2c}{15} = 8.8$ 20. $7b \div 12 = 4.2$ 21. $12.5 \cdot n = 32$ 22. $3.4m = 20.4$

23. **ERROR ANALYSIS** Describe and correct the error in solving the equation.

    ✗ $x \div 4 = 28$
    $\dfrac{x \div 4}{4} = \dfrac{28}{4}$
    $x = 7$

24. **ANOTHER WAY** Show how you can solve the equation $3x = 9$ by multiplying each side by the reciprocal of 3.

25. **BASKETBALL** Forty-five basketball players participate in a tournament. Write and solve an equation to find the number of 3-person teams that they can form.

26. **THEATER** A theater has 1200 seats. Each row has 20 seats. Write and solve an equation to find the number of rows in the theater.

Solve for x. Check your answer.

27. rectangle

Area = 45 square units

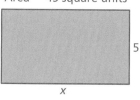

28. rectangle

Area = 176 square units

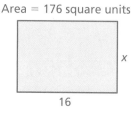

29. parallelogram

Area = 104 square units

30. TEST SCORE On a test, you correctly answer six 5-point questions and eight 2-point questions. You earn 92% of the possible points on the test. How many points p is the test worth?

31. CARD GAME You use index cards to play a homemade game. The object is to be the first to get rid of all your cards. How many cards are in your friend's stack?

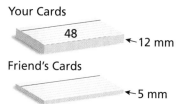

32. SLUSH DRINKS A slush drink machine fills 1440 cups in 24 hours.

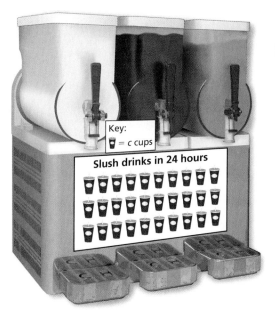

  a. Write and solve an equation to find the number c of cups each symbol represents.

  b. To lower costs, you replace the cups with paper cones that hold 20% less. Write and solve an equation to find the number n of paper cones that the machine can fill in 24 hours.

STRUCTURE Solve the equation. Explain how you found your answer.

33. $5x + 3x = 5x + 18$ **34.** $8y + 2y = 2y + 40$

35. Number Sense The area of the picture is 100 square inches. The length is 4 times the width. Find the length and width of the picture.

Fair Game Review What you learned in previous grades & lessons

Write the word sentence as an equation. *(Section 7.1)*

36. The sum of a number b and 8 is 17.   **37.** A number t divided by 3 is 7.

38. MULTIPLE CHOICE What is the value of a^3 when $a = 4$? *(Section 3.1)*

  Ⓐ 12   Ⓑ 43   Ⓒ 64   Ⓓ 81

7.4 Writing Equations in Two Variables

Essential Question How can you write an equation in two variables?

1 ACTIVITY: Writing an Equation in Two Variables

Work with a partner. You earn $8 per hour working part-time at a store.

a. Complete the table.

Hours Worked	Money Earned (dollars)
1	
2	
3	
4	
5	

b. Use the values from the table to complete the graph. Then answer each question below.

- What does the horizontal axis represent? What variable did you use to identify it?
- What does the vertical axis represent? What variable did you use to identify it?
- How are the ordered pairs in the graph related to the values in the table?
- How are the horizontal and vertical distances shown on the graph related to the values in the table?

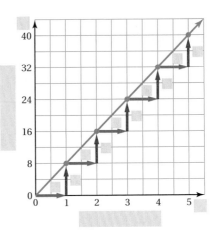

c. How can you write an equation that shows how the two variables are related?

d. What does the green line in the graph represent?

Writing Equations

In this lesson, you will
- identify independent and dependent variables.
- write equations in two variables.
- use tables and graphs to analyze the relationship between two variables.

314   Chapter 7   Equations and Inequalities

2 ACTIVITY: Describing Variables

Work with a partner. Use the equation you wrote in Activity 1.

a. How is this equation different from the equations earlier in this chapter?

b. One of the variables in this equation *depends* on the other variable. Determine which variable is which by answering the following questions:

- Does the amount of money you earn *depend* on the number of hours you work?
- Does the number of hours you work *depend* on the amount of money you earn?

What do you think is the significance of having two types of variables? How do you think you can use these types of variables in real life?

3 ACTIVITY: Describing a Formula in Two Variables

Work with a partner. Recall that the perimeter of a square is 4 times its side length.

Math Practice

Look for Patterns

What pattern do you notice in the table for the perimeter of the square?

a. Write the formula for the perimeter of a square. Tell what each variable represents.

b. Describe how the perimeter of a square changes as its side length increases by 1 unit. Use a table and a graph to support your answer.

c. In your formula, which variable depends on which?

What Is Your Answer?

4. **IN YOUR OWN WORDS** How can you write an equation in two variables?

5. The equation $y = 7.75x$ shows how the number of movie tickets is related to the total amount of money spent. Describe what each part of the equation represents.

6. **CHOOSE TOOLS** In Activity 1, you want to know the amount of money you earn after working 30.5 hours during a week. Would you use the table, the graph, or the equation to find your earnings? What are your earnings? Explain your reasoning.

7. Give an example of another real-life situation that you can model by an equation in two variables.

Practice

Use what you learned about equations in two variables to complete Exercises 4 and 5 on page 319.

7.4 Lesson

Check It Out
Lesson Tutorials
BigIdeasMath.com

An **equation in two variables** represents two quantities that change in relationship to one another. A **solution of an equation in two variables** is an ordered pair that makes the equation true.

EXAMPLE 1 — Identifying Solutions of Equations in Two Variables

Key Vocabulary
equation in two variables, *p. 316*
solution of an equation in two variables, *p. 316*
independent variable, *p. 316*
dependent variable, *p. 316*

Tell whether the ordered pair is a solution of the equation.

a. $y = 2x$; (3, 6)

$6 \stackrel{?}{=} 2(3)$ Substitute.

$6 = 6$ ✓   Compare.

∴ So, (3, 6) is a solution.

b. $y = 4x - 3$; (4, 12)

$12 \stackrel{?}{=} 4(4) - 3$

$12 \neq 13$ ✗

∴ So, (4, 12) is *not* a solution.

You can use equations in two variables to represent situations involving two related quantities. The variable representing the quantity that can change freely is the **independent variable**. The other variable is called the **dependent variable** because its value *depends* on the independent variable.

EXAMPLE 2 — Using an Equation in Two Variables

The equation $y = 128 - 8x$ gives the amount y (in fluid ounces) of milk remaining in a gallon jug after you pour x cups.

a. **Identify the independent and dependent variables.**

∴ Because the amount y remaining depends on the number x of cups you pour, y is the dependent variable and x is the independent variable.

b. **How much milk remains in the jug after you pour 10 cups?**

Use the equation to find the value of y when $x = 10$.

$y = 128 - 8x$ Write the equation.

$ = 128 - 8(10)$ Substitute 10 for x.

$ = 48$ Simplify.

∴ There are 48 fluid ounces remaining.

On Your Own

Now You're Ready
Exercises 6–11 and 13–17

Tell whether the ordered pair is a solution of the equation.

1. $y = 7x$; (2, 21)
2. $y = 5x + 1$; (3, 16)

3. The equation $y = 10x + 25$ gives the amount y (in dollars) in your savings account after x weeks.

   a. Identify the independent and dependent variables.
   b. How much is in your savings account after 8 weeks?

Key Idea

Tables, Graphs, and Equations

You can use tables and graphs to represent equations in two variables. The table and graph below represent the equation $y = x + 2$.

> **Study Tip**
> When you draw a line through the points, you graph *all* the solutions of the equation.

Independent Variable, x	Dependent Variable, y	Ordered Pair, (x, y)
1	3	(1, 3)
2	4	(2, 4)
3	5	(3, 5)

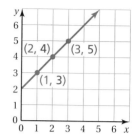

EXAMPLE 3 — Writing and Graphing an Equation in Two Variables

An athlete burns 200 calories weight lifting. The athlete then works out on an elliptical trainer and burns 10 calories for every minute. Write and graph an equation in two variables that represents the total number of calories burned during the workout.

Words   The total number of calories burned equals calories burned weight lifting plus calories burned per minute times the number of minutes.

Variables   Let c be the total number of calories burned, and let m be the number of minutes on the elliptical trainer.

Equation $c = 200 + 10 \cdot m$

> **Reading**
> Make sure you read and understand the context of the problem. Because you cannot have a negative number of minutes, use only whole number values of m.

To graph the equation, first make a table. Then plot the ordered pairs and draw a line through the points.

Minutes, m	$c = 200 + 10m$	Calories, c	Ordered Pair, (m, c)
10	$c = 200 + 10(10)$	300	(10, 300)
20	$c = 200 + 10(20)$	400	(20, 400)
30	$c = 200 + 10(30)$	500	(30, 500)

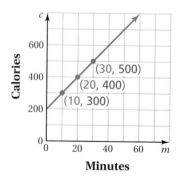

On Your Own

Now You're Ready
Exercises 22 and 23

4. It costs $25 to rent a kayak plus $8 for each hour. Write and graph an equation in two variables that represents the total cost of renting the kayak.

Section 7.4   Writing Equations in Two Variables

OPEN-ENDED Complete the table by describing possible independent or dependent variables.

	Independent Variable	Dependent Variable
18.	The number of hours you study for a test	
19.	The speed you are pedaling a bike	
20.		Your monthly cell phone bill
21.		The amount of money you earn

22. PIZZA A cheese pizza costs $5. Additional toppings cost $1.50 each. Write and graph an equation in two variables that represents the total cost of a pizza.

23. GYM MEMBERSHIP It costs $35 to join a gym. The monthly fee is $25. Write and graph an equation in two variables that represents the total cost of a gym membership.

24. TEXTING The maximum size of a text message is 160 characters. A space counts as one character.

   a. Write an equation in two variables that represents the remaining (unused) characters in a text message as you type.

   b. Identify the independent and dependent variables.

   c. How many characters remain in the message shown?

25. CHOOSE TOOLS A car averages 60 miles per hour on a road trip. Use a graph to show the relationship between the time and the distance traveled. What method did you use to create your graph?

Write and graph an equation in two variables that shows the relationship between the time and the distance traveled.

26.
Moves 2 meters every 3 hours.

27.
Rises 5 stories every 6 seconds.

28.
Moves 660 feet every 10 seconds.

29.
Moves 960 kilometers every 4 minutes.

Fill in the blank so that the ordered pair is a solution of the equation.

30. $y = 8x + 3$; (1, ___)

31. $y = 12x + 2$; (___, 14)

32. $y = 22 - 9x$; (___, 4)

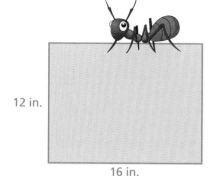

12 in.

16 in.

33. CRITICAL THINKING Can the dependent variable cause a change in the independent variable? Explain.

34. OPEN-ENDED Write an equation in two variables that has (3, 4) as a solution.

35. WALKING You walk 5 city blocks in 12 minutes. How many city blocks can you walk in 2 hours?

36. ANT How fast should the ant walk to go around the rectangle in 4 minutes?

37. LIGHTNING To estimate how far you are from lightning (in miles), count the number of seconds between a lightning flash and the thunder that follows. Then divide the number of seconds by 5. Use a graph to show the relationship between the time and the distance. Describe the method you used to create your graph.

38. PROBLEM SOLVING You and a friend start biking in opposite directions from the same point. You travel 108 feet every 8 seconds. Your friend travels 63 feet every 6 seconds.

   a. How far apart are you and your friend after 15 minutes?

   b. After 20 minutes, you take a 5-minute rest, but your friend does not. How far apart are you and your friend after 40 minutes? Explain your reasoning.

39. Reasoning The graph represents the cost c (in dollars) of buying n tickets to a baseball game.

   a. Should the points be connected with a line to show all the solutions? Explain your reasoning.

   b. Write an equation in two variables that represents the graph.

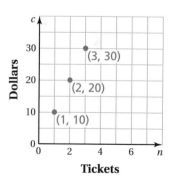

Fair Game Review *What you learned in previous grades & lessons*

Write the fraction as a percent. *(Section 5.5)*

40. $\dfrac{3}{10}$

41. $\dfrac{4}{5}$

42. $\dfrac{9}{20}$

43. $\dfrac{17}{25}$

44. MULTIPLE CHOICE What is the area of the triangle? *(Section 4.2)*

   Ⓐ 36 cm²   **Ⓑ** 68 cm²
   Ⓒ 72 cm²   **Ⓓ** 76.5 cm²

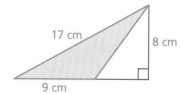

7 Study Help

You can use an **example and non-example chart** to list examples and non-examples of a vocabulary word or term. Here is an example and non-example chart for equations.

Equations

Examples	Non-Examples
$x = 5$	5
$2a = 16$	$2a$
$x + 4 = 19$	$x + 4$
$5 = x + 3$	$x + 3$
$12 - 7 = 5$	$12 - 7$
$\frac{3}{4}y = 6$	$\frac{3}{4}$

On Your Own

Make example and non-example charts to help you study these topics.

1. inverse operations
2. equations solved using addition or subtraction
3. equations solved using multiplication or division
4. equations in two variables

After you complete this chapter, make example and non-example charts for the following topics.

5. inequalities
6. graphs of inequalities
7. inequalities solved using addition or subtraction
8. inequalities solved using multiplication or division

"I need a good non-example of a cool animal for my example and non-example chart."

7.1–7.4 Quiz

Write the word sentence as an equation. *(Section 7.1)*

1. A number x decreased by 3 is 5.
2. A number a divided by 7 equals 14.

Solve the equation. Check your solution. *(Section 7.2 and Section 7.3)*

3. $4 + k = 14$
4. $3.5 = m - 2.2$
5. $8 = \dfrac{4w}{3}$
6. $31 = 6.2 \cdot y$

Tell whether the ordered pair is a solution of the equation. *(Section 7.4)*

7. $y = 6x$; $(3, 24)$
8. $y = 3x + 4$; $(4, 16)$

Write and graph an equation in two variables that shows the relationship between the time and the distance traveled. *(Section 7.4)*

9.  Rises 4 feet in 9 seconds.

10.  Moves 900 feet every 10 seconds.

11. **RIBBON** The length of the blue ribbon is two-thirds the length of the red ribbon. Write an equation you can use to find the length r of the red ribbon. *(Section 7.1)*

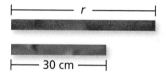

12. **BRIDGES** The main span of the Sunshine Skyway Bridge is 360 meters long. The Skyway's main span is 30 meters shorter than the main span of the Dames Point Bridge. Write and solve an equation to find the length ℓ of the main span of the Dames Point Bridge. *(Section 7.2)*

13. **SHOPPING** At a farmer's market, you buy 4 pounds of tomatoes and 2 pounds of sweet potatoes. You spend 80% of the money in your wallet. Write and solve an equation to find how much money is in your wallet before you pay. *(Section 7.3)*

 Tomatoes $3.00/pound

 Sweet Potatoes $2.00/pound

14. **SUNDAE** A sundae costs $2. Additional toppings cost $0.50 each. Write and graph an equation in two variables that represents the total cost of a sundae. *(Section 7.4)*

7.5 Writing and Graphing Inequalities

Essential Question How can you use a number line to represent solutions of an inequality?

1 ACTIVITY: Understanding Inequality Statements

Work with a partner. Read the statement. Circle each number that makes the statement true, and then answer the questions.

a. "Your friend is more than 3 minutes late."

$$-3 \quad -2 \quad -1 \quad 0 \quad 1 \quad 2 \quad 3 \quad 4 \quad 5 \quad 6$$

- What do you notice about the numbers that you circled?
- Is the number 3 included? Why or why not?
- Write four other numbers that make the statement true.

b. "The temperature is at most 2 degrees."

$$-5 \quad -4 \quad -3 \quad -2 \quad -1 \quad 0 \quad 1 \quad 2 \quad 3 \quad 4$$

- What do you notice about the numbers that you circled?
- Can the temperature be exactly 2 degrees? Explain.
- Write four other numbers that make the statement true.

c. "You need at least 4 pieces of paper for your math homework."

$$-3 \quad -2 \quad -1 \quad 0 \quad 1 \quad 2 \quad 3 \quad 4 \quad 5 \quad 6$$

- What do you notice about the numbers that you circled?
- Can you have exactly 4 pieces of paper? Explain.
- Write four other numbers that make the statement true.

d. "After playing a video game for 20 minutes, you have fewer than 6 points."

$$-2 \quad -1 \quad 0 \quad 1 \quad 2 \quad 3 \quad 4 \quad 5 \quad 6 \quad 7$$

- What do you notice about the numbers that you circled?
- Is the number 6 included? Why or why not?
- Write four other numbers that make the statement true.

Writing Inequalities
In this lesson, you will
- write word sentences as inequalities.
- use a number line to graph the solution set of inequalities.
- use inequalities to represent real-life situations.

2 ACTIVITY: Understanding Inequality Symbols

Work with a partner.

a. Consider the statement "x is a number such that $x < 2$."
- Can the number be exactly 2? Explain.
- Circle each number that makes the statement true.

    $-5 \quad -4 \quad -3 \quad -2 \quad -1 \quad 0 \quad 1 \quad 2 \quad 3 \quad 4$

- Write four other numbers that make the statement true.

b. Consider the statement "x is a number such that $x \geq 1$."
- Can the number be exactly 1? Explain.
- Circle each number that makes the statement true.

    $-5 \quad -4 \quad -3 \quad -2 \quad -1 \quad 0 \quad 1 \quad 2 \quad 3 \quad 4$

- Write four other numbers that make the statement true.

Math Practice

State the Meaning of Symbols
What do the symbols $<$ and $\geq$ mean?

3 ACTIVITY: How Close Can You Come to 0?

Work with a partner.

a. Which number line shows $x > 0$? Which number line shows $x \geq 0$? Explain your reasoning.

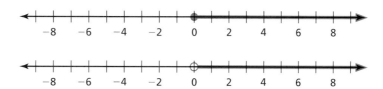

b. Write the least positive number you can think of that is still a solution of the inequality $x > 0$. Explain your reasoning.

What Is Your Answer?

4. IN YOUR OWN WORDS How can you use a number line to represent solutions of an inequality?

5. Write an inequality. Graph all solutions of your inequality on a number line.

6. Graph the inequalities $x > 9$ and $9 < x$ on different number lines. What do you notice?

Practice — Use what you learned about graphing inequalities to complete Exercises 17–20 on page 329.

Section 7.5   Writing and Graphing Inequalities   **325**

The **graph of an inequality** shows all the solutions of the inequality on a number line. An open circle ○ is used when a number is *not* a solution. A closed circle ● is used when a number is a solution. An arrow to the left or right shows that the graph continues in that direction.

EXAMPLE 3 Graphing an Inequality

Graph $g > 2$.

Use an open circle because 2 is *not* a solution.

Test a number to the left of 2. $g = 0$ is *not* a solution.

Test a number to the right of 2. $g = 3$ is a solution.

Shade the number line on the side where you found the solution. The graph shows there are *infinitely many* solutions.

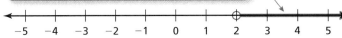

EXAMPLE 4 Real-Life Application

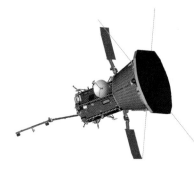

The NASA Solar Probe Plus can withstand temperatures up to and including 2600°F. Write and graph an inequality that represents the temperatures the probe can withstand.

Words   temperatures   up to and including   2600°F

Variable   Let t be the temperatures the probe can withstand.

Inequality t $\leq$ 2600

∴ An inequality is $t \leq 2600$.

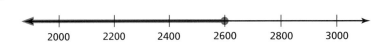

On Your Own

Graph the inequality on a number line.

8. $a < 4$ 9. $f \leq 7$ 10. $n > 0$ 11. $p \geq -3$

Write and graph an inequality for the situation.

12. A cruise ship can carry at most 3500 passengers.

13. A board game is designed for ages 12 and up.

Reading

The inequality $g > 2$ is the same as $2 < g$.

7.5 Exercises

Vocabulary and Concept Check

1. **VOCABULARY** How are *greater than* and *greater than or equal to* similar? How are they different?

2. **DIFFERENT WORDS, SAME QUESTION** Which is different? Write "both" inequalities.

   | A number n is at most 3. | A number n is at least 3. |

   | A number n is less than or equal to 3. | A number n is no more than 3. |

3. **WRITING** Explain how the graph of $x \leq 6$ is different from the graph of $x < 6$.

4. **WRITING** Are the graphs of $x \leq 5$ and $5 \geq x$ the same or different? Explain.

Practice and Problem Solving

Write the word sentence as an inequality.

5. A number k is less than 10.

6. A number a is more than 6.

7. A number z is fewer than $\dfrac{3}{4}$.

8. A number b is at least -3.

9. One plus a number y is no more than -13.

10. A number x divided by 3 is at most 5.

Tell whether the given value is a solution of the inequality.

11. $x - 1 \leq 7;\ x = 6$

12. $y + 5 < 13;\ y = 17$

13. $3z > 6;\ z = 3$

14. $\dfrac{b}{2} \geq 6;\ b = 10$

15. $c + 2.5 < 4.3;\ c = 1.8$

16. $a \leq 0;\ a = -5$

Match the inequality with its graph.

17. $x \geq 2$

18. $x < 2$

19. $x > -2$

20. $x \leq -2$

7.6 Solving Inequalities Using Addition or Subtraction

Essential Question How can you use addition or subtraction to solve an inequality?

1 ACTIVITY: Writing an Inequality

Work with a partner. In 3 years, your friend will still not be old enough to vote.

a. Which of the following represents your friend's situation? What does x represent? Explain your reasoning.

$x + 3 < 18$ $x + 3 \le 18$

$x + 3 > 18$ $x + 3 \ge 18$

b. Graph the possible ages of your friend on a number line. Explain how you decided what to graph.

2 ACTIVITY: Writing an Inequality

Work with a partner. Baby manatees are about 4 feet long at birth. They grow to a maximum length of 13 feet.

a. Which of the following can represent a baby manatee's growth? What does x represent? Explain your reasoning.

$x + 4 < 13$ $x + 4 \le 13$

$x - 4 > 13$ $x - 4 \ge 13$

b. Graph the solution on a number line. Explain how you decided what to graph.

Solving Inequalities
In this lesson, you will
- use addition or subtraction to solve inequalities.
- use a number line to graph the solution set of inequalities.
- solve real-life problems.

332   Chapter 7   Equations and Inequalities

3 ACTIVITY: Solving Inequalities

Work with a partner. Complete the following steps for Activity 1. Then repeat the steps for Activity 2.

Math Practice

Interpret Results
What does the solution of the inequality represent?

- Use your inequality from part (a). Replace the inequality symbol with an equal sign.
- Solve the equation.
- Replace the equal sign with the original inequality symbol.
- Graph this new inequality.
- Compare the graph with your graph in part (b). What do you notice?

4 ACTIVITY: The Triangle Inequality

Work with a partner. Draw different triangles whose sides have lengths 10 cm, 6 cm, and x cm.

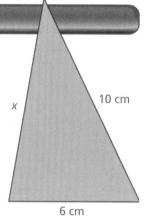

a. Which of the following describes how *small* x can be? Explain your reasoning.

$6 + x < 10$ $6 + x \leq 10$

$6 + x > 10$ $6 + x \geq 10$

b. Which of the following describes how *large* x can be? Explain your reasoning.

$x - 6 < 10$ $x - 6 \leq 10$ $x - 6 > 10$ $x - 6 \geq 10$

c. Graph the possible values of x on a number line.

What Is Your Answer?

5. **IN YOUR OWN WORDS** How can you use addition or subtraction to solve an inequality?

6. Describe a real-life situation that you can represent with an inequality. Write the inequality. Graph the solution on a number line.

Practice → Use what you learned about solving inequalities to complete Exercises 5–7 on page 336.

7.6 Lesson

Study Tip

You can solve inequalities the same way you solve equations. Use inverse operations to get the variable by itself.

🔑 Key Ideas

Addition Property of Inequality

Words When you add the same number to each side of an inequality, the inequality remains true.

Numbers
$$\begin{array}{r} 3 < 5 \\ +2 \ +2 \\ \hline 5 < 7 \end{array}$$

Algebra
$$\begin{array}{r} x - 4 > 5 \\ +4 \ +4 \\ \hline x > 9 \end{array}$$

Graph

Subtraction Property of Inequality

Words When you subtract the same number from each side of an inequality, the inequality remains true.

Numbers
$$\begin{array}{r} 3 < 5 \\ -2 \ -2 \\ \hline 1 < 3 \end{array}$$

Algebra
$$\begin{array}{r} x + 4 > 5 \\ -4 \ -4 \\ \hline x > 1 \end{array}$$

Graph

These properties are also true for ≤ and ≥.

EXAMPLE 1 Solving an Inequality Using Addition

Solve $x - 3 > 1$. Graph the solution.

$$\begin{array}{rl} x - 3 > 1 & \text{Write the inequality.} \\ +3 \ +3 & \text{Addition Property of Inequality} \\ \hline x > 4 & \text{Simplify.} \end{array}$$

Undo the subtraction.

Check:
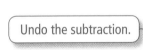

∴ The solution is $x > 4$.

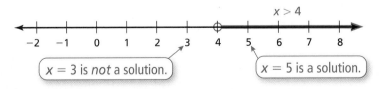

● **On Your Own**

Solve the inequality. Graph the solution.

1. $x - 2 < 3$ **2.** $x - 6 \geq 4$ **3.** $10 \geq x - 1$

EXAMPLE 2 Solving an Inequality Using Subtraction

Solve $15 \geq 6 + x$. Graph the solution.

$$15 \geq 6 + x \quad \text{Write the inequality.}$$
$$\underline{-6 \quad -6} \quad \text{Subtraction Property of Inequality}$$
$$9 \geq x \quad \text{Simplify.}$$

Undo the addition.

Reading
The inequality $x \leq 9$ is the same as $9 \geq x$.

∴ The solution is $x \leq 9$.

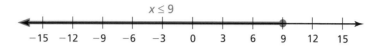

On Your Own

Now You're Ready
Exercises 5–16

Solve the inequality. Graph the solution.

4. $x + 3 > 7$
5. $y + 2 < 17$
6. $16 \leq m + 9$

EXAMPLE 3 Real-Life Application

A flea market advertises that it has more than 250 vending booths. Of these, 184 are currently filled. Write and solve an inequality to represent the number of vending booths still available.

| Words | The number of booths filled | plus | the number of remaining booths | is greater than | the total number of booths. |

Variable Let b be the number of remaining booths.

Inequality 184 + b > 250

$$184 + b > 250 \quad \text{Write the inequality.}$$
$$\underline{-184 \quad\quad -184} \quad \text{Subtraction Property of Inequality}$$
$$b > 66 \quad \text{Simplify.}$$

∴ More than 66 vending booths are still available.

On Your Own

7. You have already spent $24 shopping online for clothes. Write and solve an inequality to represent the additional amount you must spend to get free shipping.

7.6 Exercises

Vocabulary and Concept Check

1. **OPEN-ENDED** Write an inequality that can be solved by subtracting 7 from each side.
2. **WRITING** Explain how to solve the inequality $x - 6 > 3$.
3. **WRITING** Describe the graph of the solution of $x + 3 \leq 4$.
4. **OPEN-ENDED** Write an inequality that the graph represents. Then use the Subtraction Property of Inequality to write another inequality that the graph represents.

Practice and Problem Solving

Solve the inequality. Graph the solution.

5. $x - 4 < 5$
6. $5 + h > 7$
7. $3 \geq y - 2$
8. $9 \leq c + 1$
9. $18 > 12 + x$
10. $37 + z \leq 54$
11. $y - 21 < 85$
12. $g - 17 \geq 17$
13. $7.2 < x + 4.2$
14. $12.7 \geq s - 5.3$
15. $\dfrac{3}{4} \leq \dfrac{1}{2} + n$
16. $\dfrac{1}{3} + b > \dfrac{3}{4}$

17. **ERROR ANALYSIS** Describe and correct the error in solving the inequality.

$$\begin{array}{r} 28 \geq t - 9 \\ -9 \quad -9 \\ \hline 19 \geq t \end{array}$$

18. **AIR TRAVEL** Your carry-on bag can weigh at most 40 pounds. Write and solve an inequality to represent how much more weight you can add to the bag and still meet the requirement.

19. **SHOPPING** It costs $x for a round-trip bus ticket to the mall. You have $24. Write and solve an inequality to represent the greatest amount of money you can spend for the bus fare and still have enough to buy the baseball cap.

Write the word sentence as an inequality. Then solve the inequality.

20. Five more than a number is less than 17.
21. Three less than a number is more than 15.

Solve the inequality. Graph the solution.

22. $x + 9 - 3 \leq 14$ **23.** $44 > 7 + s + 26$ **24.** $6.1 - 0.3 \geq c + 1$

25. VIDEO GAME The high score for a video game is 36,480. Your current score is 34,280. Each dragonfly you catch is worth 1 point. You also get a 1000-point bonus for reaching 35,000 points. Write and solve an inequality to represent the number of dragonflies you must catch to earn a new high score.

26. PICKUP TRUCKS You can register a pickup truck as a passenger vehicle if the truck is not used for commercial purposes and the weight of the truck with its contents does not exceed 8500 pounds.

   a. Your pickup truck weighs 4200 pounds. Write an inequality to represent the number of pounds your truck can carry and still qualify as a passenger vehicle. Then solve the inequality.

   b. A cubic yard of sand weighs about 1600 pounds. How many cubic yards of sand can you haul in your truck and still qualify as a passenger vehicle? Explain your reasoning.

27. TRIATHLON You complete two events of a triathlon. Your goal is to finish with an overall time of less than 100 minutes.

   a. Write and solve an inequality to represent how many minutes you can take to finish the running event and still meet your goal.

   b. The running event is 3.1 miles long. Estimate how many minutes it would take you to run 3.1 miles. Would this time allow you to reach your goal? Explain your reasoning.

Triathlon	
Event	Your Time (minutes)
Swimming	18.2
Biking	45.4
Running	?

28. Number Sense The possible values of x are given by $x - 3 \geq 2$. What is the least possible value of $5x$?

Fair Game Review *What you learned in previous grades & lessons*

Solve the equation. Check your solution. *(Section 7.3)*

29. $\dfrac{t}{12} = 4$ **30.** $6 = \dfrac{2s}{9}$ **31.** $8x = 72$ **32.** $9 = 1.5z$

33. MULTIPLE CHOICE Which brand of turkey is the best buy? *(Section 5.4)*

   Ⓐ Brand A   **Ⓑ** Brand B
   Ⓒ Brand C   **Ⓓ** Brand D

Brand	A	B	C	D
Cost (dollars)	10.38	13.47	21.45	34.93
Pounds	2	3	5	7

7.7 Solving Inequalities Using Multiplication or Division

Essential Question How can you use multiplication or division to solve an inequality?

1 ACTIVITY: Writing an Inequality

Work with a partner. A store has a clearance rack of shirts that each cost the same amount. You buy 2 shirts and have money left after paying with a $20 bill.

a. Which of the following represents your purchase? What does x represent? Explain your reasoning.

$$2x < 20 \qquad 2x \leq 20$$

$$2x > 20 \qquad 2x \geq 20$$

b. Graph the possible values of x on a number line. Explain how you decided what to graph.

c. Can you buy a third shirt? Explain your reasoning.

2 ACTIVITY: Writing an Inequality

Work with a partner. One of your favorite stores is having a 75% off sale. You have $20. You want to buy a pair of jeans.

a. Which of the following represents your ability to buy the jeans with $20? What does x represent? Explain your reasoning.

$$\frac{1}{4}x < 20 \qquad \frac{1}{4}x \leq 20$$

$$\frac{1}{4}x > 20 \qquad \frac{1}{4}x \geq 20$$

Solving Inequalities
In this lesson, you will
- use multiplication or division to solve inequalities.
- use a number line to graph the solution set of inequalities.
- solve real-life problems.

b. Graph the possible values of x on a number line. Explain how you decided what to graph.

c. Can you afford a pair of jeans that originally costs $100? Explain your reasoning.

338 Chapter 7 Equations and Inequalities

3 ACTIVITY: Solving Inequalities

Work with a partner. Complete the following steps for Activity 1. Then repeat the steps for Activity 2.

- Use your inequality from part (a). Replace the inequality symbol with an equal sign.
- Solve the equation.
- Replace the equal sign with the original inequality symbol.
- Graph this new inequality.
- Compare the graph with your graph in part (b). What do you notice?

4 ACTIVITY: Matching Inequalities

Work with a partner. Match the inequality with its graph. Explain your method.

a. $3x < 9$ b. $3x \leq 9$ c. $\dfrac{x}{2} \geq 1$

d. $6 < 2x$ e. $12 \leq 4x$ f. $\dfrac{x}{2} < 2$

Math Practice

Make a Plan
What strategy will you use to choose the correct graph?

A.

B.

C.

D.

E.

F.

What Is Your Answer?

5. **IN YOUR OWN WORDS** How can you use multiplication or division to solve an inequality?

Practice

Use what you learned about solving inequalities to complete Exercises 8–11 on page 342.

Section 7.7   Solving Inequalities Using Multiplication or Division   **339**

7.7 Lesson

Check It Out
Lesson Tutorials
BigIdeasMath.com

Key Ideas

Multiplication and division are inverse operations.

Multiplication Property of Inequality

Words When you multiply each side of an inequality by the same *positive* number, the inequality remains true.

Numbers $8 > 6$ **Algebra** $\dfrac{x}{4} < 2$

$8 \times 2 > 6 \times 2$ $\dfrac{x}{4} \cdot 4 < 2 \cdot 4$

$16 > 12$ $x < 8$

Division Property of Inequality

Words When you divide each side of an inequality by the same *positive* number, the inequality remains true.

Numbers $8 > 6$ **Algebra** $4x < 8$

$8 \div 2 > 6 \div 2$ $\dfrac{4x}{4} < \dfrac{8}{4}$

$4 > 3$ $x < 2$

These properties are also true for ≤ and ≥.

EXAMPLE 1   Solving an Inequality Using Multiplication

Solve $\dfrac{x}{5} \leq 2$. Graph the solution.

$\dfrac{x}{5} \leq 2$ Write the inequality.

Undo the division. ⟶ $\dfrac{x}{5} \cdot 5 \leq 2 \cdot 5$ Multiplication Property of Inequality

$x \leq 10$ Simplify.

∴ The solution is $x \leq 10$.

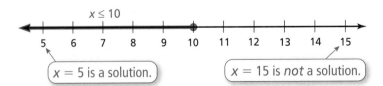

On Your Own

Now You're Ready
Exercises 6–9

Solve the inequality. Graph the solution.

1. $p \div 3 > 2$ 2. $\dfrac{3}{5}q \leq 6$ 3. $1 < \dfrac{s}{7}$

EXAMPLE 2 Solving an Inequality Using Division

Solve $4n > 32$. Graph the solution.

$4n > 32$ Write the inequality.

Undo the multiplication. → $\dfrac{4n}{4} > \dfrac{32}{4}$ Division Property of Inequality

$n > 8$ Simplify.

∴ The solution is $n > 8$.

EXAMPLE 3 Real-Life Application

A one-way bus ride costs $1.75. A 30-day bus pass costs $42.

a. Write and solve an inequality to find the least number of one-way rides you must take for the 30-day pass to be a better deal.

b. You ride the bus an average of 20 times each month. Is the pass a better deal? Explain.

a. **Words** The price of a one-way ride **times** the number of one-way rides **is more than** $42.

Variable Let r be the number of one-way rides.

Inequality $1.75 \cdot r > 42$

$1.75r > 42$ Write the inequality.

$\dfrac{1.75r}{1.75} > \dfrac{42}{1.75}$ Division Property of Inequality

$r > 24$ Simplify.

∴ So, you need to take more than 24 one-way rides for the pass to be a better deal.

b. No. The cost of 20 one-way rides is less than $42. So, the pass is not a better deal.

On Your Own

Now You're Ready
Exercises 10–13

Solve the inequality. Graph the solution.

4. $11k \le 33$ **5.** $5 \cdot j > 20$ **6.** $50 \le 2m$

7. The sign shows the toll for driving on Alligator Alley. Write and solve an inequality to represent the number of times someone can drive on Alligator Alley with $15.

Passenger Cars Toll $2.50

7.7 Exercises

Vocabulary and Concept Check

1. **REASONING** How is the graph of the solution of $2x \geq 10$ different from the graph of the solution of $2x = 10$?

Name the property you should use to solve the inequality.

2. $3x \leq 27$
3. $7x > 49$
4. $\dfrac{x}{2} < 36$

5. **OPEN-ENDED** Write two inequalities that have the same solution set: one that you can solve using division and one that you can solve using multiplication.

Practice and Problem Solving

Solve the inequality. Graph the solution.

6. $\dfrac{m}{8} < 4$
7. $n \div 6 > 2$
8. $\dfrac{t}{3} \geq 15$
9. $\dfrac{1}{11}c \geq 9$

10. $12x < 96$
11. $5x \geq 25$
12. $8 \cdot w \leq 72$
13. $7p \leq 42$

14. $\dfrac{3}{4}b > 15$
15. $6x < 90$
16. $3s \geq 36$
17. $\dfrac{5}{9}v \leq 45$

18. $4t > 72$
19. $\dfrac{3}{4}w \leq 24$
20. $12m < 132$
21. $\dfrac{5x}{8} \geq 30$

22. **ERROR ANALYSIS** Describe and correct the error in solving the inequality.

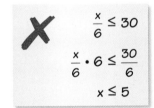

23. **GEOMETRY** The length of a rectangle is 8 feet, and its area is less than 168 square feet. Write and solve an inequality to represent the width of the rectangle.

24. **PLAYGROUND** Students at a playground are divided into 5 equal groups with at least 6 students in each group. Write and solve an inequality to represent the number of students at the playground.

Write the word sentence as an inequality. Then solve the inequality.

25. Eight times a number n is less than 72.

26. A number t divided by 32 is at most 4.25.

27. 225 is no less than 12 times a number w.

Graph the numbers that are solutions to both inequalities.

28. $x + 7 > 9$ and $8x \leq 64$

29. $x - 3 \leq 8$ and $6x < 72$

30. THRILL RIDE A thrill ride at an amusement park holds a maximum of 12 people per ride.

  a. Write and solve an inequality to find the least number of rides needed for 15,000 people.

  b. Do you think it is possible for 15,000 people to ride the thrill ride in 1 day? Explain.

31. FOOTBALL A winning football team more than doubled the offensive yards gained by its opponent. The opponent gained 272 offensive yards. The winning team had 80 offensive plays. Write and solve an inequality to find the possible number of yards per play for the winning team.

Park Hours
10:00 A.M.–10:00 P.M.

32. LOGIC Explain how you know that $7x < 7x$ has no solution.

33. OPEN-ENDED Give an example of a real-life situation in which you can list all the solutions of an inequality. Give an example of a real-life situation in which you cannot list all the solutions of an inequality.

34. FUNDRAISER You are selling items from a catalog for a school fundraiser. Write and solve two inequalities to find the range of sales that will earn you between $40 and $50.

 Let $a > b$ and $x > y$. Tell whether the statement is *always* true. Explain your reasoning.

35. $a + x > b + y$

36. $a - x > b - y$

37. $ax > by$

38. $\dfrac{a}{x} > \dfrac{y}{b}$

Fair Game Review What you learned in previous grades & lessons

Classify the quadrilateral. *(Skills Review Handbook)*

39.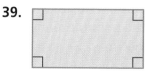

40.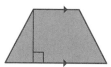

41.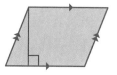

42. MULTIPLE CHOICE On a normal day, 12 airplanes arrive at an airport every 15 minutes. Which rate does not represent this situation? *(Section 5.3)*

  Ⓐ 24 airplanes every 30 minutes

  Ⓑ 4 airplanes every 5 minutes

  Ⓒ 6 airplanes every 5 minutes

  Ⓓ 48 airplanes each hour

7.5–7.7 Quiz

Write the word sentence as an inequality. *(Section 7.5)*

1. A number x is greater than 0.
2. Twice a number c is at least -8.

Tell whether the given value is a solution of the inequality. *(Section 7.5)*

3. $2n > 16$; $n = 9$
4. $x - 1 \leq 9$; $x = 10$

Graph the inequality on a number line. *(Section 7.5)*

5. $y > -4$
6. $m \leq \dfrac{3}{5}$

Solve the inequality. Graph the solution. *(Section 7.6)*

7. $x + 4 \leq 8$
8. $18 > 16 + g$

Write the word sentence as an inequality. Then solve the inequality. *(Section 7.6)*

9. Two less than a number is more than 15.
10. Seven more than a number is less than or equal to 27.

Solve the inequality. Graph the solution. *(Section 7.7)*

11. $\dfrac{3a}{2} < 24$
12. $121 \geq 11s$

Write the word sentence as an inequality. Then solve the inequality. *(Section 7.7)*

13. Three times a number x is more than 18.
14. 84 is no less than 7 times a number k.

15. **WATER PARK** Each visit to a water park costs $19.95. An annual pass to the park costs $89.95. Write an inequality to represent the number of times you would need to visit the park for the pass to be a better deal. *(Section 7.5)*

16. **GARDEN** You want to use a square section of your yard for a garden. You have at most 52 feet of fencing to surround the garden. Write and solve an inequality to represent the possible lengths of each side of the garden. *(Section 7.7)*

17. **DELIVERY** You were planning to spend $12 on a pizza. Write and solve an inequality to represent the additional amount you must spend to get free delivery. *(Section 7.6)*

7 Chapter Review

Review Key Vocabulary

equation, p. 296
solution, p. 302
inverse operations, p. 303
equation in two variables, p. 316
solution of an equation in two variables, p. 316
independent variable, p. 316
dependent variable, p. 316
inequality, p. 326
solution of an inequality, p. 327
solution set, p. 327
graph of an inequality, p. 328

Review Examples and Exercises

7.1 Writing Equations in One Variable (pp. 294–299)

Write the word sentence "The quotient of a number b and 6 is 9" as an equation.

The quotient of a number b and 6 is 9.

$b \div 6 = 9$ *Quotient of* means *division.*

An equation is $b \div 6 = 9$.

Exercises

Write the word sentence as an equation.

1. The product of a number m and 2 is 8.
2. 6 less than a number t is 7.
3. A number m increased by 5 is 7.
4. 8 is the quotient of a number g and 3.

7.2 Solving Equations Using Addition or Subtraction (pp. 300–307)

Solve $z + 5 = 13$.

$z + 5 = 13$ — Write the equation.
Undo the addition. → $-5 \quad -5$ — Subtraction Property of Equality
$z = 8$ — Simplify.

The solution is $z = 8$.

Check
$z + 5 = 13$
$8 + 5 \stackrel{?}{=} 13$
$13 = 13$ ✓

Exercises

Solve the equation. Check your solution.

5. $x - 1 = 8$
6. $m + 7 = 11$
7. $21 = p - 12$

7.3 Solving Equations Using Multiplication or Division (pp. 308–313)

Solve $4c = 32$.

Undo the multiplication.

$4c = 32$ Write the equation.

$\dfrac{4c}{4} = \dfrac{32}{4}$ Division Property of Equality

$c = 8$ Simplify.

Check
$4c = 32$
$4(8) \stackrel{?}{=} 32$
$32 = 32$ ✓

Exercises

Solve the equation. Check your solution.

8. $7 \cdot q = 42$
9. $7k \div 3 = 21$
10. $\dfrac{5a}{7} = 25$

7.4 Writing Equations in Two Variables (pp. 314–321)

Tell whether (6, 16) is a solution of the equation $y = 3x - 4$.

$16 \stackrel{?}{=} 3(6) - 4$ Substitute.

$16 \neq 14$ ✗   Compare.

So, (6, 16) is *not* a solution.

Exercises

Tell whether the ordered pair is a solution of the equation.

11. $y = 3x + 1$; (2, 7)
12. $y = 7x - 4$; (4, 22)

13. **TAXI** A taxi ride costs $3 plus $2.50 per mile. Write and graph an equation in two variables that represents the total cost of a taxi ride.

7.5 Writing and Graphing Inequalities (pp. 324–331)

Write the word sentence as an inequality.

a. A number x is more than -9.

A number x is more than -9.

x -9

So, An inequality is $x > -9$.

b. A number r divided by 2 is at most 4.

A number r divided by 2 is at most 4.

$\dfrac{r}{2}$ 4

So, An inequality is $\dfrac{r}{2} \leq 4$.

Exercises

Write the word sentence as an inequality.

14. A number m is less than 5.
15. A number h is at least -12.

Graph the inequality on a number line.

16. $x < 0$
17. $a \geq 3$
18. $n \leq -1$

7.6 Solving Inequalities Using Addition or Subtraction (pp. 332–337)

Solve $1 \leq x - 4$. Graph the solution.

$$\begin{aligned} 1 &\leq x - 4 &&\text{Write the inequality.} \\ +4 & +4 &&\text{Addition Property of Inequality} \\ \hline 5 &\leq x &&\text{Simplify.} \end{aligned}$$

Undo the subtraction.

The inequality $5 \leq x$ is the same as $x \geq 5$.

∴ The solution is $x \geq 5$.

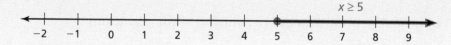

Exercises

Solve the inequality. Graph the solution.

19. $x + 1 > 3$
20. $k - 7 \leq 0$
21. $y + 8 \geq 9$
22. $24 < 11 + x$
23. $4 \leq n - 4$
24. $x - 20 > 24$
25. $b + 12 \leq 26$
26. $s - 1.5 < 2.5$
27. $\dfrac{1}{4} + m \leq \dfrac{1}{2}$

7.7 Solving Inequalities Using Multiplication or Division (pp. 338–343)

Solve $7n < 42$. Graph the solution.

$$\begin{aligned} 7n &< 42 &&\text{Write the inequality.} \\ \dfrac{7n}{7} &< \dfrac{42}{7} &&\text{Division Property of Inequality} \\ n &< 6 &&\text{Simplify.} \end{aligned}$$

Undo the multiplication.

∴ The solution is $n < 6$.

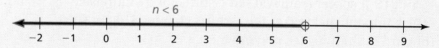

Exercises

Solve the inequality. Graph the solution.

28. $x \div 2 < 4$
29. $9n \geq 63$
30. $\dfrac{5}{3}x \leq 10$
31. $9 \geq 3b$
32. $10p > 40$
33. $\dfrac{3}{11}k < 15$

34. **TICKETS** The cost of three tickets to a movie is at least $20. Write and solve an inequality that represents the situation.

7 Chapter Test

Write the word sentence as an equation.

1. 7 times a number s is 84.
2. 13 is one-third of a number m.

Solve the equation. Check your solution.

3. $15 = 7 + b$
4. $v - 6 = 16$
5. $5x = 70$
6. $3b = 45$
7. $\dfrac{6m}{7} = 30$
8. $\dfrac{8k}{3} = 32$

Tell whether the ordered pair is a solution of the equation.

9. $y = 9x$; (3, 27)
10. $y = 4x + 2$; (8, 36)

Write an inequality for the situation.

11. An MP3 player holds up to 300 songs.
12. Riders must be at least 48 inches tall.

Graph the inequality on a number line.

13. $x \geq 5$
14. $m \leq -2$

Solve the inequality. Graph the solution.

15. $x - 3 < 7$
16. $12 \geq n + 6$
17. $\dfrac{4}{3}b \leq 12$
18. $72 > 12p$

19. **SCHOOL DANCE** Each ticket to a school dance is $4. The total amount collected in ticket sales is $332. Write and solve an equation to find the number of students attending the dance.

20. **T-SHIRTS** A soccer team will sell T-shirts for a fundraiser. The company that makes the T-shirts charges $10 per shirt plus a $20 shipping fee per order.

    a. Write and graph an equation in two variables that represents the total cost of ordering the shirts.
    b. Choose an ordered pair that lies on your graph in part (a). Interpret it in the context of the problem.

21. **HURRICANE** A hurricane has wind speeds that are greater than or equal to 74 miles per hour. Write an inequality to represent the possible wind speeds during a hurricane.

7 Cumulative Assessment

1. What is the area of the balcony shown below?

   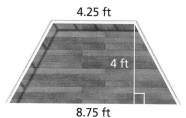

   A. 9 ft²   C. 26 ft²
   B. 18 ft²   D. 52 ft²

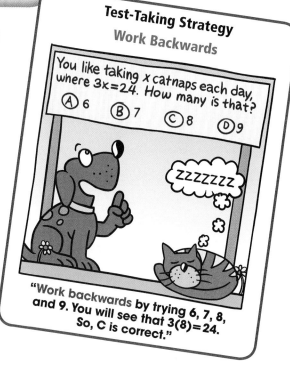

2. You are making identical fruit baskets using 16 apples, 24 pears, and 32 bananas. What is the greatest number of baskets you can make using all the fruit?

   F. 2   H. 8
   G. 4   I. 16

3. Which equation represents the word sentence below?

   The sum of 18 and 5 is equal to 9 less than a number y.

   A. $18 - 5 = 9 - y$ C. $18 + 5 = y - 9$
   B. $18 + 5 = 9 - y$ D. $18 - 5 = y - 9$

4. Which number line is a graph of the solution of the inequality below?

   $x \geq 5$

   F.

   G.

   H.

   I.

5. The steps your friend took to divide two mixed numbers are shown below.

$$3\frac{3}{5} \div 1\frac{1}{2} = \frac{18}{5} \times \frac{3}{2}$$
$$= \frac{27}{5}$$
$$= 5\frac{2}{5}$$

What should your friend change in order to divide the two mixed numbers correctly?

A. Find a common denominator of 5 and 2.

B. Multiply by the reciprocal of $\frac{18}{5}$.

C. Multiply by the reciprocal of $\frac{3}{2}$.

D. Rename $3\frac{3}{5}$ as $2\frac{8}{5}$.

6. An inequality is graphed on the number line below.

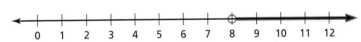

What is the least whole number value that is a solution of the inequality?

7. A company ordering parts receives a charge of $25 for shipping and handling plus $20 per part. Which equation represents the cost c of ordering p parts?

F. $c = 25 + 20p$

G. $c = 20 + 25p$

H. $p = 25 + 20c$

I. $p = 20 + 25c$

8. Which property is illustrated by the statement below?

$$5(3 + 6) = 5(3) + 5(6)$$

A. Associative Property of Multiplication

B. Commutative Property of Multiplication

C. Commutative Property of Addition

D. Distributive Property

9. What is the value of the expression below?

    $46.8 \div 0.156$

10. In a fish tank, 75% of the fish are goldfish. How many fish are in the tank if there are 24 goldfish?

    F. 6

    G. 18

    H. 32

    I. 96

11. What are the coordinates of point P in the coordinate plane below?

    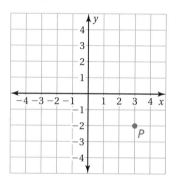

    A. $(-3, -2)$

    B. $(3, -2)$

    C. $(-2, -3)$

    D. $(-2, 3)$

12. What is the first step in evaluating the expression below?

    $3 \cdot (5 + 2)^2 \div 7$

    F. Multiply 3 and 5.

    G. Add 5 and 2.

    H. Evaluate 5^2.

    I. Evaluate 2^2.

13. Jeff wants to save $4000 to buy a used car. He has already saved $850. He plans to save an additional $150 each week.

    Part A Write and solve an equation to represent the number of weeks remaining until he can afford the car.

    Jeff saves $150 per week by saving $\frac{3}{4}$ of what he earns at his job each week. He works 20 hours per week.

    Part B Write an equation to represent the amount per hour that Jeff must earn to save $150 per week. Explain your reasoning.

    Part C What is the amount per hour that Jeff must earn? Show your work and explain your reasoning.

8 Surface Area and Volume

8.1 **Three-Dimensional Figures**

8.2 **Surface Areas of Prisms**

8.3 **Surface Areas of Pyramids**

8.4 **Volumes of Rectangular Prisms**

"I petitioned my owner for a doghouse with greater volume."

"And this is what he built for me."

"I want to paint my doghouse. To make sure I buy the correct amount of paint, I want to calculate the lateral surface area."

"Then, because I want to paint the inside and the outside, I will multiply by 2. Does this seem right to you?"

What You Learned Before

● **Classifying Figures**

Example 1

Identify the figure.

∴ Because the figure has a right angle and three sides of different lengths, it is a right scalene triangle.

Example 2

Identify the figure.

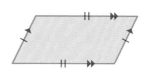

∴ Because the figure is a quadrilateral with opposite sides that are parallel, it is a parallelogram.

Try It Yourself

Identify the figure.

1.
2.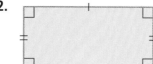
3.

● **Finding Volumes of Rectangular Prisms**

Example 3

Find the volume of the rectangular prism.

There are $4 \times 7 = 28$ unit cubes in each layer.

Because there are 5 layers, there are $5 \times 28 = 140$ unit cubes in the prism.

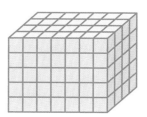

∴ So, the volume is 140 cubic units.

Try It Yourself

Find the volume of the rectangular prism.

4.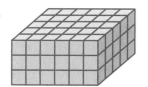
5.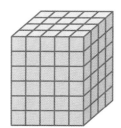
6.

8.1 Three-Dimensional Figures

Essential Question How can you draw three-dimensional figures?

Dot paper can help you draw three-dimensional figures, or *solids*.

Square Dot Paper

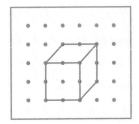

Face-On view

Isometric Dot Paper

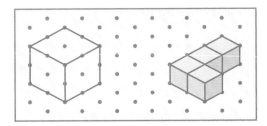

Corner view

1 ACTIVITY: Drawing Views of a Solid

Work with a partner. Draw the front, side, and top views of each stack of cubes. Then find the number of cubes in the stack.

a. Sample:

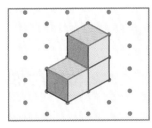 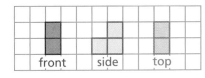

Number of cubes: 3

b.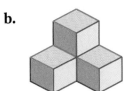

c.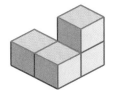

d.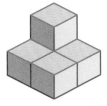

e.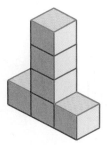

f.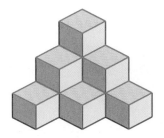

g.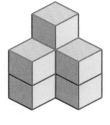

Geometry

In this lesson, you will
- draw three-dimensional figures.
- find the number of faces, edges, and vertices of solids.

354   Chapter 8   Surface Area and Volume

2 ACTIVITY: Drawing Solids

Work with a partner.

a. Use isometric dot paper to draw three different solids that use the same number of cubes as the solid at the right.

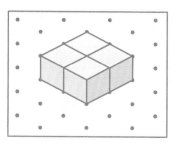

b. Use square dot paper to draw a different solid that uses the same number of *prisms* as the solid at the right.

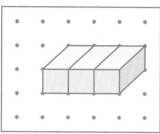

3 ACTIVITY: Exploring Faces, Edges, and Vertices

Work with a partner. Use the solid shown.

Math Practice

View as Components
What are the different parts of a three-dimensional object? How can dot paper help you draw the parts of the object?

a. Match each word to the figure. Then write a definition for each word.

    face   edge   vertex

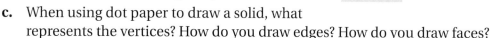

b. Identify the number of faces, edges, and vertices in a rectangular prism.

c. When using dot paper to draw a solid, what represents the vertices? How do you draw edges? How do you draw faces?

d. What do you think it means for lines or planes to be parallel or perpendicular in three dimensions? Use drawings to identify one pair of each of the following:

- parallel faces
- parallel edges
- edge parallel to a face
- perpendicular faces
- perpendicular edges
- edge perpendicular to a face

What Is Your Answer?

4. **IN YOUR OWN WORDS** How can you draw three-dimensional figures?

Practice

Use what you learned about three-dimensional figures to complete Exercises 7–9 on page 358.

Section 8.1 Three-Dimensional Figures 355

8.1 Lesson

Check It Out
Lesson Tutorials
BigIdeasMath.com

Key Vocabulary
solid, *p. 356*
polyhedron, *p. 356*
face, *p. 356*
edge, *p. 356*
vertex, *p. 356*
prism, *p. 356*
pyramid, *p. 356*

A **solid** is a three-dimensional figure that encloses a space. A **polyhedron** is a solid whose *faces* are all polygons.

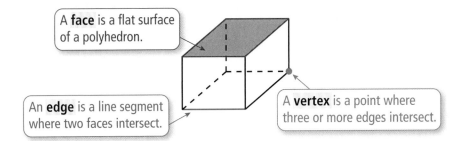

A **face** is a flat surface of a polyhedron.

An **edge** is a line segment where two faces intersect.

A **vertex** is a point where three or more edges intersect.

EXAMPLE 1 Finding the Number of Faces, Edges, and Vertices

Find the number of faces, edges, and vertices of the solid.

The solid has 1 face on the bottom, 1 face on the top, and 4 faces on the sides.

The faces intersect at 12 different line segments.

The edges intersect at 8 different points.

∴ So, the solid has 6 faces, 12 edges, and 8 vertices.

On Your Own

Exercises 10–12

1. Find the number of faces, edges, and vertices of the solid.

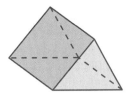

Key Ideas

Prisms

A **prism** is a polyhedron that has two parallel, identical *bases*. The *lateral faces* are parallelograms.

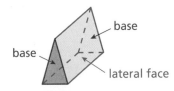

base
base
lateral face

Triangular Prism

Pyramids

A **pyramid** is a polyhedron that has one base. The lateral faces are triangles.

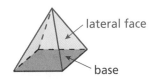

lateral face
base

Rectangular Pyramid

The shape of the base tells the name of the prism or the pyramid.

356 Chapter 8   Surface Area and Volume   Multi-Language Glossary at BigIdeasMath.com

EXAMPLE 2 Drawing Solids

a. Draw a rectangular prism.

Step 1:
Draw identical rectangular bases.

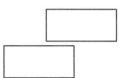

Step 2:
Connect corresponding vertices.

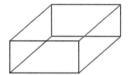

Step 3:
Change any *hidden* lines to dashed lines.

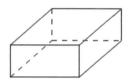

b. Draw a triangular pyramid.

Step 1:
Draw a triangular base and a point.

Step 2:
Connect the vertices of the triangle to the point.

Step 3:
Change any *hidden* lines to dashed lines.

EXAMPLE 3 Drawing Views of a Solid

Draw the front, side, and top views of the eraser.

The front view is a parallelogram.

The side view is a rectangle.

The top view is a rectangle.

On Your Own

Now You're Ready
Exercises 13–22

Draw the solid.

2. square prism

3. pentagonal pyramid

Draw the front, side, and top views of the solid.

4.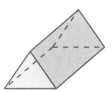

5.

Section 8.1 Three-Dimensional Figures 357

8.2 Surface Areas of Prisms

Essential Question How can you find the area of the entire surface of a prism?

1 ACTIVITY: Identifying Prisms

Work with a partner. Label one of the faces as a "base" and the other as a "lateral face." Use the shape of the base to identify the prism.

a.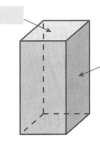

_____ Prism

b.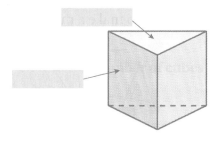

_____ Prism

2 ACTIVITY: Using Grid Paper to Construct a Prism

Work with a partner.

a. Copy the figure shown below onto grid paper.
b. Cut out the figure and fold it to form a prism. What type of prism does it form?

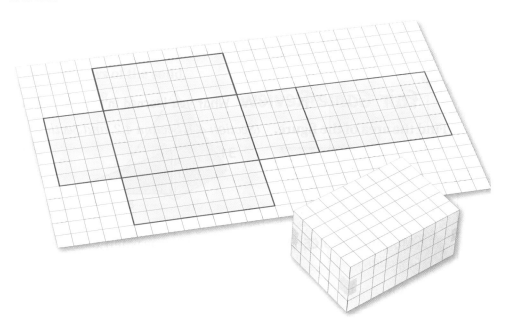

Geometry

In this lesson, you will
- use nets to represent prisms.
- find the surface area of prisms.
- solve real-life problems.

3 ACTIVITY: Finding the Area of the Entire Surface of a Prism

Work with a partner. Label each face in the two-dimensional representation of the prism as a "base" or a "lateral face." Then find the area of the entire surface of each prism.

Math Practice

Repeat Calculations
When finding the areas of the faces, what calculations do you repeat?

a.

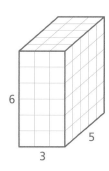 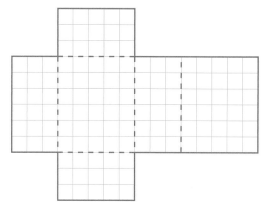

b.

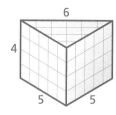 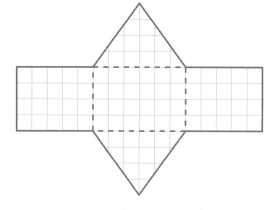

4 ACTIVITY: Drawing Two-Dimensional Representations of Prisms

Work with a partner. Draw a two-dimensional representation of each prism. Then find the area of the entire surface of each prism.

a.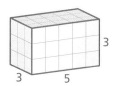

b.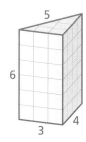

What Is Your Answer?

5. **IN YOUR OWN WORDS** How can you find the area of the entire surface of a prism?

Practice

Use what you learned about the area of the entire surface of a prism to complete Exercises 3–5 on page 364.

Section 8.2 Surface Areas of Prisms **361**

8.2 Lesson

Key Vocabulary
surface area, p. 362
net, p. 362

The **surface area** of a solid is the sum of the areas of all of its faces. You can use a two-dimensional representation of a solid, called a **net**, to find the surface area of the solid. Surface area is measured in *square units*.

Key Idea

Net of a Rectangular Prism

A *rectangular prism* is a prism with rectangular bases.

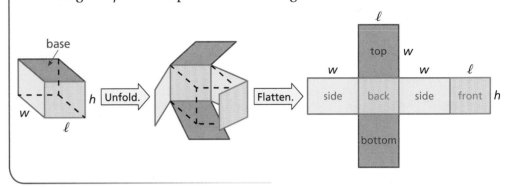

EXAMPLE 1 — Finding the Surface Area of a Rectangular Prism

Find the surface area of the rectangular prism.

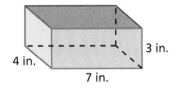

Use a net to find the area of each face.

Top: $7 \cdot 4 = 28$
Bottom: $7 \cdot 4 = 28$
Front: $7 \cdot 3 = 21$
Back: $7 \cdot 3 = 21$
Side: $4 \cdot 3 = 12$
Side: $4 \cdot 3 = 12$

Find the sum of the areas of the faces.

Surface Area = Area of top + Area of bottom + Area of front + Area of back + Area of a side + Area of a side

$S = 28 + 28 + 21 + 21 + 12 + 12$

$ = 122$

∴ So, the surface area is 122 square inches.

Key Idea

Remember

The area A of a triangle with base b and height h is $A = \frac{1}{2}bh$.

Net of a Triangular Prism

A *triangular prism* is a prism with triangular bases.

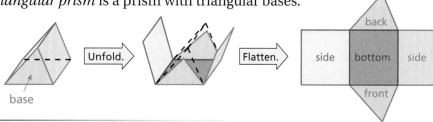

EXAMPLE 2 Finding the Surface Area of a Triangular Prism

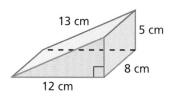

Find the surface area of the triangular prism.

Use a net to find the area of each face.

Bottom: $12 \cdot 8 = 96$

Front: $\frac{1}{2} \cdot 12 \cdot 5 = 30$

Back: $\frac{1}{2} \cdot 12 \cdot 5 = 30$

Side: $13 \cdot 8 = 104$

Side: $8 \cdot 5 = 40$

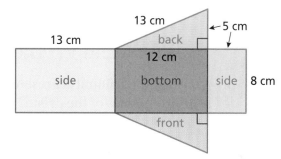

Find the sum of the areas of the faces.

Surface Area = Area of bottom + Area of front + Area of back + Area of a side + Area of a side

$S = 96 + 30 + 30 + 104 + 40 = 300$

∴ So, the surface area is 300 square centimeters.

On Your Own

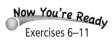
Exercises 6–11

Find the surface area of the rectangular prism.

1.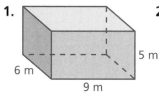

2.

3.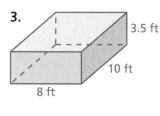

Find the surface area of the triangular prism.

4.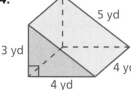

5.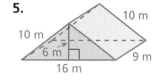

6.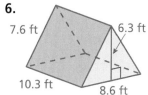

8.2 Exercises

Vocabulary and Concept Check

1. **VOCABULARY** Explain how to find the surface area of a prism.

2. **DIFFERENT WORDS, SAME QUESTION** Which is different? Find "both" answers.

   What is the sum of the areas of the faces of the prism?

   What is the area of the entire surface of the prism?

   What is the area of the triangular faces of the prism?

   What is the surface area of the prism?

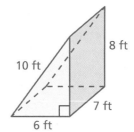

Practice and Problem Solving

Draw a two-dimensional representation of the prism. Then find the area of the entire surface of the prism.

3.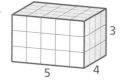

4.

5.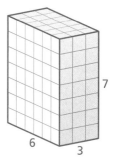

Find the surface area of the prism.

6.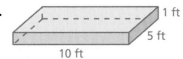

7.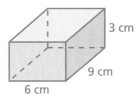

8.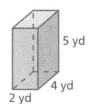

9.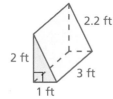

10.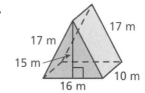

11.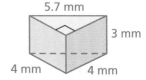

12. **GIFT BOX** A gift box in the shape of a rectangular prism measures 8 inches by 8 inches by 10 inches. What is the least amount of wrapping paper needed to wrap the gift box? Explain.

13. **TENT** What is the least amount of fabric needed to make the tent?

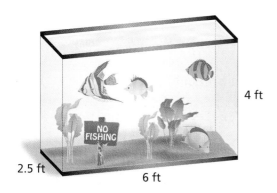

14. AQUARIUM A public library has an aquarium in the shape of a rectangular prism. The base is 6 feet by 2.5 feet. The height is 4 feet. How many square feet of glass were used to build the aquarium? (The top of the aquarium is open.)

15. STORAGE BOX The material used to make a storage box costs $1.25 per square foot. The boxes have the same volume. How much does a company save by choosing to make 50 of Box 2 instead of 50 of Box 1?

	Length	Width	Height
Box 1	20 in.	6 in.	4 in.
Box 2	15 in.	4 in.	8 in.

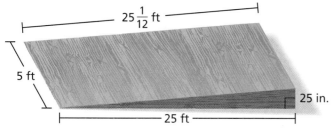

16. RAMP A quart of stain covers 100 square feet. How many quarts should you buy to stain the wheelchair ramp? (Assume you do not have to stain the bottom of the ramp.)

17. **Critical Thinking** A cube is removed from a rectangular prism. Find the surface area of the figure after removing the cube.

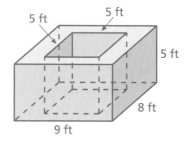

Fair Game Review *What you learned in previous grades & lessons*

Find the area of the triangle. *(Section 4.2)*

18.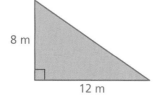

19.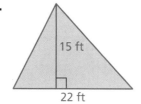

20.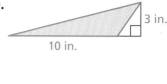

21. MULTIPLE CHOICE Which value is *not* a solution of the inequality $x - 4 \geq 2$? *(Section 7.5)*

　Ⓐ $x = 10$　　Ⓑ $x = 6$　　Ⓒ $x = 4$　　Ⓓ $x = 14$

8 Study Help

You can use a **process diagram** to show the steps involved in a procedure. Here is an example of a process diagram for drawing a prism.

Example

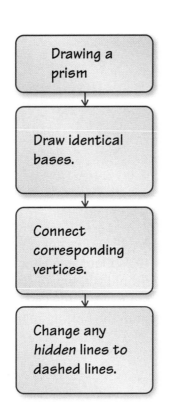
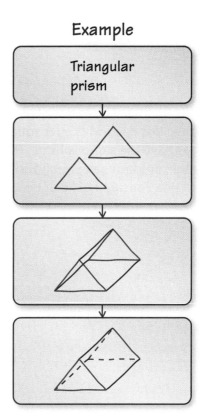

On Your Own

Make process diagrams with examples to help you study these topics.

1. drawing a pyramid
2. finding the surface area of a prism

After you complete this chapter, make process diagrams with examples for the following topics.

3. finding the surface area of a pyramid
4. finding the volume of a rectangular prism

"Descartes, you should use my process diagram when you eat your treats."

366 Chapter 8 Surface Area and Volume

8.1–8.2 Quiz

Find the number of faces, edges, and vertices of the solid. *(Section 8.1)*

1.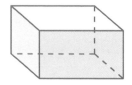
2.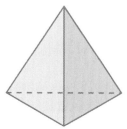

Draw the solid. *(Section 8.1)*

3. trapezoidal prism
4. octagonal pyramid

Draw the front, side, and top views of the solid. *(Section 8.1)*

5.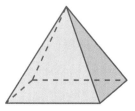
6.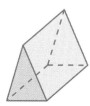

Find the surface area of the prism. *(Section 8.2)*

7.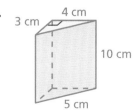
8.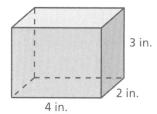

9. **CEREAL** A cereal box has the dimensions shown. *(Section 8.2)*

    a. Find the surface area of the cereal box.

    b. The manufacturer decides to decrease the size of the box by reducing each of the dimensions by 1 inch. Find the decrease in surface area.

10. **GIFT BOX** Find the surface area of the gift box. *(Section 8.2)*

8.3 Surface Areas of Pyramids

Essential Question How can you use a net to find the surface area of a pyramid?

1 ACTIVITY: Identifying Pyramids

Work with a partner. Label one of the faces as a "base" and the other as a "lateral face." Use the shape of the base to identify the pyramid.

a.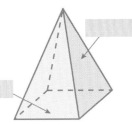

b.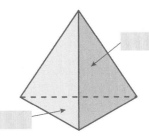

_____ Pyramid   _____ Pyramid

2 ACTIVITY: Using a Net

Work with a partner.

a. Copy the net shown below onto grid paper.
b. Cut out the net and fold it to form a pyramid. What type of rectangle is the base? Use this shape to name the pyramid.
c. Find the surface area of the pyramid.

Geometry

In this lesson, you will
- use nets to represent pyramids.
- find the surface area of pyramids.
- solve real-life problems.

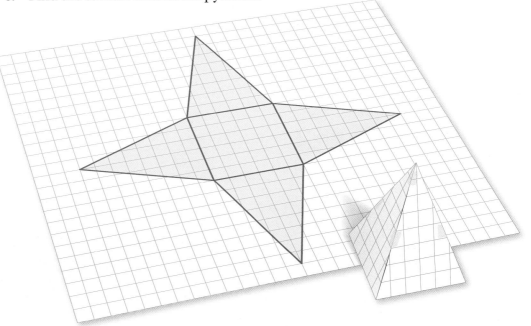

Key Idea

Net of a Triangular Pyram[...]

A *triangular pyramid* is a [...]

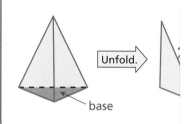

Study Tip

The base of each triangular pyramid in this section is an equilateral triangle.

EXAMPLE 2 Finding the Surface [...]

Find the surface area of th[...]

Use a net to find the area of [...]

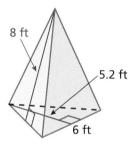

Bottom: $\dfrac{1}{2} \cdot 6 \cdot 5.2 = 15.6$

Side: $\dfrac{1}{2} \cdot 6 \cdot 8 = 24$

Side: $\dfrac{1}{2} \cdot 6 \cdot 8 = 24$

Side: $\dfrac{1}{2} \cdot 6 \cdot 8 = 24$

Find the sum of the areas of [...]

$$\text{Surface Area} = \text{Area of bottom} + \text{Area a sid[...]}$$

$S = 15.6 + 24$ [...]

$= 87.6$

∴ So, the surface area is 87[...]

On Your Own

Now You're Ready
Exercises 9–11

Find the surface area of th[...]

4.   **5.**

3 cm, 1.7 cm, 2 cm

---

3 ACTIVITY: Estimating the Surface Area of a Triangular Pyramid

Work with a partner. Label each face in the net of the triangular pyramid as a "base" or a "lateral face." Then estimate the surface area of the pyramid.

Math Practice

Analyze Givens
What information is given in the diagram? How does this help you estimate the surface area of the pyramid?

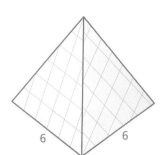

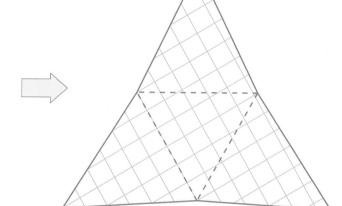

4 ACTIVITY: Finding the Surface Area of a Square Pyramid

Work with a partner. Draw a net for each square pyramid. Use the net to find the surface area of the pyramid.

a. 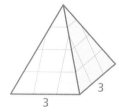 **b.**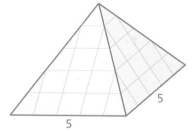

What Is Your Answer?

5. IN YOUR OWN WORDS How can you use a net to find the surface area of a pyramid?

6. CONJECTURE Make a conjecture about the lateral faces of a pyramid when the side lengths of the base have the same measure. Explain.

Practice — Use what you learned about the surface area of a pyramid to complete Exercises 3–5 on page 372.

8.3 Lesson

Key Idea

Net of a Square Pyramid

A *square pyramid* is a pyram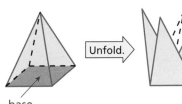

EXAMPLE 1 Finding the Surface A

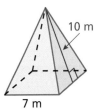

Find the surface area of the s

Use a net to find the area of e

Bottom: 7 • 7 = 49

Side: $\frac{1}{2}$ • 7 • 10 = 35

Side: $\frac{1}{2}$ • 7 • 10 = 35

Side: $\frac{1}{2}$ • 7 • 10 = 35

Side: $\frac{1}{2}$ • 7 • 10 = 35

Find the sum of the areas of t

Surface Area = Area of bottom + Area o a side

S = 49 + 35

∴ So, the surface area is 189

On Your Own

Now You're Ready Exercises 6–8

Find the surface area of the

1.
2.

8.3 Exercises

Vocabulary and Concept Check

1. **PRECISION** Explain how to find the surface area of a pyramid.
2. **WHICH ONE DOESN'T BELONG?** Which figure does *not* belong with the other three? Explain your reasoning.

 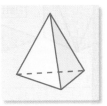 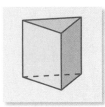 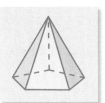

Practice and Problem Solving

Draw a net of the square pyramid. Then find the surface area of the pyramid.

3.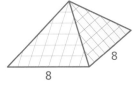
4.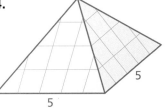
5.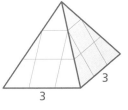

Find the surface area of the pyramid. The side lengths of the base are equal.

6.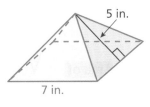
7.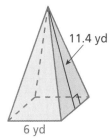
8.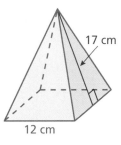

9.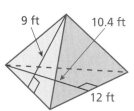
10.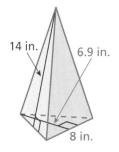
11.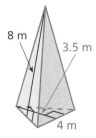

12. **PAPERWEIGHT** A paperweight is shaped like a triangular pyramid. The base is an equilateral triangle. Find the surface area of the paperweight.

13. **LOUVRE** The entrance to the Louvre Museum in Paris, France, is a square pyramid. The side length of the base is 116 feet, and the height of one of the triangular faces is 91.7 feet. Find the surface area of the four triangular faces of the entrance to the Louvre Museum.

14. **LIGHT COVER** A hanging light cover made of glass is shaped like a square pyramid. The cover does not have a bottom. One square foot of the glass weighs 2.45 pounds. The chain can support 35 pounds. Will the chain support the light cover? Explain.

15. **GEOMETRY** The surface area of a square pyramid is 84 square inches. The side length of the base is 6 inches. What is the value of x?

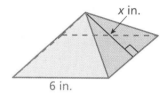

16. **STRUCTURE** In the diagram of the base of the hexagonal pyramid, all the triangles are the same. Find the surface area of the hexagonal pyramid.

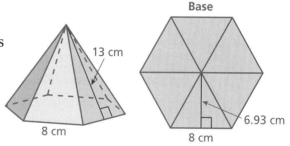

17. **Critical Thinking** Can you form a square pyramid using four of the triangles shown? Explain your reasoning.

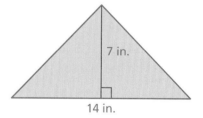

Fair Game Review *What you learned in previous grades & lessons*

Find the missing values in the ratio table. Then write the equivalent ratios. *(Section 5.2)*

18.
Frogs	7		28
Turtles	3	6	

19.
Apples	10	5	
Oranges	4		12

20. **MULTIPLE CHOICE** Which ordered pair is in Quadrant III? *(Section 6.5)*

    Ⓐ (5, −1)   Ⓑ (−2, −3)   Ⓒ (2, 4)   Ⓓ (−7, 1)

8.4 Volumes of Rectangular Prisms

Essential Question How can you find the volume of a rectangular prism with fractional edge lengths?

Recall that the **volume** of a three-dimensional figure is a measure of the amount of space that it occupies. Volume is measured in *cubic units*.

A *unit cube* is a cube with an edge length of 1 unit.

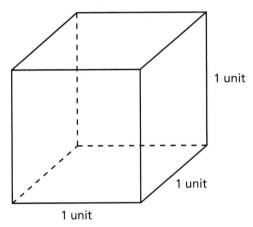

1 ACTIVITY: Using a Unit Cube

Work with a partner. The parallel edges of the unit cube have been divided into 2, 3, and 4 equal parts to create smaller rectangular prisms that are identical.

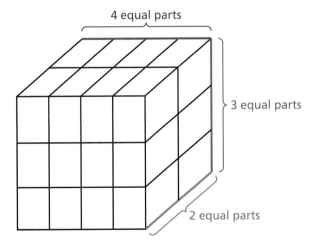

Geometry

In this lesson, you will
- find the volume of prisms with fractional edge lengths by using models.
- find the volume of prisms by using formulas.

a. Draw one of these identical prisms and label its dimensions.

b. What fraction of the volume of the unit cube does one of these identical prisms represent? Use this value to find the volume of one of the identical prisms. Explain your reasoning.

2 ACTIVITY: Finding the Volume of a Rectangular Prism

Work with a partner.

a. How many of the identical prisms in Activity 1(a) does it take to fill the rectangular prism below? Support your answer with a drawing.

Math Practice

Analyze Relationships
What is the relationship between the solid shown here and the solid in the previous activity?

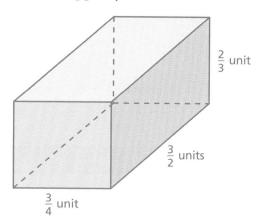

b. Use the volume of one of the identical prisms in Activity 1(a) to find the volume of the rectangular prism above. Explain your reasoning.

3 ACTIVITY: Finding the Volumes of Rectangular Prisms

Work with a partner. Explain how you can use the procedure in Activities 1 and 2 to find the volume of each rectangular prism. Then find the volume of each prism.

a.
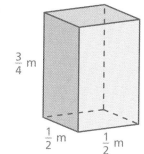

b.
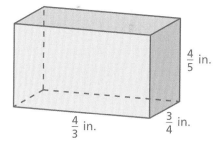

What Is Your Answer?

4. You have used the formulas $V = Bh$ and $V = \ell wh$ to find the volume V of a rectangular prism with whole number edge lengths. Do you think the formulas work for rectangular prisms with fractional edge lengths? Give examples with your answer.

5. **IN YOUR OWN WORDS** How can you find the volume of a rectangular prism with fractional edge lengths?

Practice — Use what you learned about the volume of a rectangular prism to complete Exercises 4–6 on page 378.

8.4 Lesson

Key Vocabulary
volume, p. 374

Key Idea

Volume of a Rectangular Prism

Words The volume V of a rectangular prism is the product of the area of the base and the height of the prism.

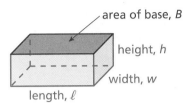

Algebra $V = Bh$ or $V = \ell wh$

EXAMPLE 1 Finding Volumes of Rectangular Prisms

Find the volume of each prism.

a. (prism with dimensions $\frac{7}{8}$ m, $\frac{1}{2}$ m, $\frac{5}{8}$ m)

b. (cube with edge $\frac{3}{2}$ in.)

Study Tip
In Example 1(b), the rectangular prism is a cube. You can use the formula $V = s^3$ to find the volume V of a cube with an edge length of s.

a.
$V = \ell wh$ Write formula.
$= \frac{7}{8}\left(\frac{1}{2}\right)\left(\frac{5}{8}\right)$ Substitute values.
$= \frac{35}{128}$ Multiply.

∴ So, the volume is $\frac{35}{128}$ cubic meter.

b.
$V = \ell wh$
$= \frac{3}{2}\left(\frac{3}{2}\right)\left(\frac{3}{2}\right)$
$= \frac{27}{8}$
$= 3\frac{3}{8}$

∴ So, the volume is $3\frac{3}{8}$ cubic inches.

On Your Own

Now You're Ready
Exercises 4–9

Find the volume of the prism.

1. (prism: 1 ft, $1\frac{1}{3}$ ft, $\frac{1}{2}$ ft)

2. (cube: $\frac{3}{4}$ yd, $\frac{3}{4}$ yd, $\frac{3}{4}$ yd)

EXAMPLE 2 — Using the Volume of a Rectangular Prism

One cubic foot of dirt weighs about 70 pounds. How many pounds of dirt can the dump truck haul when it is full?

Find the volume of dirt that the dump truck can haul when it is full.

$V = \ell w h$ Write formula for volume.

$= 17(8)\left(4\dfrac{3}{4}\right)$ Substitute values.

$= 646$ Multiply.

So, the dump truck can haul 646 cubic feet of dirt when it is full. To find the weight of the dirt, multiply by $\dfrac{70 \text{ lb}}{1 \text{ ft}^3}$.

$$646 \text{ ft}^3 \times \dfrac{70 \text{ lb}}{1 \text{ ft}^3} = 45{,}220 \text{ lb}$$

∴ The dump truck can haul about 45,220 pounds of dirt when it is full.

EXAMPLE 3 — Finding a Missing Dimension of a Rectangular Prism

Write and solve an equation to find the height of the computer tower.

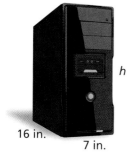

Volume = 1792 in.³

$V = \ell w h$ Write formula for volume.

$1792 = 16(7)h$ Substitute values.

$1792 = 112h$ Simplify.

$\dfrac{1792}{112} = \dfrac{112h}{112}$ Division Property of Equality

$16 = h$ Simplify.

∴ So, the height of the computer tower is 16 inches.

On Your Own

3. **WHAT IF?** In Example 2, the length of the dump truck is 20 feet. How many pounds of dirt can the dump truck haul when it is full?

Now You're Ready
Exercises 10–12

Write and solve an equation to find the missing dimension of the prism.

4. Volume = 72 in.³

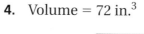

5. Volume = 1375 cm³

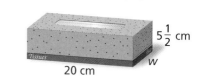

8.4 Exercises

Vocabulary and Concept Check

1. **CRITICAL THINKING** Explain how volume and surface area are different.

2. **REASONING** Will the formulas for volume work for rectangular prisms with decimal edge lengths? Explain.

3. **DIFFERENT WORDS, SAME QUESTION** Which is different? Find "both" answers.

   How much does it take to fill the rectangular prism?

   What is the capacity of the rectangular prism?

   How much does it take to cover the rectangular prism?

   How much does the rectangular prism contain?

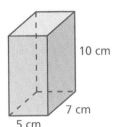

Practice and Problem Solving

Find the volume of the prism.

4.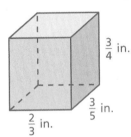

   $\frac{3}{4}$ in., $\frac{3}{5}$ in., $\frac{2}{3}$ in.

5.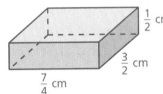

   $\frac{1}{2}$ cm, $\frac{3}{2}$ cm, $\frac{7}{4}$ cm

6.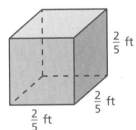

   $\frac{2}{5}$ ft, $\frac{2}{5}$ ft, $\frac{2}{5}$ ft

7.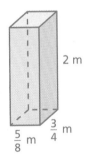

   2 m, $\frac{3}{4}$ m, $\frac{5}{8}$ m

8.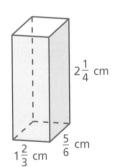

   $2\frac{1}{4}$ cm, $\frac{5}{6}$ cm, $1\frac{2}{3}$ cm

9.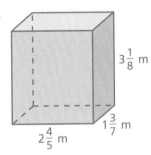

   $3\frac{1}{8}$ m, $1\frac{3}{7}$ m, $2\frac{4}{5}$ m

Write and solve an equation to find the missing dimension of the prism.

10. Volume = 1620 cm³

    9 cm, 9 cm, h

11. Volume = 220.5 cm³

    7 cm, 7 cm, w

12. Volume = 532 in.³

    $1\frac{3}{4}$ in., 19 in., w

13. FISH TANK One cubic foot of water weighs about 62.4 pounds. How many pounds of water can the fish tank hold when it is full?

14. CUBE How many $\frac{3}{4}$-centimeter cubes do you need to create a cube with an edge length of 12 centimeters?

15. REASONING How many 1-inch cubes do you need to fill a cube that has an edge length of 1 foot? How can this result help you convert a volume from cubic inches to cubic feet? from cubic feet to cubic inches?

16. FOOD STORAGE

a. Estimate the amount of casserole left in the dish.

b. Will the casserole fit in the storage container? Explain your reasoning.

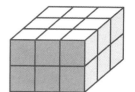

17. PROBLEM SOLVING The area of the shaded face is 96 square centimeters. What is the volume of the rectangular prism?

18. Project You have 1400 square feet of boards to use for a new tree house.

a. Design a tree house that has a volume of at least 250 cubic feet. Include sketches of your tree house.

b. Are your dimensions reasonable? Explain your reasoning.

Fair Game Review   *what you learned in previous grades & lessons*

Tell whether the given value is a solution of the equation. *(Section 7.2)*

19. $x + 17 = 24$; $x = 7$

20. $\frac{x}{5} = 6$; $x = 35$

21. $x - 19 = 42$; $x = 21$

22. MULTIPLE CHOICE Which set of integers is ordered from least to greatest? *(Section 6.2)*

- Ⓐ −1, 3, −5, −8, 12
- Ⓑ −1, 3, −5, −8, 12
- Ⓒ −4, −2, 1, 7, 10
- Ⓓ −14, −9, 6, −4, 2

8.3–8.4 Quiz

Find the surface area of the pyramid. The side lengths of the base are equal. *(Section 8.3)*

1.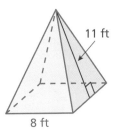

2.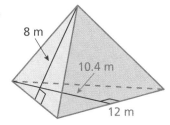

Find the volume of the prism. *(Section 8.4)*

3.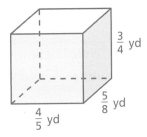

4.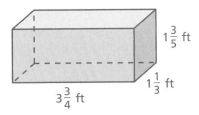

Write and solve an equation to find the missing dimension of the prism. *(Section 8.4)*

5. Volume = 1620 in.³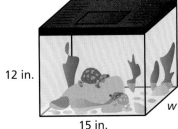

6. Volume = 154 in.³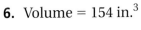

7. Volume = 4250 in.³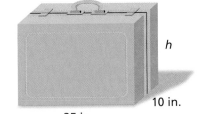

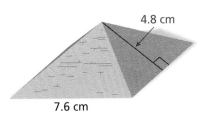

8. **GREAT PYRAMID** The Great Pyramid of Giza is a square pyramid. A gift shop sells miniature models of this pyramid. Find the surface area of the model shown at the left. *(Section 8.3)*

9. **CUBE** How many 1-inch cubes do you need to create a cube with an edge length of 7 inches? *(Section 8.4)*

10. **TOY CHEST** A toy company sells two different toy chests. The toy chests have different dimensions, but the same volume. What is the width w of Toy Chest 2? *(Section 8.4)*

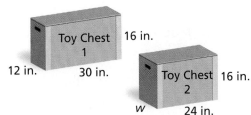

8 Chapter Review

Review Key Vocabulary

solid, *p. 356*
polyhedron, *p. 356*
face, *p. 356*
edge, *p. 356*

vertex, *p. 356*
prism, *p. 356*
pyramid, *p. 356*
surface area, *p. 362*

net, *p. 362*
volume, *p. 374*

Review Examples and Exercises

Three-Dimensional Figures *(pp. 354–359)*

a. Find the number of faces, edges, and vertices of the solid.

The solid has 1 face on the bottom and 4 faces on the sides.

The faces intersect at 8 different line segments.

The edges intersect at 5 different points.

∴ So, the solid has 5 faces, 8 edges, and 5 vertices.

b. Draw a triangular prism.

| Draw identical triangular bases. | Connect corresponding vertices. | Change any *hidden* lines to dashed lines. |

Exercises

Find the number of faces, edges, and vertices of the solid.

1.

2.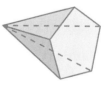

Draw the solid.

3. square pyramid

4. hexagonal prism

Chapter Review **381**

8 Chapter Test

Find the number of faces, edges, and vertices of the solid.

1.

2.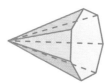

Find the surface area of the prism.

3.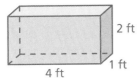

4.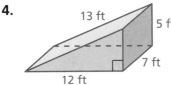

Find the surface area of the pyramid. The side lengths of the base are equal.

5.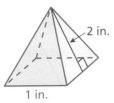

6.

Find the volume of the prism.

7.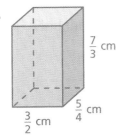

8.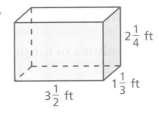

9. **DRAWING A SOLID** Draw an octagonal prism.

10. **DVD COLLECTION** You are wrapping the boxed DVD collection as a present. What is the least amount of wrapping paper needed to wrap the box?

11. **SKATEBOARD RAMP** A quart of paint covers 80 square feet. How many quarts should you buy to paint the ramp with two coats? (Assume you will not paint the bottom of the ramp.)

12. **CUBE** A cube has an edge length of 4 inches. You double the side lengths. How many times greater is the volume of the new cube?

384   Chapter 8   Surface Area and Volume

8 Cumulative Assessment

1. The temperature in a town has never been above 38 degrees Fahrenheit. Let t represent the temperature, in degrees Fahrenheit. Which inequality represents the temperature in the town?

   A. $t < 38$ **C.** $t > 38$

   B. $t \leq 38$ **D.** $t \geq 38$

2. Which number is equivalent to the expression below?

   $$3 \cdot 4^2 + 6 \div 2$$

   F. 27   **H.** 51

   G. 33   **I.** 75

3. What is the volume of the package shown below?

   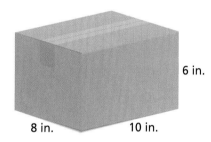

   A. 240 in.³   **C.** 480 in.³

   B. 376 in.³   **D.** 960 in.³

4. A housing community started with 60 homes. In each of the following years, 8 more homes were built. Let y represent the number of years that have passed since the first year, and let n represent the number of homes. Which equation describes the relationship between n and y?

   F. $n = 8y + 60$ **H.** $n = 60y + 8$

   G. $n = 68y$ **I.** $n = 60 + 8 + y$

Cumulative Assessment

5. What is the value of *m* that makes the equation below true?

$$4m = 6$$

6. A square pyramid is shown below.

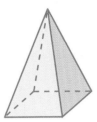

The square base and one of the triangular faces of the square pyramid are shown below with their dimensions.

 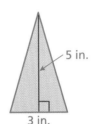

3 in.   3 in.

Square Base   **A Triangular Face**

What is the total surface area of the square pyramid?

A. 16.5 in.^{2}

B. 31.5 in.^{2}

C. 39 in.^{2}

D. 69 in.^{2}

7. A wooden box has a length of 12 inches, a width of 6 inches, and a height of 8 inches.

Part A Draw and label a rectangular prism with the dimensions of the wooden box.

Part B What is the surface area, in square inches, of the wooden box? Show your work.

Part C You have a 2-ounce sample of wood stain that covers 900 square inches. Is this enough to give the entire box two coats of stain? Show your work and explain your reasoning.

8. A biologist measures the lengths of a crazy ant and a green anole that he has in his laboratory. His measurements are shown below.

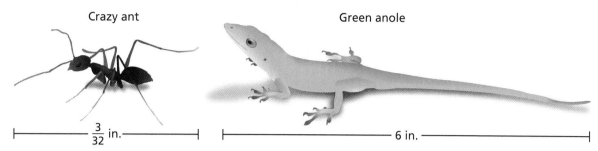

Not drawn to scale

The length of the green anole is how many times greater than the length of the crazy ant?

F. $\dfrac{9}{16}$

G. $5\dfrac{29}{32}$

H. 16

I. 64

9. What is the missing value in the ratio table?

Castles	1	2		12
Towers	4	8	24	48

10. What is the area of the shaded figure shown below?

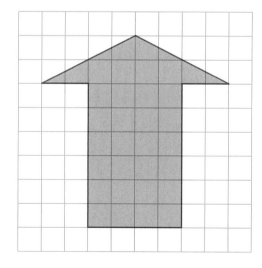

A. 32 units^2

B. 36 units^2

C. 40 units^2

D. 64 units^2

Cumulative Assessment 387

9 Statistical Measures

9.1 Introduction to Statistics
9.2 Mean
9.3 Measures of Center
9.4 Measures of Variation
9.5 Mean Absolute Deviation

"Mom, my owner, and Fluffy have agreed to participate in my survey. Will you be my fourth participant?"

"Please hold still. I am trying to find the mean of 6, 8, and 10 by dividing their sum into three equal piles."

What You Learned Before

"Our mean weight is 18 pounds."

Ordering Decimals

Example 1 Use a number line to order 8, 6.5, 7.25, 5.5, 4.25, and 7 from least to greatest.

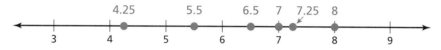

Try It Yourself

Use a number line to order the numbers from least to greatest.

1. 7.25, 4.5, 6.5, 6, 5.5, 8.75
2. 4, 2.5, 3.25, 5.5, 4.5, 6.75
3. 6.25, 3, 2.5, 3.5, 5.75, 5
4. 1.25, 5.5, 4.75, 4.5, 3.5, 2.25

Analyzing Double Bar Graphs

Example 2 How many more male athletes than female athletes participated in the 1992 Summer Olympics?

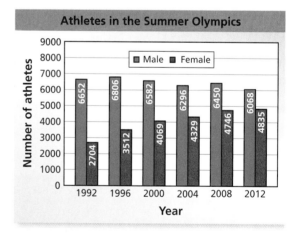

$6652 - 2704 = 3948$

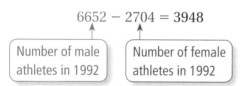

∴ 3948 more male athletes participated.

Example 3 How many athletes participated in the 2000 Summer Olympics?

∴ $6582 + 4069 = 10{,}651$ athletes participated in the 2000 Summer Olympics.

Try It Yourself

5. How many more female athletes participated in the 2012 Summer Olympics than in the 1992 Summer Olympics?

6. Describe the relationship between the number of athletes in the 2000 Summer Olympics and the number of athletes in the 2004 Summer Olympics.

9.1 Introduction to Statistics

Essential Question How can you tell whether a question is a statistical question?

Your heart rate is the number of times your heart beats in a certain time period, such as 1 minute. To measure your heart rate, you can check your pulse. The illustration shows how to check your pulse by pressing lightly on your wrist.

Here are other places to check your pulse:
- inside your elbow
- side of your neck
- top of your foot

1 ACTIVITY: Using Data to Answer Questions

Work with a partner.

a. Find your pulse by counting the number of beats in 10 seconds. Have your partner keep track of the time. Write a rate to describe your result.

b. Complete the ratio table. What is your heart rate in beats per minute?

Time (seconds)	10	30	60
Number of Beats			

c. Collect the recorded heart rates (in beats per minute) of the students in your class, including yourself. Compare the heart rates.

d. **MODELING** Make a *line plot* of your data. Then answer the following questions:
   - How many values are in your data set?
   - Do the heart rates *cluster* around a particular value or values?
   - Are there any *peaks* or *gaps* in the data?
   - Are there any unusual heart rates that are far removed from the other values?

e. **REASONING** How would you answer the following question by using only one value? Explain your reasoning.

   "What is the heart rate of sixth grade students?"

f. **REASONING** Read and compare the following questions. How did you answer each question? Could the answer be the same for both questions? Explain.
   - *What is your heart rate?*
   - *What is the heart rate of sixth grade students?*

Statistics

In this lesson, you will
- recognize statistical questions.
- use dot plots to display numerical data.

390 Chapter 9 Statistical Measures

2 ACTIVITY: Identifying Types of Questions

Work with a partner.

Math Practice

Build Arguments
How can comparing your answers help you support your conjecture?

a. Answer each question below on your own. Then compare your answers with your partner's answers. For which questions should your answers be the same? For which questions might your answers be different?

1. What is your shoe size?
2. How many states are in the United States?
3. How many brothers and sisters do you have?
4. How many U.S. presidents have been in office?
5. What is your favorite type of movie?
6. How tall are you?

b. **CONJECTURE** Some of the questions above are considered *statistical* questions. Which ones do you think they are? Why?

3 ACTIVITY: Analyzing a Question in a Survey

Work with a partner. A student asks the following question in a survey:

"Do you prefer salty potato chips or healthy granola bars to be sold in the school's vending machines?"

a. Do you think this is a fair question to ask in a survey? Explain.

b. **LOGIC** Identify the words in the question that may influence someone's response. Then explain how you can reword the question.

c. How might the results of the survey differ when the student asks the original question and your reworded question in part (b)?

What Is Your Answer?

4. **REASONING** What do you think "statistics" means?

5. **IN YOUR OWN WORDS** How can you tell whether a question is a statistical question? Give examples to support your explanation.

6. Find the least and the greatest heart rates in your class. How can you use these two values to answer the question in Activity 1(e)?

7. Create a one-question survey. Explain why your question is a statistical question. Then conduct your survey and organize your results in a line plot. Make three observations about your data set.

Practice

Use what you learned about different types of questions to complete Exercises 4–7 on page 394.

9.1 Lesson

Key Vocabulary
statistics, *p. 392*
statistical question, *p. 392*

Statistics is the science of collecting, organizing, analyzing, and interpreting data. A **statistical question** is one for which you do not expect to get a single answer. Instead, you expect a variety of answers, and you are interested in the distribution and tendency of those answers.

Recall that a dot plot uses a number line to show the number of times each value in a data set occurs. Dot plots show the *spread* and the *distribution* of a data set.

EXAMPLE 1 Answering a Statistical Question

You conduct a science experiment on house mice. Your teacher asks you, "What is the weight of a mouse?"

a. Is this a statistical question? Explain.

∴ Because you can anticipate that the weights of mice will vary, it is a statistical question.

Weights (grams)			
20	19	21	20
18	20	27	21
28	23	20	19
20	21	18	27
19	22	21	20

b. You weigh some mice and record the weights (in grams) in the table. Display the data in a dot plot. Identify any clusters, peaks, or gaps in the data.

Draw a number line that includes the least value, 18, and greatest value, 28. Then place a dot above the number line for each data value.

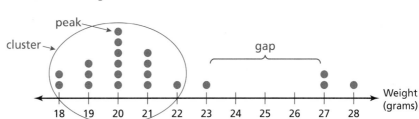

Most of the data are clustered around 20. There is a peak at 20 and a gap between 23 and 27.

Study Tip
Dot plots are sometimes called *line plots*. It is easy to see clusters, peaks, and gaps in a dot plot.

c. Use the distribution of the data to answer the question.

∴ Most mice weigh about 20 grams.

On Your Own

Now You're Ready
Exercises 8–16

1. The table shows the ages of some people who retired early. You are asked, "How old are people who retire early?"

Ages			
60	61	59	60
62	56	64	59
58	60	61	60
59	60	58	61

   a. Is this a statistical question? Explain.
   b. Display the data in a dot plot. Identify any clusters, peaks, or gaps in the data.
   c. Use the distribution of the data to answer the question.

EXAMPLE 2 **Using a Dot Plot**

You record the high temperature every day while at summer camp in August. Then you create the vertical dot plot.

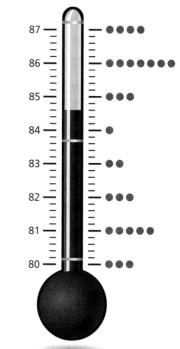

a. **How many weeks were you at summer camp?**

Because there are 28 data values on the dot plot, you were at camp 28 days.

$$28 \text{ days} \cdot \frac{1 \text{ week}}{7 \text{ days}} = 4 \text{ weeks}$$

∴ So, you were at summer camp for 4 weeks.

b. **How can you collect these data? What are the units?**

∴ You can collect these data with a thermometer. The units are degrees Fahrenheit (°F).

c. **Write a statistical question that you can answer using the dot plot. Then answer the question.**

One possible statistical question is:

What is the daily high temperature in August?

∴ The high temperatures are spread out with about half of the temperatures around 81°F and half of the temperatures around 86°F.

On Your Own

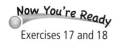
Exercises 17 and 18

2. The dot plot shows the times of sixth grade students in a 100-meter race.

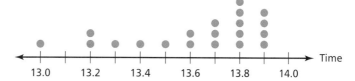

a. How many students ran in the race?

b. How can you collect these data? What are the units?

c. Write a statistical question that you can answer using the dot plot. Then answer the question.

9.2 Mean

Essential Question How can you find an average value of a data set?

1 ACTIVITY: Finding a Balance Point

Work with a partner. Discuss the distribution of the data. Where on the number line do you think the data set is *balanced*? Is this a good representation of the average? Explain.

a. number of quarters brought to a batting cage

b. annual income of recent graduates (in thousands of dollars)

c. hybrid fuel economy (miles per gallon)

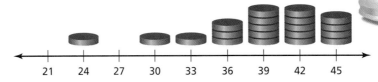

2 ACTIVITY: Finding a Fair Share

Statistics
In this lesson, you will
- understand the concept of the mean of data sets.
- find the mean of data sets.
- compare and interpret the means of data sets.

Work with a partner. It costs $0.25 to hit 12 baseballs in a batting cage. The table shows the numbers of quarters six friends bring to the batting cage. They want to group the quarters so that everyone has the same amount.

Quarters					
John	Lisa	Miguel	Matt	Cheryl	Jean
6	3	4	5	2	4

Use counters to represent each number in the table. How can you use the counters to determine how many times each friend can use the batting cage? Explain how this procedure results in a "fair share."

396   Chapter 9   Statistical Measures

3 ACTIVITY: Finding an Average

Work with a partner. Use the information in Activity 2.

a. What is the total number of quarters the group of friends brought to the batting cage?

b. **REASONING** How can you use math to find the average number of quarters that each friend brought to the batting cage? Find the average number of quarters. Why do you think this average represents a fair share?

4 ACTIVITY: Answering a Statistical Question

Work with a partner. The table shows the numbers of quarters several people bring to a batting cage. You want to answer the question:

"How many quarters do people bring to a batting cage?"

a. Explain why this question is a statistical question.

b. **MODELING** Make a dot plot of the data. Use the distribution of the data to answer the question. Explain your reasoning.

c. **REASONING** Use an average to answer the question. Explain your reasoning.

> **Math Practice**
>
> **Use Clear Definitions**
> What does it mean for data to have an average? How does this help you answer the question?

What Is Your Answer?

5. **IN YOUR OWN WORDS** How can you find an average value of a data set?

6. Give two real-life examples of averages.

7. Explain what it means to say the average of a data set is the point on a number line where the data set is balanced.

8. There are 5 students in the cartoon. Four of the students are 66 inches tall. One is 96 inches tall.

   a. How do you think the students decided that their average height is 6 feet?

   b. Does a height of 6 feet seem like a good representation of the average height of the 5 students? Explain why or why not.

"Yup, the average height in our class is 6 feet."

Practice — Use what you learned about averages to complete Exercises 4 and 5 on page 400.

Section 9.2   Mean   397

9.2 Lesson

Key Vocabulary
mean, p. 398
outlier, p. 399

A mean is a type of average.

Key Idea

Mean

Words The **mean** of a data set is the sum of the data divided by the number of data values.

Numbers Data: 8, 5, 6, 9 → 4 data values   Mean: $\dfrac{8 + 5 + 6 + 9}{4} = \dfrac{28}{4} = 7$

EXAMPLE 1 Finding the Mean

Text Messages Sent
Mark: 120
Laura: 95
Stacy: 101
Josh: 125
Kevin: 82
Maria: 108
Manny: 90

The table shows the number of text messages sent by a group of friends over 1 week. What is the mean number of messages sent?

Ⓐ 100   Ⓑ 102   Ⓒ 103   Ⓓ 104

$$\text{mean} = \dfrac{120 + 95 + 101 + 125 + 82 + 108 + 90}{7}$$ ← Sum of the data / Number of values

$$= \dfrac{721}{7}, \text{ or } 103 \quad \text{Simplify.}$$

∴ The mean number of text messages sent is 103. The correct answer is Ⓒ.

EXAMPLE 2 Comparing Means

The double bar graph shows the monthly rainfall amounts for two cities over a six-month period. Compare the mean monthly rainfalls.

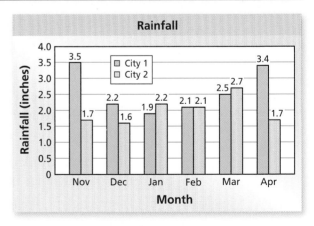

City 1 mean: $\dfrac{3.5 + 2.2 + 1.9 + 2.1 + 2.5 + 3.4}{6} = \dfrac{15.6}{6}$, or 2.6

City 2 mean: $\dfrac{1.7 + 1.6 + 2.2 + 2.1 + 2.7 + 1.7}{6} = \dfrac{12}{6}$, or 2

∴ Because 2.6 is greater than 2, City 1 averaged more rainfall.

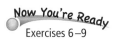

On Your Own

Find the mean of the data.

1. 49, 62, 52, 54, 61, 70, 55, 53
2. 7.2, 8.5, 7.0, 8.1, 6.7

An **outlier** is a data value that is much greater or much less than the other values. When included in a data set, it can affect the mean.

EXAMPLE 3 Finding the Mean With and Without an Outlier

Shetland Pony Heights (inches)				
40	37	39	40	42
38	38	37	28	40

The table shows the heights of several Shetland ponies.

a. Identify the outlier.
b. Find the mean with and without the outlier.
c. Describe how the outlier affects the mean.

a. Display the data in a dot plot.

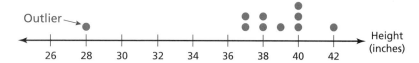

The height of 28 inches is much less than the other heights. So, it is an outlier.

b. **Mean with outlier:**

$$\frac{40 + 37 + 39 + 40 + 42 + 38 + 38 + 37 + 28 + 40}{10} = \frac{379}{10}, \text{ or } 37.9$$

Mean without outlier:

$$\frac{40 + 37 + 39 + 40 + 42 + 38 + 38 + 37 + 40}{9} = \frac{351}{9}, \text{ or } 39$$

c. With the outlier, the mean is less than all but three of the heights. Without the outlier, the mean better represents the heights.

On Your Own

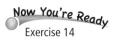

For each data set, identify the outlier. Then describe how it affects the mean.

3. Weights (in pounds) of dogs at a kennel:

    48, 50, 55, 60, 8, 37, 50

4. Prices for flights from Miami, Florida, to San Juan, Puerto Rico:

    $456, $512, $516, $900, $436, $516

Section 9.2 Mean 399

9.2 Exercises

Vocabulary and Concept Check

1. **VOCABULARY** Arrange the words to explain how to find a mean.

   | the data values | divide by | the number of data values | add | then |

2. **NUMBER SENSE** Is the mean always equal to a value in the data set? Explain.

3. **REASONING** Can you use the mean to answer a statistical question? Explain.

Practice and Problem Solving

Describe an average value of the data.

4. Ages in a class: 11, 12, 12, 12, 12, 12, 13

5. Movies seen this week: 0, 0, 0, 1, 1, 2, 3

Find the mean of the data.

6.
Pets Owned	
Brandon	I
Jill	III
Mark	II
Nicole	IIII
Steve	0

7.
Brothers and Sisters	
Amanda	♀
Eve	♀♀♀♀♀
Joseph	♀♀♀♀
Michael	♀♀

8.
Sit-ups		
108	85	94
103	112	115
98	119	126
105	82	89

9. Visits to Your Website (bar graph): Day 1: 12, Day 2: 16, Day 3: 0, Day 4: 8, Day 5: 31, Day 6: 28, Day 7: 17

10. **GOLF** The table shows tournament finishes for a golfer.

    a. What was the golfer's mean finish?
    b. Identify two outliers for the data.

Tournament Finishes							
1	1	2	1	1	12	6	2
15	37	1	2	1	26	9	1

Time (minutes)				
4.2	3.5	4.55	2.75	2.25

11. **COMMERCIALS** You and your friends are watching a television show. One of your friends asks, "How long are the commercial breaks during this show?"

    a. Is this a statistical question? Explain.
    b. Use the mean of the values in the table to answer the question.

Month	Rainfall (inches)	Month	Rainfall (inches)
Jan	2.22	Jul	3.27
Feb	1.51	Aug	5.40
Mar	1.86	Sep	5.45
Apr	2.06	Oct	4.34
May	3.48	Nov	2.64
Jun	4.57	Dec	2.14

12. **RAINFALL** The table shows the monthly rainfall at a measuring station. What is the mean monthly rainfall?

13. **OPEN-ENDED** Create two different sets of data that have six values and a mean of 21.

3 14. **CELL PHONE** The bar graph shows your cell phone usage for five months.

   a. Which data value is an outlier? Explain.

   b. Find the mean with and without the outlier. Then describe how the outlier affects the mean.

   c. Describe a situation that could have caused the outlier in this problem.

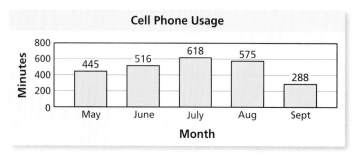

15. **HEIGHT** The table shows the heights of the volleyball players from two schools. What is the difference between the mean heights of the two teams? Do outliers affect either mean? Explain.

Player Height (inches)														
Dolphins	59	65	53	56	58	61	64	68	51	56	54	57		
Tigers	63	68	66	58	54	55	61	62	53	70	64	64	62	67	69

16. **REASONING** Make a dot plot of the data set 11, 13, 17, 15, 12, 18, and 12. Use the dot plot to explain how the mean is the point where the data set is balanced.

17. **ALLOWANCE** In your class, 7 students do not receive a weekly allowance, 5 students receive $3, 7 students receive $5, 3 students receive $6, and 2 students receive $8. What is the mean weekly allowance? Explain how you found your answer.

18. **Precision** A collection of 8 backpacks has a mean weight of 14 pounds. A different collection of 12 backpacks has a mean weight of 9 pounds. What is the mean weight of the 20 backpacks? Explain how you found your answer.

Fair Game Review What you learned in previous grades & lessons

Evaluate the expression. *(Section 1.3)*

19. $\dfrac{8 + 10}{2}$ 20. $\dfrac{26 + 34}{2}$ 21. $\dfrac{18 + 19}{2}$ 22. $\dfrac{14 + 17}{2}$

23. **MULTIPLE CHOICE** 60% of what number is 105? *(Section 5.6)*

   Ⓐ 63   Ⓑ 175   Ⓒ 630   Ⓓ 1750

9.3 Measures of Center

Essential Question In what other ways can you describe an average of a data set?

1 ACTIVITY: Finding a Median

Work with a partner.

a. Write the total number of letters in the first and last names of 19 celebrities, historical figures, or people you know. Organize your data in a table. One person is already listed for you.

Person	Number of letters in first and last name
Abraham Lincoln	14

b. Order the values in your data set from least to greatest. Then write the data on a strip of grid paper with 19 boxes.

c. Place a finger on the square at each end of the strip. Move your fingers toward the center of the ordered data set until your fingers touch. On what value do your fingers touch?

Statistics

In this lesson, you will
- understand the concept of measures of center.
- find the median and mode of data sets.

d. Now take your strip of grid paper and fold it in half. On what number is the crease? What do you notice? This value is called the *median*. How would you describe to another student what the median of a data set represents?

e. How many values are greater than the median? How many are less than the median?

f. Why do you think the median is considered an average of a data set?

2 ACTIVITY: Adding a Value to a Data Set

Work with a partner.

a. How many total letters are in your first name and last name? Add this value to the ordered data set in Activity 1. How many values are now in your data set?

b. Write the ordered data, including your new value from part (a), on a strip of grid paper.

c. Repeat parts (c) and (d) from Activity 1. Explain your findings. How do you think you can find the median of this data set?

d. Compare the medians in Activities 1 and 2. Then answer the following questions. Explain your reasoning.
   - Do you think the median always has to be a value in the data set?
   - Do you think the median always has to be a whole number?

3 ACTIVITY: Finding a Mode

Work with a partner.

Math Practice

Use a Graph
How can you use the dot plot to find the mode?

a. Make a dot plot for the data set in Activity 2. Describe the distribution of the data.

b. Which value occurs most often in the data set? This value is called the *mode*.

c. Do you think a data set can have no mode or more than one mode? Explain.

d. Do you think the mode always has to be a value in the data set? Explain.

e. Why do you think the mode is considered an average of a data set?

What Is Your Answer?

4. **IN YOUR OWN WORDS** In what other ways can you describe an average of a data set?

5. Find the mean of your data set in Activity 2. Then compare the mean, median, and mode. Is there one measure that you think best represents your data set? Explain your reasoning.

Practice Use what you learned about the median of a data set to complete Exercises 5 and 6 on page 407.

Section 9.3   Measures of Center   403

9.3 Lesson

Check It Out
Lesson Tutorials
BigIdeasMath.com

Key Vocabulary
measure of center, p. 404
median, p. 404
mode, p. 404

A **measure of center** is a measure that describes the typical value of a data set. The mean is one type of measure of center. Here are two others.

Key Ideas

Median

Words Order the data. For a set with an odd number of values, the **median** is the middle value. For a set with an even number of values, the **median** is the mean of the two middle values.

Numbers Data: 5, 8, 9, 12, 14   The median is 9.

Data: 2, 3, 5, 7, 10, 11

The median is $\frac{5+7}{2}$, or 6.

Mode

Words The **mode** of a data set is the value or values that occur most often. Data can have one mode, more than one mode, or no mode. When all values occur only once, there is no mode.

Numbers Data: 11, 13, 15, 15, 18, 21, 24, 24

The modes are 15 and 24.

Study Tip
The mode is the only measure of center that you can use to describe a set of data that is *not* made up of numbers.

EXAMPLE 1 Finding the Median and Mode

Bowling Scores

120	135	160	125	90
205	160	175	105	145

Find the median and mode of the bowling scores.

90, 105, 120, 125, 135, 145, 160, 160, 175, 205   Order the data.

Median: $\frac{135+145}{2} = \frac{280}{2}$, or 140   Add the two middle values and divide by 2.

Mode: 90, 105, 120, 125, 135, 145, 160, 160, 175, 205

The value 160 occurs most often.

∴ The median is 140. The mode is 160.

On Your Own

Now You're Ready
Exercises 7–12

Find the median and mode of the data.

1. 20, 4, 17, 8, 12, 9, 5, 20, 13
2. 100, 75, 90, 80, 110, 102

EXAMPLE 2 Finding the Mode

The list shows the favorite types of movies for students in a class. Organize the data in a frequency table. Then find the mode.

Favorite Types of Movies

Comedy	Drama	Horror
Horror	Drama	Horror
Comedy	Comedy	Action
Action	Comedy	Action
Horror	Drama	Comedy
Comedy	Comedy	Horror
Horror	Comedy	Action
Horror	Action	Drama

Type	Tally	Frequency
Action	||||	5
Comedy	|||| |||	8
Drama	||||	4
Horror	|||| ||	7

The number of tally marks is the frequency.

Make a tally for each vote.

Comedy received the most votes.

∴ So, the mode is comedy.

On Your Own

Now You're Ready Exercises 14–15

3. One member of the class was absent and ends up voting for horror. Does this change the mode? Explain.

EXAMPLE 3 Choosing the Best Measure of Center

Find the mean, median, and mode of the sneaker prices. Which measure best represents the data?

Mean: $\dfrac{20 + 31 + 122 + 48 + 37 + 20 + 45 + 65}{8} = \dfrac{388}{8}$, or 48.5

Median: 20, 20, 31, 37, 45, 48, 65, 122   Order from least to greatest.

$$\dfrac{37 + 45}{2} = \dfrac{82}{2}, \text{ or } 41$$

Mode: 20, 20, 31, 37, 45, 48, 65, 122   The value 20 occurs most often.

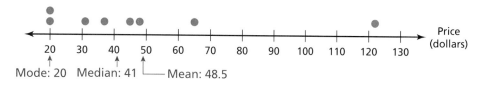

∴ The median best represents the data. The mode is less than most of the data, and the mean is greater than most of the data.

On Your Own

Now You're Ready Exercises 17–20

Find the mean, median, and mode of the data. Choose the measure that best represents the data. Explain your reasoning.

4. 1, 93, 46, 48, 34, 194, 67, 55   **5.** 96, 150, 102, 87, 150, 75

EXAMPLE 4 Removing an Outlier

Identify the outlier in Example 3. Find the mean, median, and mode without the outlier. Which measure does the outlier affect the most?

The price of $122 is much greater than any other price. So, it is the outlier.

	Mean	Median	Mode
With Outlier (Example 3)	48.5	41	20
Without Outlier	38	37	20

∴ The mean is affected the most by the outlier.

On Your Own

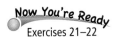
Exercises 21–22

6. The times (in minutes) it takes six students to travel to school are 8, 10, 10, 15, 20, and 45. Identify the outlier. Find the mean, median, and mode with and without the outlier. Which measure does the outlier affect the most?

EXAMPLE 5 Changing the Values of a Data Set

The prices of six video games at an online store are shown in the table. The price of each game increases by $4.98 when a shipping charge is included. How does this increase affect the mean, median, and mode?

Video Game Prices

$53.42	$35.69
$18.99	$25.13
$27.97	$53.42

Make a new table by adding $4.98 to each price. Then find the mean, median, and mode of both data sets.

Video Game Prices with Shipping Charge

$58.40	$40.67
$23.97	$30.11
$32.95	$58.40

	Mean	Median	Mode
Original Price	35.77	31.83	53.42
Price with Shipping Charge	40.75	36.81	58.4

Compare:

Mean: $40.75 - 35.77 = 4.98$

Median: $36.81 - 31.83 = 4.98$

Mode: $58.4 - 53.42 = 4.98$

∴ By increasing each video game price by $4.98 for shipping, the mean, median, and mode all increase by $4.98.

On Your Own

7. **WHAT IF?** The store decreases the price of each video game by $3. How does this decrease affect the mean, median, and mode?

9.3 Exercises

Vocabulary and Concept Check

1. **NUMBER SENSE** Give an example of a data set that has no mode.

2. **WRITING** Which is affected most by an outlier: the mean, median, or mode? Explain.

3. **WHICH ONE DOESN'T BELONG** Which word does *not* belong with the other three? Explain.

    | median | outlier | mode | mean |

4. **NUMBER SENSE** A data set has a mean of 7, a median of 5, and a mode of 8. Which of the numbers 7, 5, and 8 *must* be in the data set? Explain.

Practice and Problem Solving

Use grid paper to find the median of the data.

5. 9, 7, 2, 4, 3, 5, 9, 6, 8, 0, 3, 8

6. 16, 24, 13, 36, 22, 26, 22, 28, 25

Find the median and mode(s) of the data.

7. 3, 5, 7, 9, 11, 3, 8

8. 14, 19, 16, 13, 16, 14

9. 93, 81, 94, 71, 89, 92, 94, 99

10. 44, 13, 36, 52, 19, 27, 33

11. 12, 33, 18, 28, 29, 12, 17, 4, 2

12. 55, 44, 40, 55, 48, 44, 58, 67

13. **ERROR ANALYSIS** Describe and correct the error in finding the median of the data.

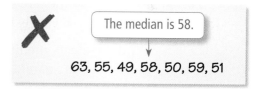

Find the mode(s) of the data.

14.
Shirt Color		
Black	Blue	Red
Pink	Black	Black
Gray	Green	Blue
Blue	Blue	Red
Yellow	Blue	Blue
Black	Orange	Black
Black		

15.
Talent Show Acts		
Singing	Dancing	Comedy
Singing	Singing	Dancing
Juggling	Dancing	Singing
Singing	Poetry	Dancing
Comedy	Magic	Dancing
Poetry	Singing	Singing

16. **REASONING** In Exercises 14 and 15, can you find the mean and median of the data? Explain.

Find the mean, median, and mode(s) of the data. Choose the measure that best represents the data. Explain your reasoning.

17. 48, 12, 11, 45, 48, 48, 43, 32

18. 12, 13, 40, 95, 88, 7, 95

19. 2, 8, 10, 12, 56, 9, 5, 2, 4

20. 126, 62, 144, 81, 144, 103

Find the mean, median, and mode(s) of the data with and without the outlier. Describe the effect of the outlier on the measures of center.

21. 45, 52, 17, 63, 57, 42, 54, 58

22. 85, 77, 211, 88, 91, 84, 85

Find the mean, median, and mode(s) of the data.

23. 4.7, 8.51, 6.5, 7.42, 9.64, 7.2, 9.3

24. $8\frac{1}{2}, 6\frac{5}{8}, 3\frac{1}{8}, 5\frac{3}{4}, 6\frac{5}{8}, 5\frac{1}{4}, 10\frac{5}{8}, 4\frac{1}{2}$

25. WEATHER The weather forecast for a week is shown.

	Sun	Mon	Tue	Wed	Thu	Fri	Sat
High	90° F	91° F	89° F	97° F	101° F	99° F	91° F
Low	74° F	78° F	77° F	77° F	83° F	78° F	72° F

a. Find the mean, median, and mode(s) of the high temperatures. Which measure best represents the data? Explain your reasoning.

b. Repeat part (a) for the low temperatures.

26. RESEARCH Find the unit costs of 10 different kinds of cereal. Choose one cereal whose unit cost will be an outlier.

a. Find the mean, median, and mode(s) of the data. Which measure best represents the data? Explain your reasoning.

b. Identify the outlier in the data set. Find the mean, median, and mode(s) of the data set without the outlier. Which measure does the outlier affect the most?

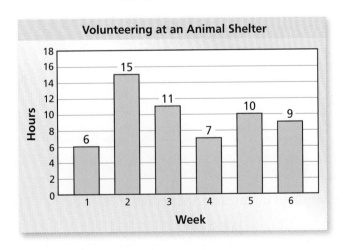

27. PROBLEM SOLVING The bar graph shows the numbers of hours you volunteered at an animal shelter. What is the minimum number of hours you need to work in the seventh week to justify that you worked an average of 10 hours for the 7 weeks? Explain your answer using measures of center.

28. REASONING Why do you think the mode is the least frequently used measure to describe a data set? Explain.

29. **MOTOCROSS** The ages of the racers in a bicycle motocross race are 14, 22, 20, 25, 26, 17, 21, 30, 27, 25, 14, and 29. The 30-year-old drops out of the race and is replaced with a 15-year-old. How are the mean, median, and mode of the ages affected?

30. **CAMERAS** The data are the prices of several digital cameras at a store.

    $130 $170 $230 $130
    $250 $275 $130 $185

    a. Does the price shown in the advertisement represent the prices well? Explain.
    b. Why might the store use this advertisement?
    c. In this situation, why might a person want to know the mean? the median? the mode? Explain.

31. **SALARIES** The table shows the monthly salaries for employees at a company.

Monthly Salaries (dollars)				
1940	1660	1860	2100	1720
1540	1760	1940	1820	1600

    a. Find the mean, median, and mode of the data.
    b. Each employee receives a 5% raise. Find the mean, median, and mode of the data with the raise. How does this increase affect the mean, median, and mode of the data?
    c. Use the original monthly salaries to calculate the annual salaries. Find the mean, median, and mode of the annual salaries. How are these values related to the mean, median, and mode of the monthly salaries?

32. **Critical Thinking** Consider the algebraic expressions $3x$, $9x$, $4x$, $23x$, $6x$, and $3x$. Assume $x > 0$.

    a. Find the mean, median, and mode.
    b. Is there an outlier? If so, what is it?

Fair Game Review What you learned in previous grades & lessons

Find the value of the expression. *(Section 1.1)*

33. $48 - 35$
34. $188 - 123$
35. $416 - 297$
36. $6249 - 3374$

37. **MULTIPLE CHOICE** A shelf in your room can hold at most 30 pounds. There are 12 pounds of books already on it. Which inequality represents the number of pounds you can add to the shelf? *(Section 7.6)*

    Ⓐ $x < 18$ Ⓑ $x \geq 18$ Ⓒ $x \leq 42$ Ⓓ $x \leq 18$

9.4 Measures of Variation

Essential Question How can you describe the spread of a data set?

1 ACTIVITY: Interpreting Statements

Work with a partner. There are 24 students in your class. Your teacher makes the following statements:

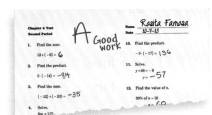

- "The exam scores range from 75% to 96%."
- "Most of the students received high scores."

a. What do you think the first statement means? Explain.

b. In the first statement, is your teacher describing the center of the data set? If not, what do you think your teacher is describing?

c. What do you think the scores are for most of the students in the class? Explain your reasoning.

d. Use your teacher's statements to make a dot plot that can represent the distribution of the exam scores of the class.

2 ACTIVITY: Grouping Data

Work with a partner. The numbers of U.S. states visited by each student in a sixth grade class are shown.

Number of States Visited

1	7	5	2
11	6	3	20
4	18	1	6
2	7	1	8
10	2	12	5
	3	21	

a. Between what values do the data range?

b. Write the ordered data values on a strip of grid paper and fold it to find the median. How many values are greater than the median? How many are less than the median?

c. **REPEATED REASONING** Fold the strip in half again. On what values are the two new creases? What do you think these values represent?

d. Into how many parts did you divide the data set? How many data values are in each part?

e. Graph the median and the values you found in parts (a) and (c) on a number line. Are the distances the same between these points?

f. How can you use these values to describe the spread of the data?

Statistics
In this lesson, you will
- find the range of data sets.
- find the interquartile range of data sets.
- check for outliers in data sets.

412   Chapter 9   Statistical Measures

3 ACTIVITY: Adding a Value to a Data Set

Work with a partner. A new student joins the class in Activity 2. The new student has visited 41 states.

a. Add this value to the ordered data set in Activity 2. Does your answer to part (a) change? Explain.

b. How does the distribution of the data change when this value is added? Explain your reasoning.

c. How does adding this value affect the values on your number line in part (e) of Activity 2?

4 ACTIVITY: Analyzing Data Sets

Work with a partner. Identify the data set that is the least spread out and the data set that is the most spread out. Explain your reasoning.

a.

b.

c.

d.

Math Practice

Analyze Givens
How can you use the given information to determine how spread out the data are?

What Is Your Answer?

5. **IN YOUR OWN WORDS** How can you describe the spread of a data set?

6. Make a dot plot of the data set in Activity 2. Describe any similarities between the dot plot and the number line in part (e).

Practice
Use what you learned about variation to complete Exercises 4 and 5 on page 416.

Section 9.4 Measures of Variation

9.4 Lesson

A **measure of variation** is a measure that describes the distribution of a data set. A simple measure of variation to find is the *range*. The **range** of a data set is the difference between the greatest value and the least value.

EXAMPLE 1 Finding the Range

Key Vocabulary
measure of variation, *p. 414*
range, *p. 414*
quartiles, *p. 414*
first quartile, *p. 414*
third quartile, *p. 414*
interquartile range, *p. 414*

The table shows the lengths of several Burmese pythons captured for a study. Find and interpret the range of their lengths.

Lengths (feet)	
18.5	8
11	10
14	15.5
12.5	6.25
16.25	5

To find the least and the greatest values, order the lengths from least to greatest.

5, 6.25, 8, 10, 11, 12.5, 14, 15.5, 16.25, 18.5

The least value is 5. The greatest value is 18.5.

∴ So, the range of the lengths is 18.5 − 5, or 13.5 feet. This means that the lengths vary by no more than 13.5 feet.

On Your Own

Exercises 6–9

1. The ages of people in line for a roller coaster are 15, 17, 21, 32, 41, 30, 25, 52, 16, 39, 11, and 24. Find and interpret the range of their ages.

Key Ideas

Quartiles

The **quartiles** of a data set divide the data into four equal parts. Recall that the median (second quartile) divides the data set into two halves.

Reading

The first quartile can also be called the *lower quartile*. The third quartile can also be called the *upper quartile*.

```
     lower half          Median = 29     upper half
 18  21  22  24  28  ↓  30  31  32  36  37
         ↑                       ↑
The median of the lower half      The median of the upper half
is the first quartile, $Q_1$.      is the third quartile, $Q_3$.
```

Interquartile Range (IQR)

The difference between the third quartile and the first quartile is called the **interquartile range**. The IQR represents the range of the middle half of the data and is another measure of variation.

```
 18  21  22  24  28  30  31  32  36  37
         IQR = $Q_3$ − $Q_1$
             =  32  −  22
             =  10
```

EXAMPLE 2 Finding the Interquartile Range

The dot plot shows the top speeds of 12 sports cars. Find and interpret the interquartile range of the data.

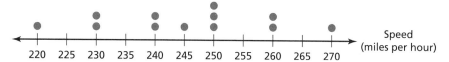

Order the speeds from slowest to fastest. Find the quartiles.

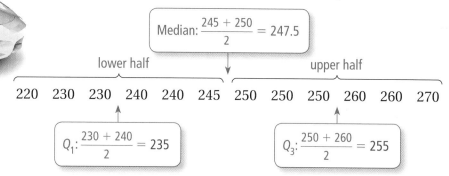

∴ So, the interquartile range is $255 - 235 = 20$. This means that the middle half of the speeds vary by no more than 20 miles per hour.

You can use the quartiles and the interquartile range to check for outliers.

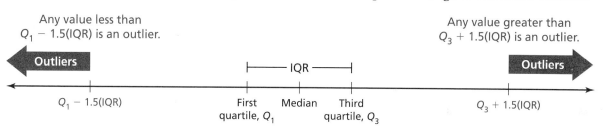

Any value less than $Q_1 - 1.5(IQR)$ is an outlier.

Any value greater than $Q_3 + 1.5(IQR)$ is an outlier.

EXAMPLE 3 Checking for Outliers

Check for outliers in the data set in Example 2.

$Q_1 - 1.5(IQR)$	Outlier boundaries	$Q_3 + 1.5(IQR)$
$235 - 1.5(20)$	Substitute values.	$255 + 1.5(20)$
205	Simplify.	285

∴ There are no speeds less than 205 miles per hour or greater than 285 miles per hour. So, the data set has no outliers.

On Your Own

2. The number of pages in each of an author's novels is shown.

   356, 364, 390, 468, 400, 382, 376, 396, 350

   a. Find and interpret the interquartile range of the data.

   b. Does this data set contain any outliers? Justify your answer.

9.4 Exercises

Vocabulary and Concept Check

1. **VOCABULARY** How are measures of center different from measures of variation?

2. **VOCABULARY** How many quartiles does a data set have?

3. **DIFFERENT WORDS, SAME QUESTION** Which is different? Find "both" answers.

   53, 47, 60, 45, 62, 59, 65, 50, 56, 48

   What is the interquartile range of the data?

   What is the range of the data?

   What is the range of the middle half of the data?

   What is the difference between the third quartile and the first quartile?

Practice and Problem Solving

Use grid paper to find the median of the data. Then find the median of the lower half and the median of the upper half of the data. Describe the spread of the data.

4. 5, 8, 10, 1, 7, 6, 15, 8, 6

5. 82, 62, 95, 81, 89, 51, 72, 56, 97, 98, 79, 85

Find the range of the data.

6. 26, 21, 27, 33, 24, 29

7. 52, 40, 49, 48, 62, 54, 44, 58, 39

8. 133, 117, 152, 127, 168, 146, 174

9. 4.8, 5.5, 4.2, 8.9, 3.4, 7.5, 1.6, 3.8

10. **ERROR ANALYSIS** Describe and correct the error in finding the range of the data.

    49, 48, 51, 41, 35, 44, 38
    The range is 49 − 38, or 11.

Find the median, first quartile, third quartile, and interquartile range of the data.

11. 40, 33, 37, 54, 41, 34, 27, 39, 35

12. 84, 75, 90, 87, 99, 91, 85, 88, 76, 92, 94

13. 132, 127, 106, 140, 158, 135, 129, 138

14. 38, 55, 61, 56, 46, 67, 59, 75, 65, 58

15. **PAPER AIRPLANE** The table shows the distances traveled by a paper airplane. Find and interpret the range and the interquartile range of the distances.

Distances (feet)			
$13\frac{1}{2}$	$21\frac{1}{2}$	21	$16\frac{3}{4}$
$10\frac{1}{4}$	19	32	$26\frac{1}{2}$
29	$16\frac{1}{4}$	$28\frac{1}{2}$	$18\frac{1}{2}$

416 Chapter 9 Statistical Measures

16. **WRITING** Consider a data set that has no mode. Which measure of variation is greater, the range or the interquartile range? Explain your reasoning.

17. **OUTLIERS** Use the interquartile range to identify any outliers in Exercises 11–14.

18. **REASONING** How does an outlier affect the range of a data set? Explain.

19. **BASKETBALL** The table shows the numbers of points scored by players on a basketball team.

Points Scored					
21	53	74	82	84	93
103	108	116	122	193	

   a. Find the range and the interquartile range of the data.

   b. Use the interquartile range to identify the outlier(s) in the data set. Find the range and the interquartile range of the data set without the outlier(s). Which measure did the outlier(s) affect more?

20. **STRUCTURE** Two data sets have the same range. Can you assume that the interquartile ranges of the two data sets are about the same? Give an example to justify your answer.

21. **SINGING** The tables show the ages of the finalists for two reality singing competitions.

   a. Find the mean, median, range, and interquartile range of the ages for each show. Compare the results.

   b. A 21-year-old is voted off Show A, and the 36-year-old is voted off Show B. How do these changes affect the measures in part (a)? Explain.

Ages for Show A	
18	17
15	21
22	16
18	28
24	21

Ages for Show B	
21	20
23	13
15	18
17	22
36	25

22. Create a set of data with 7 values that has a mean of 30, a median of 26, a range of 50, and an interquartile range of 36.

Fair Game Review What you learned in previous grades & lessons

Find the mean of the data. *(Section 9.2)*

23. 8, 14, 22, 7, 2, 11, 25, 7, 5, 9

24. 55, 64, 58, 43, 49, 67

25. **MULTIPLE CHOICE** What is the surface area of the rectangular prism? *(Section 8.2)*

   Ⓐ 62 m²   Ⓑ 72 m²
   Ⓒ 88 m²   Ⓓ 124 m²

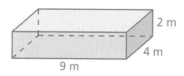

9.5 Lesson

Key Vocabulary
mean absolute deviation, *p. 420*

Another measure of variation is the *mean absolute deviation*. The **mean absolute deviation** is an average of how much data values differ from the mean.

Key Idea

Finding the Mean Absolute Deviation (MAD)

Step 1: Find the mean of the data.

Step 2: Find the distance between each data value and the mean.

Step 3: Find the sum of the distances in Step 2.

Step 4: Divide the sum in Step 3 by the total number of data values.

EXAMPLE 1 — Finding the Mean Absolute Deviation

You record the numbers of raisins in 8 scoops of cereal. Find and interpret the mean absolute deviation of the data.

$$1, 2, 2, 2, 4, 4, 4, 5$$

Step 1: Mean $= \dfrac{1+2+2+2+4+4+4+5}{8} = \dfrac{24}{8} = 3$

Step 2: You can use a dot plot to organize the data. Replace each dot with its distance from the mean.

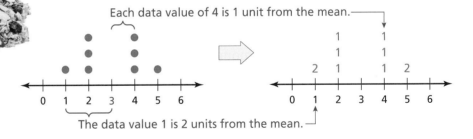

Each data value of 4 is 1 unit from the mean.
The data value 1 is 2 units from the mean.

Step 3: The sum of the distances is $2 + 1 + 1 + 1 + 1 + 1 + 1 + 2 = 10$.

Step 4: The mean absolute deviation is $\dfrac{10}{8} = 1.25$.

∴ So, the data values differ from the mean by an average of 1.25 raisins.

On Your Own

Now You're Ready Exercises 5–8

1. Find and interpret the mean absolute deviation of the data.

$$5, 8, 8, 10, 13, 14, 16, 22$$

EXAMPLE 2 Real-Life Application

The smartphones show the numbers of runs allowed by two pitchers in their last 10 starts.

a. Find the mean, median, and mean absolute deviation of the numbers of runs allowed for each pitcher.

Order the runs allowed for Mendoza:
0, 0, 0, 2, 4, 4, 5, 6, 6, 8.

$$\text{Mean} = \frac{35}{10} = 3.5 \qquad \text{Median} = \frac{4+4}{2} = 4$$

Mean absolute deviation:

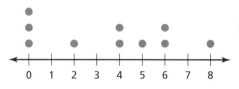 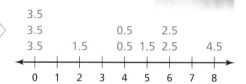

The mean absolute deviation is $\frac{24}{10} = 2.4$

Order the runs allowed for Rodriguez: 0, 2, 2, 3, 4, 4, 4, 5, 5, 6.

$$\text{Mean} = \frac{35}{10} = 3.5 \qquad \text{Median} = \frac{4+4}{2} = 4$$

Mean absolute deviation:

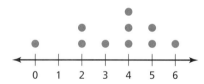 

The mean absolute deviation is $\frac{14}{10} = 1.4$.

b. Which measure can you use to distinguish the data? What can you conclude about the pitchers from this measure?

You cannot use the measures of center to distinguish the data because they are the same for each data set. The measure of variation, MAD, is 2.4 for Mendoza and 1.4 for Rodriguez. This indicates that the data for Rodriguez has less variation.

> Using the MAD to distinguish the data, you can conclude that Rodriguez is more consistent than Mendoza.

> **Study Tip**
> The greater the mean absolute deviation, the greater the variation of the data.

On Your Own

2. **WHAT IF?** Mendoza allows 4 runs in the next game. How would you expect the mean absolute deviation to change? Explain.

9.4–9.5 Quiz

Find the range of the data. *(Section 9.4)*

1. 35, 76, 43, 58, 34, 67
2. 19, 21, 22, 22, 19, 25, 24, 23, 24

Find the median, first quartile, third quartile, and interquartile range of the data. *(Section 9.4)*

3. 56, 48, 72, 37, 35, 42, 48, 33, 28
4. 95, 14, 86, 62, 55, 46, 28, 37, 33, 70, 31

Find and interpret the mean absolute deviation of the data. Round your answer to the nearest tenth if necessary. *(Section 9.5)*

5.
Ages of Television Show Viewers (years)			
29	18	26	33
33	22	34	26

6.
Prices of Houses (thousands of dollars)				
80	120	95	240	140
75	135	110	90	125

7. **AMUSEMENT PARKS** The data set shows the admission prices at several amusement parks.

    $65, $70, $40, $55, $35, $40, $60

    Find and interpret the range, interquartile range, and mean absolute deviation of the data. *(Section 9.4 and Section 9.5)*

8. **TEACHING EXPERIENCE** The tables show the years of teaching experience of faculty members at two schools. *(Section 9.4)*

    a. Find the mean, median, range, and interquartile range of the years of experience for each school. Compare the results.

    b. The teacher with 11 years of experience leaves School A, and the teacher with 33 years of experience retires from School B. How does this affect the measures in part (a)? Explain.

School A: Teaching Experience (years)		School B: Teaching Experience (years)	
5	11	4	15
10	22	6	12
7	8	10	33
8	6	12	20
10	35	8	7

9. **BOOK CLUB** You survey the students in your book club about the number of books they read last summer. The results are shown in the table. *(Section 9.4 and Section 9.5)*

Books Read			
8	14	15	9
6	12	9	13
11	11	7	5
12	6	10	8

a. Find the measures of center and the measures of variation for the data.

b. A new student who read 18 books last summer joins the club. Is 18 an outlier? How does adding this value to the data set affect the measures of center and variation? Explain.

9 Chapter Review

Review Key Vocabulary

statistics, *p. 392*
statistical question, *p. 392*
mean, *p. 398*
outlier, *p. 399*
measure of center, *p. 404*

median, *p. 404*
mode, *p. 404*
measure of variation, *p. 414*
range, *p. 414*
quartiles, *p. 414*

first quartile, *p. 414*
third quartile, *p. 414*
interquartile range, *p. 414*
mean absolute deviation, *p. 420*

Review Examples and Exercises

9.1 Introduction to Statistics *(pp. 390–395)*

Display the data in a dot plot. Identify any clusters, peaks, or gaps in the data.

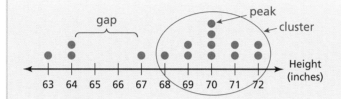

Heights (inches)				
70	71	70	72	69
68	69	71	64	70
64	63	72	70	67

Exercises

Display the data in a dot plot. Identify any clusters, peaks, or gaps in the data.

1.
Distance (feet)			
56	55	56	57
58	54	51	55
51	56	49	56

2.
Weight (pounds)				
83	88	89	90	89
91	89	84	90	92
90	88	89	83	88

9.2 Mean *(pp. 396–401)*

Find the mean of 5, 9, 10, 6, 6, and 12.

$$\text{mean} = \frac{5 + 9 + 10 + 6 + 6 + 12}{6} \quad \begin{array}{l}\leftarrow \text{sum of the data} \\ \leftarrow \text{number of values}\end{array}$$

$$= \frac{48}{6}, \text{ or } 8 \qquad \text{Simplify.}$$

Exercises

Find the mean of the data.

3. 4, 5, 7, 14, 17, 12, 18

4. 15, 5, 8, 12, 5, 9, 4, 10, 2, 11

9.3 Measures of Center (pp. 402–409)

Find the median and the mode of the movie lengths in the table.

Order the data from least to greatest.

Movie Lengths (minutes)		
91	112	126
142	122	112
92	144	

Median:
91, 92, 112, 112, 122, 126, 142, 144

$$\frac{112 + 122}{2} = \frac{234}{2}, \text{ or } 117$$

Mode:
91, 92, 112, 112, 122, 126, 142, 144

The value 112 occurs most often.

∴ The median is 117 minutes, and the mode is 112 minutes.

Exercises

Find the median and the mode(s) of the data.

5. 8, 8, 6, 8, 4, 5, 6

6. 24, 74, 61, 29, 38, 27, 68, 54

9.4 Measures of Variation (pp. 412–417)

The table shows the weights of several adult emperor penguins. (a) Find and interpret the range. (b) Find and interpret the interquartile range. (c) Check for outliers.

Weights (kilograms)	
25	27
36	23.5
33.5	31.25
30.75	32
24	29.25

a. Ordered from least to greatest, the weights are 23.5, 24, 25, 27, 29.25, 30.75, 31.25, 32, 33.5, and 36.

∴ So, the range of the weights is 36 − 23.5, or 12.5 kilograms. The weights vary by no more than 12.5 kilograms.

b. Find the quartiles.

Median: $\frac{29.25 + 30.75}{2} = 30$

lower half: 23.5, 24, 25, 27, 29.25, Q_1: 25

upper half: 30.75, 31.25, 32, 33.5, 36 Q_3: 32

∴ So, the interquartile range is 32 − 25 = 7. This means that the middle half of the weights vary by no more than 7 kilograms.

c. Calculate the outlier boundaries.

$Q_1 - 1.5(\text{IQR}) = 25 - 1.5(7) = 14.5$

$Q_3 + 1.5(\text{IQR}) = 32 + 1.5(7) = 42.5$

∴ There are no weights less than 14.5 kilograms or greater than 42.5 kilograms. So, the data set has no outliers.

Exercises

Find the range of the data.

7. 45, 76, 98, 21, 52, 39

8. 95, 63, 52, 8, 93, 16, 42, 37, 62

Find the median, first quartile, third quartile, and interquartile range of the data.

9. 28, 46, 25, 76, 18, 25, 47, 83, 44

10. 14, 25, 97, 55, 66, 28, 92, 38, 94

9.5 Mean Absolute Deviation *(pp. 418–423)*

You record the prices of 8 printers. Find and interpret the mean absolute deviation of the data.

$120, $150, $90, $110,

$140, $120, $140, $90

Step 1: Mean $= \dfrac{120 + 150 + 90 + 110 + 140 + 120 + 140 + 90}{8} = \dfrac{960}{8} = 120$

Step 2: Use a dot plot to organize the data. Replace each dot with its distance from the mean.

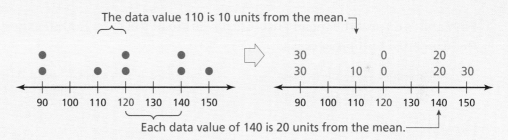

Step 3: The sum of the distances is $30 + 30 + 10 + 0 + 0 + 20 + 20 + 30 = 140$.

Step 4: The mean absolute deviation is $\dfrac{140}{8} = 17.50$.

∴ The data values differ from the mean by an average of $17.50.

Exercises

Find and interpret the mean absolute deviation of the data. Round your answer to the nearest tenth if necessary.

11.
Shoe Sizes			
6	8.5	6	9
10	7	8	9.5

12.
Prices of Monitors (dollars)				
130	150	190	100	175
120	165	140	180	190

Chapter Review 427

9 Chapter Test

Display the data in a dot plot. Identify any clusters, peaks, or gaps in the data.

1.
Time (minutes)			
33	40	32	40
39	38	40	39
38	39	39	33

2.
Temperature (°F)				
81	81	80	82	81
83	76	83	76	80
75	83	82	82	81

Find the mean, median, and mode(s) of the data.

3. 2, 7, 7, 12, 4

4. 4, 5, 7, 5, 9, 9, 10, 6

Find the mean, median, and mode(s) of the data. Choose the measure that best represents the data. Explain your reasoning.

5. 5, 6, 4, 24, 18

6. 46, 27, 94, 56, 53, 65, 43

Find the range of the data.

7. 24, 56, 9, 83, 77, 14

8. 43, 12, 55, 91, 25, 86, 84, 23, 1

Find the median, first quartile, third quartile, and interquartile range of the data.

9. 32, 58, 19, 36, 44, 57, 11, 26, 74

10. 36, 24, 49, 32, 37, 28, 38, 40, 39

Find and interpret the mean absolute deviation of the data. Round your answer to the nearest tenth if necessary.

11.
Distances Driven (miles)			
312	286	196	201
158	225	206	192

12.
Prices of Sunglasses (dollars)				
15	8	19	20	18
20	22	14	10	15

13. **HOTEL** The table shows the numbers of guests at a hotel on different days.

Numbers of Guests					
66	58	90	57	63	55
60	62	56	54	72	

   a. Find the range and the interquartile range of the data.

   b. Use the interquartile range to identify the outlier(s) in the data set. Find the range and the interquartile range of the data set without the outlier(s). Which measure did the outlier(s) affect more?

14. **JOBS** The data sets show the numbers of hours worked each week by two friends for several weeks.

   Greg's hours: 9, 18, 12, 6, 9, 21, 3, 12
   Tom's hours: 12, 18, 15, 16, 14, 12, 15, 18

   Find the measures of center and the measures of variation for each data set. Compare the measures. What can you conclude?

9 Cumulative Assessment

1. What is the value of the expression below?

$$8\frac{4}{9} \div 4\frac{2}{3}$$

   A. $1\frac{17}{21}$

   B. $2\frac{2}{3}$

   C. $32\frac{8}{27}$

   D. $39\frac{11}{27}$

Test-Taking Strategy
Use Intelligent Guessing

What is the mean length of these hyena fangs: 4 in., 3 in., 3 in., 4 in., 5 in., 5 in.?
Ⓐ 6 in.   Ⓑ $\frac{1}{3}$ ft   Ⓒ 2 in.   Ⓓ 5 in.

"The mean can't be 6 or 2 or 5 inches. So, you can use intelligent guessing to find that the answer is $\frac{1}{3}$ ft, or 4 in."

2. What is the value of the expression below?

$$4.18 + 6.225 + 5.7$$

   F. 15.005

   G. 15.105

   H. 16.005

   I. 16.105

3. One number is missing from the data set in the box below.

   18, 24, 22, 30, 26, ____, 25

   The median of the data set is 24. What is the greatest possible value of the missing number?

4. The number of hours that each of 6 students spent reading last week is shown in the bar graph below.

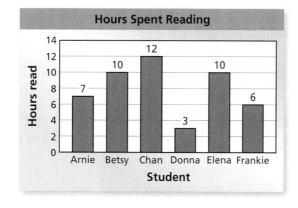

   For the data in the bar graph, which measure is the *least*?

   A. mean

   B. median

   C. mode

   D. range

Cumulative Assessment   429

10 Data Displays

10.1 **Stem-and-Leaf Plots**

10.2 **Histograms**

10.3 **Shapes of Distributions**

10.4 **Box-and-Whisker Plots**

What You Learned Before

"Okay, I have the box. But I need your help to complete my box-and-whisker plot."

Analyzing Bar Graphs

Example 1 The bar graph shows the favorite colors of the students in a class. How many students said their favorite color is blue?

The height of the bar labeled "Blue" is 8.

So, 8 students said their favorite color is blue.

Try It Yourself

1. What color was chosen the least?
2. How many students said green or red is their favorite color?
3. How many students did *not* choose yellow as their favorite color?
4. How many students are in the class?

Finding Percents

Example 2 The circle graph shows the favorite fruits of the students in a class. There are 20 students in the class. How many students said their favorite fruit is an orange?

Find 25% of 20.

$$25\% \text{ of } 20 = \frac{1}{4} \cdot 20 = \frac{1 \cdot \overset{5}{\cancel{20}}}{\cancel{4}} = 5$$

So, 5 students said their favorite fruit is an orange.

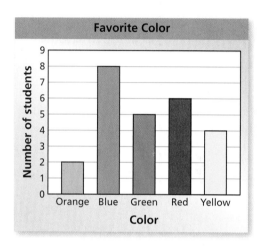

Try It Yourself

5. How many students said their favorite fruit is an apple?
6. How many students said their favorite fruit is a banana?

10.1 Stem-and-Leaf Plots

Essential Question How can you use place values to represent data graphically?

1 ACTIVITY: Making a Data Display

Work with a partner. The list below gives the ages of these women when they became first ladies of the United States.

THE WHITE HOUSE — WASHINGTON

Frances Cleveland – 21
Caroline Harrison – 56
Ida McKinley – 49
Edith Roosevelt – 40
Helen Taft – 48
Ellen Wilson – 52
Florence Harding – 60
Grace Coolidge – 44
Lou Hoover – 54
Eleanor Roosevelt – 48
Elizabeth Truman – 60
Mamie Eisenhower – 56
Jacqueline Kennedy – 31
Claudia Johnson – 50
Patricia Nixon – 56
Elizabeth Ford – 56
Rosalynn Carter – 49
Nancy Reagan – 59
Barbara Bush – 63
Hillary Clinton – 45
Laura Bush – 54
Michelle Obama – 45

Data Displays

In this lesson, you will
- make and interpret stem-and-leaf plots.

a. The incomplete data display shows the ages of the first ladies in the left column of the list above.

   What do the numbers to the left of the line represent? What do the numbers to the right of the line represent?

Ages of First Ladies

2	1
3	
4	0 4 8 8 9
5	2 4 6
6	0 0

b. This data display is called a *stem-and-leaf plot*. What numbers do you think represent the *stems*? *leaves*? Explain your reasoning.

c. Complete the stem-and-leaf plot using the remaining ages in the right column. Order the numbers to the right of the line in numerical order.

d. **REASONING** Write a question about the ages of first ladies that would be easier to answer using a stem-and-leaf plot than a dot plot.

434 Chapter 10 Data Displays

2 ACTIVITY: Making a Back-to-Back Stem-and-Leaf Plot

Work with a partner. The table below shows the ages of presidents of the United States from 1885 to 2009 on their first inauguration day.

Ages of Presidents

47	55	54	42	51	56	55	51	54	51	60
62	43	55	56	61	52	69	64	46	54	47

a. On your stem-and-leaf plot from Activity 1(c), draw a vertical line to the left of the display. Represent the ages of the presidents by including numbers to the left of the line.

b. Find the median ages of both the first ladies and the presidents of the United States.

c. Compare the distribution of each data set.

3 ACTIVITY: Conducting an Experiment

Math Practice

Interpret Results
How can you use the stem-and-leaf plot to interpret your results? Explain.

Work with a partner. Use two number cubes to conduct the following experiment.

- Toss the cubes and find the product of the resulting numbers.
- Repeat this process 30 times. Record your results.

a. Use a stem-and-leaf plot to organize your results.
b. Describe the distribution of the data.

What Is Your Answer?

4. **IN YOUR OWN WORDS** How can you use place values to represent data graphically?

5. How can you display data in a stem-and-leaf plot whose values range from 82 through 129?

Practice

Use what you learned about stem-and-leaf plots to complete Exercises 4 and 5 on page 438.

10.1 Lesson

Check It Out
Lesson Tutorials
BigIdeasMathcom

Key Vocabulary
stem-and-leaf plot, p. 436
stem, p. 436
leaf, p. 436

Key Idea

Stem-and-Leaf Plots

A **stem-and-leaf plot** uses the digits of data values to organize a data set. Each data value is broken into a **stem** (digit or digits on the left) and a **leaf** (digit or digits on the right).

A stem-and-leaf plot shows how data are distributed.

Stem	Leaf
2	0 0 1 2 5 7
3	1 4 8
4	2
5	8 9

Key: 2 | 0 = 20

The *key* explains what the stems and leaves represent.

EXAMPLE 1 Making a Stem-and-Leaf Plot

	A	B
1	DATE	MINUTES
2	JULY 9	55
3	JULY 9	3
4	JULY 9	6
5	JULY 10	14
6	JULY 10	18
7	JULY 10	5
8	JULY 10	23
9	JULY 11	30
10	JULY 11	23
11	JULY 11	10
12	JULY 11	2
13	JULY 11	36

Make a stem-and-leaf plot of the length of the 12 cell phone calls.

Step 1: Order the data.

2, 3, 5, 6, 10, 14, 18, 23, 23, 30, 36, 55

Step 2: Choose the stems and the leaves. Because the data values range from 2 to 55, use the *tens* digits for the stems and the *ones* digits for the leaves. Be sure to include the key.

Step 3: Write the stems to the *left* of the vertical line.

Step 4: Write the leaves for each stem to the *right* of the vertical line.

Cell Phone Call Lengths

Stem	Leaf
0	2 3 5 6
1	0 4 8
2	3 3
3	0 6
4	
5	5

Order the stems vertically. The stem for data values less than 10 is 0.

Write the leaves horizontally.

Include stems without leaves.

Key: 1 | 4 = 14 minutes

On Your Own

Now You're Ready
Exercises 4–9

1. Make a stem-and-leaf plot of the hair lengths.

Hair Length (centimeters)								
5	1	20	12	27	2	30	5	7	38
40	47	1	2	1	32	4	44	33	23

EXAMPLE 2 Interpreting a Stem-and-Leaf Plot

Test Scores

Stem	Leaf
6	6
7	0 5 7 8
8	1 1 3 4 4 6 8 8 9
9	0 2 9
10	0

Key: 9 | 2 = 92 points

The stem-and-leaf plot shows student test scores. (a) How many students scored less than 80 points? (b) How many students scored at least 90 points? (c) How are the data distributed?

a. There are five scores less than 80 points: 66, 70, 75, 77, and 78.

∴ Five students scored less than 80 points.

b. There are four scores of at least 90 points: 90, 92, 99, and 100.

∴ Four students scored at least 90 points.

c. There are few low test scores and few high test scores. So, most of the scores are in the middle.

On Your Own

Now You're Ready
Exercises 12–15

2. Use the grading scale at the right.

   a. How many students received a B on the test?

   b. How many students received a C on the test?

   A: 90–100
   B: 80–89
   C: 70–79
   D: 60–69
   F: 59 and below

EXAMPLE 3 Making Conclusions from a Stem-and-Leaf Plot

Which statement is *not* true?

Ⓐ Most of the plants are less than 20 inches tall.

Ⓑ The median plant height is 11 inches.

Ⓒ The range of the plant heights is 35 inches.

Ⓓ The plant height that occurs most often is 11 inches.

Plant Heights

Stem	Leaf
0	1 2 4 5 6 8 9
1	0 1 1 5 7
2	2 5
3	6

Key: 1 | 5 = 15 inches

There are 15 plant heights. So, the median is the eighth data value, 10 inches.

∴ The correct answer is Ⓑ.

On Your Own

3. You are told that three plants are taller than 20 inches. Is the statement true? Explain.

Section 10.1 Stem-and-Leaf Plots 437

10.1 Exercises

Check It Out
Help with Homework
BigIdeasMath com

Vocabulary and Concept Check

1. **VOCABULARY** The key for a stem-and-leaf plot is 3 | 4 = 34. Which number is the stem? Which number is the leaf?

2. **WRITING** Describe how to make a stem-and-leaf plot of the data values 14, 22, 9, 13, 30, 8, 25, and 29.

3. **WRITING** How does a stem-and-leaf plot show the distribution of data?

Practice and Problem Solving

Make a stem-and-leaf plot of the data.

4. **Books Read**

26	15	20	9
31	25	29	32
17	26	19	40

5. **Hours Online**

8	12	21	14
18	6	15	24
12	17	2	0

6. **Test Scores (%)**

87	82	95	91	69
88	68	87	65	81
97	85	80	90	62

7. **Points Scored**

58	50	42	71	75
45	51	43	38	71
42	70	56	58	43

8. **Bikes Sold**

78	112	105	99
86	96	115	100
79	81	99	108

9. **Minutes in Line**

4.0	2.6	1.9	3.1
3.6	2.2	2.7	3.8
1.6	2.0	3.1	2.9

10. **ERROR ANALYSIS** Describe and correct the error in making a stem-and-leaf plot of the data.

    51, 25, 47, 42, 55, 26, 50, 44, 55

    ✗
Stem	Leaf
2	5 6
4	2 4 7
5	0 1 5 5

    Key: 4 | 2 = 42

11. **PUPPIES** The weights (in pounds) of eight puppies at a pet store are 12, 24, 17, 8, 18, 31, 24, and 15. Make a stem-and-leaf plot of the data. Describe the distribution of the data.

438   Chapter 10   Data Displays

VOLLEYBALL The stem-and-leaf plot shows the number of *digs* for the top 15 players at a volleyball tournament.

Stem	Leaf
4	1 1 3 3 5
5	0 2 3 4
6	2 3 3 7
7	5
8	
9	7

Key: 5 | 0 = 50 digs

② 12. How many players had more than 60 digs?

13. Find the mean, median, mode, range, and interquartile range of the data.

14. Describe the distribution of the data.

15. Which data value is the outlier? Describe how the outlier affects the mean.

16. REASONING Each stem-and-leaf plot below has a mean of 39. Without calculating, determine which stem-and-leaf plot has the lesser mean absolute deviation. Explain your reasoning.

Stem	Leaf
2	3 7
3	0 2 6 9
4	1 2 5 8
5	1 4

Key: 4 | 1 = 41

Stem	Leaf
2	2 4 5 8 9
3	3 8
4	5
5	3 6 7 8

Key: 5 | 3 = 53

17. TEMPERATURE The stem-and-leaf plot shows the daily high temperatures (in degrees Fahrenheit) for the first 15 days of a month.

Stem	Leaf
6	7 8
7	0 0 3 4 6 8 9
8	2 3 6 7 8 9

Key: 6 | 7 = 67°F

a. Find and interpret the mean absolute deviation of the data.

b. After you include the daily high temperatures for the rest of the month in the stem-and-leaf plot, the mean absolute deviation increases. Where do you think most of the data values for the rest of the month are located in the stem-and-leaf plot? Explain.

18. The back-to-back stem-and-leaf plot shows the 9-hole golf scores for two golfers. Only one of the golfers can compete in a tournament. Use measures of center and measures of variation to give reasons why you would choose each golfer.

Rich		Will
7 5	3	
8 5 4 3 2 1	4	2 3 4 4 6 7 7 8 9
5 0	5	0

Key: 1 | 4 | 2 = 41 and 42 strokes

Fair Game Review *What you learned in previous grades & lessons*

Draw the solid. *(Section 8.1)*

19. square pyramid

20. hexagonal prism

21. MULTIPLE CHOICE In a bar graph, what determines the length of each bar? *(Skills Review Handbook)*

Ⓐ frequency Ⓑ data value Ⓒ leaf Ⓓ change in data

10.2 Histograms

Essential Question How can you use intervals, tables, and graphs to organize data?

1 ACTIVITY: Conducting an Experiment

Work with a partner.

a. Roll a number cube 20 times. Record your results in a tally chart.

b. Make a bar graph of the totals.

c. Go to the board and enter your totals in the class tally chart.

d. Make a second bar graph showing the class totals. Compare and contrast the two bar graphs.

Tally Chart

1	
2	
3	
4	
5	
6	

Key: | = 1 ||||| = 5

2 ACTIVITY: Using Intervals to Organize Data

Work with a partner. You are judging a paper airplane contest. A contestant flies a paper airplane 20 times. You record the following distances:

20.5 ft, 24.5 ft, 18.5 ft, 19.5 ft, 21.0 ft, 14.0 ft, 12.5 ft, 20.5 ft, 17.5 ft, 24.5 ft, 19.5 ft, 17.0 ft, 18.5 ft, 12.0 ft, 21.5 ft, 23.0 ft, 13.5 ft, 19.0 ft, 22.5 ft, 19.0 ft

a. Complete the tally chart and the bar graph of the distances.

Tally Chart

Interval	Tally	Total
10.0 – 12.9		
13.0 – 15.9		
16.0 – 18.9		
19.0 – 21.9		
22.0 – 24.9		

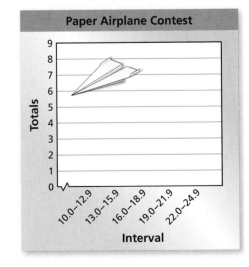

Data Displays

In this lesson, you will
- make histograms.
- use histograms to analyze data.

b. Make a different tally chart and bar graph of the distances. Use the following intervals:

10.0–11.9, 12.0–13.9, 14.0–15.9, 16.0–17.9, 18.0–19.9, 20.0–21.9, 22.0–23.9, 24.0–25.9

c. Which graph do you think represents the distances better? Explain.

440   Chapter 10   Data Displays

The tally chart in Activity 2 is also called a *frequency table*. A **frequency table** groups data values into intervals. The **frequency** is the number of values in an interval.

3 ACTIVITY: Developing an Experiment

Math Practice

Specify Units
What units will you use to measure the distance flown each time? Will the units you use affect the results in your frequency table? Explain.

Work with a partner.

a. Make the airplane shown from a single sheet of $8\frac{1}{2}$-by-11-inch paper. Then design and make your own paper airplane.

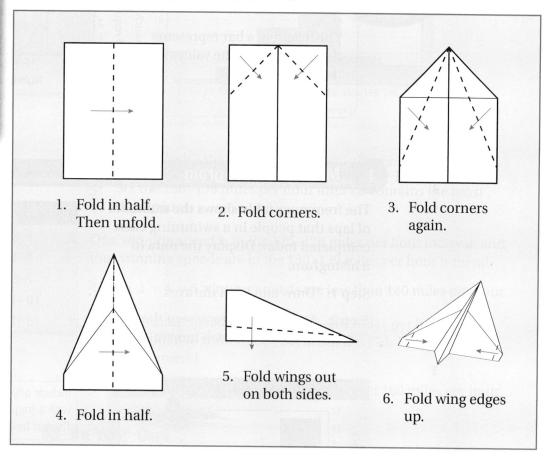

1. Fold in half. Then unfold.
2. Fold corners.
3. Fold corners again.
4. Fold in half.
5. Fold wings out on both sides.
6. Fold wing edges up.

b. **PRECISION** Fly each airplane 20 times. Keep track of the distance flown each time.

c. **MODELING** Organize the results of the flights using frequency tables and graphs. Which airplane flies farther? Explain your reasoning.

What Is Your Answer?

4. **IN YOUR OWN WORDS** How can you use intervals, tables, and graphs to organize data?

5. What intervals could you use in a graph that displays data whose values range from 40 through 59?

Practice — Use what you learned about organizing data into intervals to complete Exercises 4 and 5 on page 445.

Section 10.2 Histograms 441

EXAMPLE 3 Comparing Data Displays

The data displays show how many push-ups students in a class completed for a physical fitness test. Which data display can you use to find how many students are in the class? Explain.

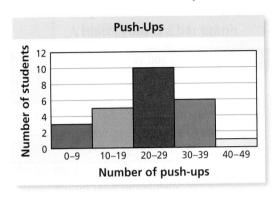

You can use the histogram because it shows the number of students in each interval. The sum of these values represents the number of students in the class. You cannot use the circle graph because it does not show the number of students in each interval.

EXAMPLE 4 Making Conclusions from Data Displays

Which statement *cannot* be made using the data displays in Example 3?

- **A** Twelve percent of the class completed less than 10 push-ups.
- **B** Five students completed at least 10 and at most 19 push-ups.
- **C** At least one student completed more than 39 push-ups.
- **D** Twenty-nine percent of the class completed 30 or more push-ups.

The circle graph shows that 12% completed 0–9 push-ups. So, Statement A can be made.

In the histogram, the bar height for the 10–19 interval is 5, and the bar height for the 40–49 interval is 1. So, Statements B and C can be made.

The circle graph shows that 24% completed 30–39 push-ups, and 4% completed 40–49 push-ups. So, 24% + 4% = 28% completed 30 or more push-ups. Statement D cannot be made.

The correct answer is **D**.

On Your Own

Now You're Ready
Exercises 14 and 15

3. In Example 3, which data display should you use to describe the portion of the entire class that completed 30–39 push-ups?

4. Make two more conclusions from the data displays in Example 3.

10.1–10.2 Quiz

Make a stem-and-leaf plot of the data. *(Section 10.1)*

1. **Cans Collected Each Month**

80	90	84	92
76	83	79	59
68	55	58	61

2. **Miles Driven Each Day**

21	18	12	16	10
16	9	15	20	28
35	50	37	20	11

3. **Ages of Tortoises**

86	99	100	124
92	85	110	130
115	129	83	104

4. **Kilometers Run Each Day**

6.0	5.6	6.2	3.0	2.5
3.5	2.0	5.0	3.9	3.1
6.2	3.1	4.5	3.8	6.1

Display the data in a histogram. *(Section 10.2)*

5. **Soccer Team Goals**

Goals per Game	Frequency
0–1	5
2–3	4
4–5	0
6–7	1

6. **Minutes Practiced**

Minutes	Frequency
0–19	8
20–39	10
40–59	11
60–79	2

7. **Poems Written for Class**

Poems	Frequency
0–4	6
5–9	16
10–14	4
15–19	2
20–24	2

8. **WEIGHTS** The weights (in ounces) of nine packages are 7, 22, 16, 12, 6, 18, 15, 13, and 25. Make a stem-and-leaf plot of the data. Describe the distribution of the data. *(Section 10.1)*

9. **REBOUNDS** The histogram shows the number of rebounds per game for a middle school basketball player this season. *(Section 10.2)*

   a. Which interval contains the most data values?

   b. How many games did the player play this season?

   c. What percent of the games did the player have 4 or more rebounds?

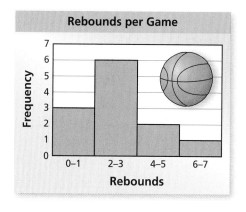

Stem	Leaf
0	6 8 8 9
1	0 1 2 3 7 8
2	0

Key: 0 | 9 = 9 hours

10. **STAGE CREW** The stem-and-leaf plot shows the number of hours 11 stage crew members spent building sets. Find the mean, median, mode, range, and interquartile range of the data. *(Section 10.1)*

10.3 Shapes of Distributions

Essential Question How can you describe the shape of the distribution of a data set?

1 ACTIVITY: Describing the Shape of a Distribution

Work with a partner. The lists at the left show the last four digits of a set of phone numbers in a phone book.

a. Create a list that represents the last digit of each phone number shown. Make a dot plot of the data.

b. In your own words, how would you describe the shape of the distribution? What single word do you think you can use to identify this type of distribution? Explain your reasoning.

-7253
-7290
-7200
-1192
-1142
-3500
-2531
-2079
-5897
-5341
-1392
-5406
-7875
-7335
-0494
-9018
-2184
-2367

-8678
-2063
-2911
-2103
-4328
-7826
-7957
-7246
-2119
-7845
-1109
-9154

2 ACTIVITY: Describing the Shape of a Distribution

Work with a partner. The lists at the right show the first three digits of a set of phone numbers in a phone book.

a. Create a list that represents the first digit of each phone number shown. Make a dot plot of the data.

b. In your own words, how would you describe the shape of the distribution? What single word do you think you can use to identify this type of distribution? Explain your reasoning.

c. In your dot plot, draw a vertical line through the middle of the data set. What do you notice?

d. Repeat part (c) for the dot plot you constructed in Activity 1. What do you notice? Compare the distributions from Activities 1 and 2.

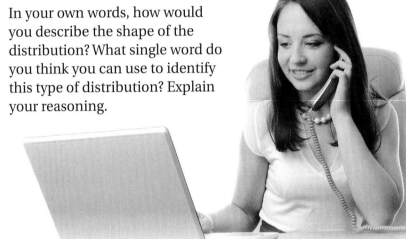

Data Displays

In this lesson, you will
- describe shapes of distributions.

The Meaning of a Word • Skewed

When something is **skewed**, it has a slanted direction or position.

3 ACTIVITY: Describing the Shape of a Distribution

Work with a partner. The table shows the ages of cellular phones owned by a group of students.

Ages of Cellular Phones (years)				
0	1	0	6	4
2	3	5	1	1
0	1	2	3	1
0	0	1	1	1
7	1	4	2	2
0	2	0	1	2

a. Make a dot plot of the data.

b. In your own words, how would you describe the shape of the distribution? Compare it to the distributions in Activities 1 and 2.

c. Why do you think this type of distribution is called a *skewed distribution*?

4 ACTIVITY: Finding Measures of Center

Math Practice

Use Prior Results

How is the distribution of the data related to the mean and the median?

Work with a partner.

a. Find the means and the medians of the data sets in Activities 1–3.

b. What do you notice about the means and the medians of the data sets and the shapes of the distributions? Explain.

c. Which measure of center do you think best describes the data set in Activity 2? in Activity 3? Explain your reasoning.

d. Using your answers to part (c), decide which measure of variation you think best describes the data set in Activity 2. Which measure of variation do you think best describes the data set in Activity 3? Explain your reasoning.

What Is Your Answer?

5. **IN YOUR OWN WORDS** How can you describe the shape of the distribution of a data set?

6. Name two other ways you can describe the distribution of a data set.

Practice — Use what you learned about shapes of distributions to complete Exercises 3 and 4 on page 454.

Section 10.3 Shapes of Distributions

10.3 Lesson

You can use dot plots and histograms to identify shapes of distributions.

Key Ideas

Symmetric and Skewed Distributions

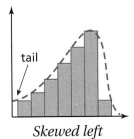
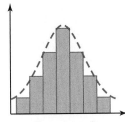
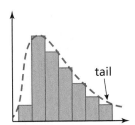

Skewed left   *Symmetric*   *Skewed right*

- The "tail" of the graph extends to the left.
- Most data are on the right.

- The left side of the graph is a mirror image of the right side of the graph.

- The "tail" of the graph extends to the right.
- Most data are on the left.

Study Tip

If all the dots of a dot plot or bars of a histogram are about the same height, then the distribution is a *flat*, or *uniform*, distribution. A uniform distribution is also symmetric.

EXAMPLE 1   Describing the Shapes of Distributions

Describe the shape of each distribution.

a. **Daily Snowfall Amounts**

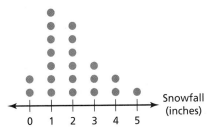

Most of the data are on the left, and the tail extends to the right.

∴ So, the distribution is skewed right.

b. **Passes Thrown**

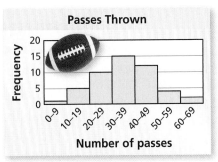

The left side of the graph is approximately a mirror image of the right side of the graph.

∴ So, the distribution is symmetric.

On Your Own

Now You're Ready
Exercises 5–8

1. Describe the shape of the distribution.

Daily Spam Emails Received

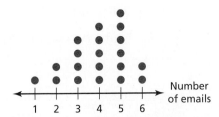

EXAMPLE 2 Describing the Shape of a Distribution

Ages	Frequency
10–13	1
14–17	3
18–21	7
22–25	12
26–29	20
30–33	18
34–37	3

The frequency table shows the ages of people watching a comedy in a theater. Display the data in a histogram. Describe the shape of the distribution.

Draw and label the axes. Then draw a bar to represent the frequency of each interval.

Most of the data are on the right, and the tail extends to the left.

∴ So, the distribution is skewed left.

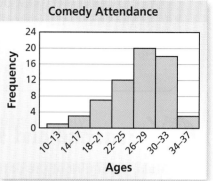

EXAMPLE 3 Comparing Shapes of Distributions

The histogram shows the ages of people watching an animated movie in the same theater as in Example 2.

a. **Describe the shape of the distribution.**

Most of the data are on the left, and the tail extends to the right.

∴ So, the distribution is skewed right.

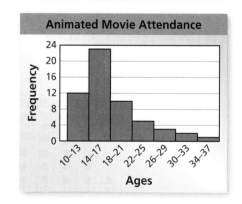

b. **Which movie has an older audience?**

The intervals in the histograms are the same. Most of the data for the animated movie are on the left, while most of the data for the comedy are on the right. This means that the people watching the comedy are generally older than the people watching the animated movie.

∴ So, the comedy has an older audience.

On Your Own

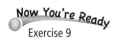
Exercise 9

2. The frequency table shows the ages of people watching a historical movie in a theater.

Ages	10–19	20–29	30–39	40–49	50–59	60–69
Frequency	3	18	36	40	14	5

a. Display the data in a histogram. Describe the shape of the distribution.

b. Compare the distribution of the data to the distributions in Examples 2 and 3. What can you conclude?

Extension 10.3 Choosing Appropriate Measures

You can use a measure of center and a measure of variation to describe the distribution of a data set. The shape of the distribution can help you choose which measures are the most appropriate to use.

Math Practice

Understand Quantities

What effect can outliers have on the mean? on the median? Explain.

Choosing Appropriate Measures

The mean absolute deviation (MAD) uses the mean in its calculation. So, when a data distribution is *symmetric*,

- use the mean to describe the center and
- use the MAD to describe the variation.

The interquartile range (IQR) uses quartiles in its calculation. So, when a data distribution is *skewed*,

- use the median to describe the center and
- use the IQR to describe the variation.

EXAMPLE 1 Choosing Appropriate Measures

The dot plot shows the average number of hours students in a class sleep each night.

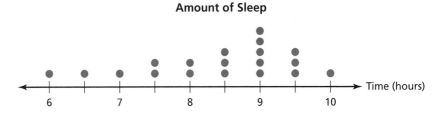

a. **What are the most appropriate measures to describe the center and the variation?**

Most of the data values are on the right clustered around 9, and the tail extends to the left. The distribution is skewed left.

∴ So, the median and the interquartile range are the most appropriate measures to describe the center and the variation.

b. **Describe the center and the variation of the data set.**

The median is 8.5 hours. The first quartile is 7.5, and the third quartile is 9. So, the interquartile range is $9 - 7.5 = 1.5$ hours.

∴ The data are centered around 8.5 hours. The middle half of the data varies by no more than 1.5 hours.

Data Displays

In this extension, you will
- choose appropriate measures of center and variation to represent data sets.

EXAMPLE 2 Choosing Appropriate Measures

Bordering States	Frequency
0–1	3
2–3	13
4–5	22
6–7	10
8–9	2

The frequency table shows the number of states that border each state in the United States.

a. Display the data in a histogram.

Draw and label the axes. Then draw a bar to represent the frequency of each interval.

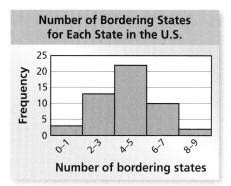

b. What are the most appropriate measures to describe the center and the variation?

The left side of the graph is approximately a mirror image of the right side of the graph. The distribution is symmetric.

∴ So, the mean and the mean absolute deviation are the most appropriate measures to describe the center and the variation.

Practice

Choose the most appropriate measures to describe the center and the variation. Find the measures you chose.

1. Prices of Jeans

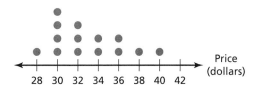

2. Weekly Biking Times

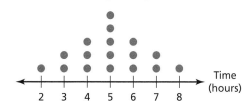

3. REASONING Can you find the exact values of the mean and the mean absolute deviation for the data in Example 2? Explain.

4. GAS MILEAGE The frequency table shows the gas mileages of several vehicles made by a company.

Mileage (miles per gallon)	Frequency
10–14	2
15–19	1
20–24	6
25–29	8
30–34	10
35–39	3

a. What are the most appropriate measures to describe the center and the variation?

b. What conclusions can you make?

5. OPEN-ENDED Construct a dot plot for which the mean is the most appropriate measure to describe the center of the distribution.

Extension 10.3 Choosing Appropriate Measures

10.4 Box-and-Whisker Plots

Essential Question How can you use quartiles to represent data graphically?

1 ACTIVITY: Drawing a Box-and-Whisker Plot

Work with a partner.

The numbers of pairs of footwear owned by each student in a sixth grade class are shown.

Numbers of Pairs of Footwear

2	5	12	3
7	2	4	6
14	10	6	28
5	3	2	4
9	25	4	10
8	15	5	8

a. Order the data set from least to greatest. Then write the data on a strip of grid paper with 24 boxes.

b. Use the strip of grid paper to find the median, the first quartile, and the third quartile. Identify the least value and the greatest value in the data set.

c. Graph the five numbers that you found in part (b) on the number line below.

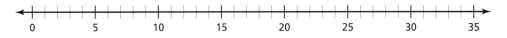

d. The data display shown below is called a *box-and-whisker plot*. Fill in the missing labels and numbers. Explain how a box-and-whisker plot uses quartiles to represent the data.

Data Displays

In this lesson, you will
- make and interpret box-and-whisker plots.
- compare box-and-whisker plots.

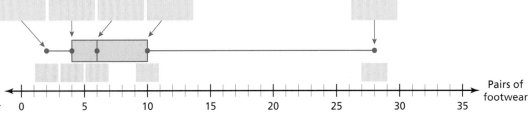

e. Using only the box-and-whisker plot, which measure(s) of center can you find for the data set? Which measure(s) of variation can you find for the data set? Explain your reasoning.

f. Why do you think this type of data display is called a box-and-whisker plot? Explain.

2 ACTIVITY: Conducting a Survey

Have your class conduct a survey. Each student will write on the chalkboard the number of pairs of footwear that he or she owns.

Now, work with a partner to draw a box-and-whisker plot of the data.

3 ACTIVITY: Reading a Box-and-Whisker Plot

Math Practice

View as Components
What do the different components of a box-and-whisker plot represent?

Work with a partner. The box-and-whisker plots show the test score distributions of two sixth grade achievement tests. The same group of students took both tests. The students took one test in the fall and the other in the spring.

a. Compare and contrast the test results.

b. Decide which box-and-whisker plot represents the results of which test. How did you make your decision?

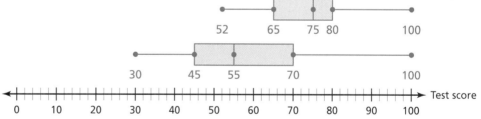

What Is Your Answer?

4. **IN YOUR OWN WORDS** How can you use quartiles to represent data graphically?

5. Describe who might be interested in test score distributions like those shown in Activity 3. Explain why it is important for such people to know test score distributions.

Practice

Use what you learned about box-and-whisker plots to complete Exercise 4 on page 463.

10.4 Lesson

Key Vocabulary
box-and-whisker plot, *p. 460*
five-number summary, *p. 460*

Box-and-Whisker Plot

A **box-and-whisker plot** represents a data set along a number line by using the least value, the greatest value, and the quartiles of the data. A box-and-whisker plot shows the *variability* of a data set.

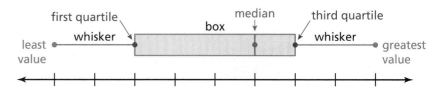

The five numbers that make up the box-and-whisker plot are called the **five-number summary** of the data set.

EXAMPLE 1 Making a Box-and-Whisker Plot

Make a box-and-whisker plot for the ages (in years) of the spider monkeys at a zoo:

15, 20, 14, 38, 30, 36, 30, 30, 27, 26, 33, 35

Step 1: Order the data. Find the median and the quartiles.

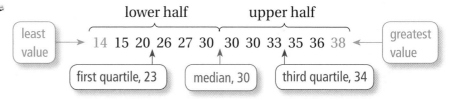

Step 2: Draw a number line that includes the least and greatest values. Graph points above the number line that represent the five-number summary.

Step 3: Draw a box using the quartiles. Draw a line through the median. Draw whiskers from the box to the least and the greatest values.

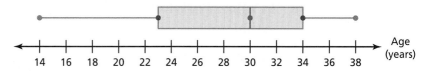

On Your Own

Now You're Ready Exercises 5–8

1. A group of friends spent 1, 0, 2, 3, 4, 3, 6, 1, 0, 1, 2, and 2 hours online last night. Make a box-and-whisker plot for the data.

460 Chapter 10 Data Displays

The figure shows how data are distributed in a box-and-whisker plot.

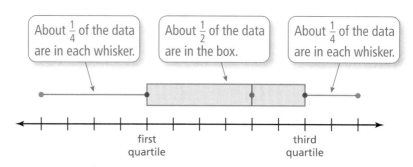

Study Tip

A long whisker or box indicates that the data are more spread out.

EXAMPLE 2   Analyzing a Box-and-Whisker Plot

The box-and-whisker plot shows the body mass index (BMI) of a sixth grade class.

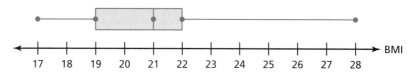

a. What fraction of the students have a BMI of at least 22?

The right whisker represents students who have a BMI of at least 22.

∴ So, about $\frac{1}{4}$ of the students have a BMI of at least 22.

b. Are the data more spread out below the first quartile or above the third quartile? Explain.

The right whisker is longer than the left whisker.

∴ So, the data are more spread out above the third quartile than below the first quartile.

c. Find and interpret the interquartile range of the data.

$$\text{interquartile range} = \text{third quartile} - \text{first quartile}$$
$$= 22 - 19 = 3$$

∴ So, the middle half of the students' BMIs varies by no more than 3.

On Your Own

Now You're Ready
Exercises 11 and 12

2. The box-and whisker plot shows the heights of the roller coasters at an amusement park. (a) What fraction of the roller coasters are between 120 feet tall and 220 feet tall? (b) Are the data more spread out below or above the median? Explain. (c) Find and interpret the interquartile range of the data.

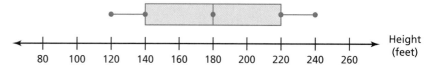

A box-and-whisker plot also shows the shape of a distribution.

Key Ideas

Shapes of Box-and-Whisker Plots

Skewed left
- Left whisker longer than right whisker
- Most data on the right

Symmetric
- Whiskers about same length
- Median in the middle of the box

Skewed right
- Right whisker longer than left whisker
- Most data on the left

Study Tip

If you can draw a line through the median of a box-and-whisker plot, and each side is a mirror image of the other, then the distribution is symmetric.

EXAMPLE 3 Comparing Box-and-Whisker Plots

The double box-and-whisker plot represents the prices of snowboards at two stores.

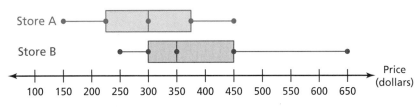

a. Identify the shape of each distribution.

For Store A, the whisker lengths are equal. The median is in the middle of the box. The data on the left are the mirror image of the data on the right. So, the distribution is symmetric.

For Store B, the right whisker is longer than the left whisker, and most of the data are on the left side of the display. So, the distribution is skewed right.

b. Which store's prices are more spread out? Explain.

Both boxes appear to be the same length. So, the interquartile range of each data set is equal. However, the range of the prices in Store B is greater than the range of the prices in Store A. So, the prices in Store B are more spread out.

On Your Own

Now You're Ready
Exercises 13–17

3. The double box-and-whisker plot represents the life spans of crocodiles and alligators at a zoo. Identify the shape of each distribution. Which reptile's life spans are more spread out? Explain.

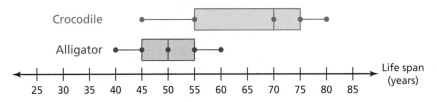

10.4 Exercises

Vocabulary and Concept Check

1. **VOCABULARY** Explain how to find the five-number summary of a data set.

2. **NUMBER SENSE** In a box-and-whisker plot, what fraction of the data is greater than the first quartile?

3. **DIFFERENT WORDS, SAME QUESTION** Which is different? Find "both" answers.

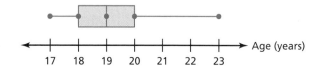

Is the distribution skewed right?

Is the left whisker longer than the right whisker?

Are the data more spread out below the first quartile than above the third quartile?

Does the lower fourth of the data vary more than the upper fourth of the data?

Practice and Problem Solving

4. The box-and-whisker plots represent the daily attendance at two beaches during July. Compare and contrast the attendances for the two beaches.

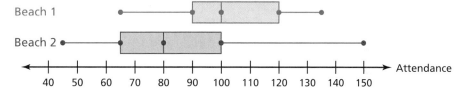

Make a box-and-whisker plot for the data.

① 5. Ages of teachers (in years): 30, 62, 26, 35, 45, 22, 49, 32, 28, 50, 42, 35

6. Quiz scores: 8, 12, 9, 10, 12, 8, 5, 9, 7, 10, 8, 9, 11

7. Donations (in dollars): 10, 30, 5, 15, 50, 25, 5, 20, 15, 35, 10, 30, 20

8. Ski lengths (in centimeters): 180, 175, 205, 160, 210, 175, 190, 205, 190, 160, 165, 195

9. **ERROR ANALYSIS** Describe and correct the error in making a box-and-whisker plot for the data.

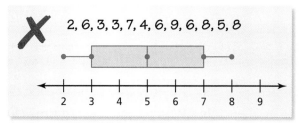

10. **CAMPING** The numbers of days 12 friends went camping during the summer are 6, 2, 0, 10, 3, 6, 6, 4, 12, 0, 6, and 2. Make a box-and-whisker plot for the data. What is the range of the data?

Section 10.4   Box-and-Whisker Plots   463

11. **DUNK TANK** The box-and-whisker plot represents the numbers of gallons of water needed to fill different types of dunk tanks offered by a company.

    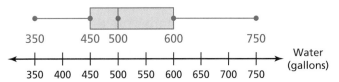

    a. What fraction of the dunk tanks require at least 500 gallons of water?
    b. Are the data more spread out below the first quartile or above the third quartile? Explain.
    c. Find and interpret the interquartile range of the data.

12. **BUILDINGS** The box-and-whisker plot represents the heights (in meters) of the tallest buildings in Chicago.

    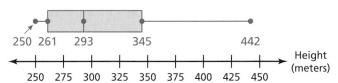

    a. What percent of the buildings are no taller than 345 meters?
    b. Is there more variability in the heights above 345 meters or below 261 meters? Explain.
    c. Find and interpret the interquartile range of the data.

Identify the shape of the distribution. Explain.

13.   14.

15.   16.

17. **RECESS** The double box-and-whisker plot represents the start times of recess for two schools.

    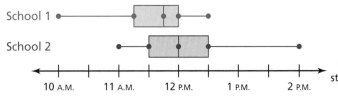

    a. Identify the shape of each distribution.
    b. Which school's start times for recess are more spread out? Explain.
    c. Which school is more likely to have recess before lunch? Explain.

Make a box-and-whisker plot for the data.

18. Temperatures (in °C): 5, 1, 4, 0, 9, 0, −8, 5, 2, 4, −1, 10, 7, −5

19. Checking account balances (in dollars): 30, 0, 50, 20, 90, −15, 40, 100, 45, −20, 70, 0

20. **REASONING** The data set in Exercise 18 has an outlier. Describe how removing the outlier affects the box-and-whisker plot.

21. **CHOOSE TOOLS** What are the most appropriate measures to describe the center and the variation of the distribution in Exercise 12?

22. **OPEN-ENDED** Write a data set with 12 values that has a symmetric box-and-whisker plot.

23. **CRITICAL THINKING** When would a box-and-whisker plot *not* have one or both whiskers?

24. **STRUCTURE** Draw a histogram that could represent the distribution shown in Exercise 15.

25. **REASONING** The double box-and-whisker plot represents the runs scored per game by two softball teams during a 32-game season.

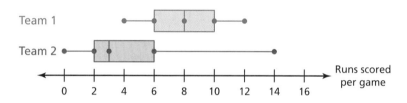

   a. Which team is more consistent at scoring runs? Explain.
   b. In how many games did Team 2 score 6 runs or less?
   c. Team 1 played Team 2 once during the season. Which team do you think won? Explain.
   d. Which team do you think has the greater mean? Explain.

26. A market research company wants to summarize the variability of the SAT scores of graduating seniors in the United States. Do you think the company should use a stem-and-leaf plot, a histogram, or a box-and-whisker plot? Explain.

Fair Game Review What you learned in previous grades & lessons

Copy and complete the statement using < or >. *(Section 6.3)*

27. $-\dfrac{2}{3}$ ▢ $-\dfrac{3}{4}$

28. $-2\dfrac{1}{5}$ ▢ $-2\dfrac{1}{6}$

29. -5.3 ▢ -5.5

30. **MULTIPLE CHOICE** Which of the following items is most likely represented by a rectangular prism with a volume of 1785 cubic inches? *(Section 8.4)*

    Ⓐ closet   Ⓑ computer tower
    Ⓒ filing cabinet   Ⓓ your math book

10.3–10.4 Quiz

Describe the shape of each distribution. *(Section 10.3)*

1.

2.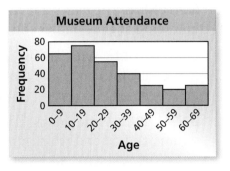

Choose the most appropriate measures to describe the center and the variation. Find the measures you chose. *(Section 10.3)*

3.
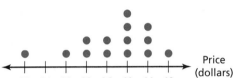

4.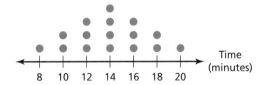

Make a box-and-whisker plot for the data. *(Section 10.4)*

5. Science test scores: 85, 76, 99, 84, 92, 95, 68, 100, 93, 88, 87, 85

6. Shoe sizes: 12, 8.5, 9, 10, 9, 11, 11.5, 9, 9, 10, 10, 10.5, 8

7. **MOVIES** The box-and-whisker plot represents the lengths (in minutes) of movies being shown at a theater. *(Section 10.4)*

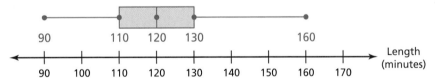

   a. What percent of the movies are no longer than 120 minutes?

   b. Is there more variability in the movie lengths longer than 130 minutes or shorter than 110 minutes? Explain.

   c. Find and interpret the interquartile range of the data.

8. **EXPERIENCE** The frequency table shows the years of experience of employees at two branches of a company. Display the data for each branch in a histogram. Describe the shape of each distribution. Which branch has less experience? Explain. *(Section 10.3)*

Years of Experience	0–2	3–6	7–10	11–14	15–18	19–22	23–26
Frequency at Branch A	10	25	14	20	8	5	2
Frequency at Branch B	3	6	8	10	15	25	8

10 Chapter Review

Review Key Vocabulary

stem-and-leaf plot, *p. 436*
stem, *p. 436*
leaf, *p. 436*

frequency table, *p. 441*
frequency, *p. 441*
histogram, *p. 442*

box-and-whisker plot, *p. 460*
five-number summary, *p. 460*

Review Examples and Exercises

10.1 Stem-and-Leaf Plots (pp. 434–439)

Make a stem-and-leaf plot of the number of DVDs rented each day at a store.

Day	DVDs Rented
Sun	50
Mon	19
Tue	25
Wed	28
Thu	39
Fri	53
Sat	50

Step 1: Order the data. 19, 25, 28, 39, 50, 50, 53

Step 2: Choose the stems and the leaves. Because the data range from 19 to 53, use the *tens* digits for the stems and the *ones* digits for the leaves. Be sure to include the key.

Step 3: Write the stems to the *left* of the vertical line.

Step 4: Write the leaves for each stem to the *right* of the vertical line.

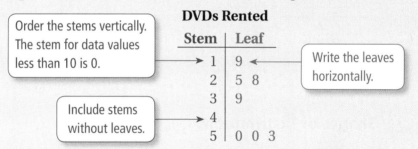

DVDs Rented

Stem	Leaf
1	9
2	5 8
3	9
4	
5	0 0 3

Order the stems vertically. The stem for data values less than 10 is 0.

Include stems without leaves.

Write the leaves horizontally.

Key: 2|5 = 25 DVDs

Exercises

Make a stem-and-leaf plot of the data.

1. **Hats Sold Each Day**

5	18	12	15
21	30	8	12
13	9	14	25

2. **Ages of Park Volunteers**

13	17	40	15
48	21	19	52
13	55	60	20

The stem-and-leaf plot shows the weights (in pounds) of yellowfin tuna caught during a fishing contest.

3. How many tuna weigh less than 90 pounds?

4. What is the median weight of the tuna?

Stem	Leaf
7	6
8	0 2 5 7 9
9	5 6
10	2

Key: 8|5 = 85 pounds

10.2 Histograms (pp. 440–447)

The frequency table shows the number of crafts each member of a craft club made for a fundraiser. Display the data in a histogram.

Crafts	Frequency
0–2	10
3–5	8
6–8	5
9–11	0
12–14	2

Step 1: Draw and label the axes.

Step 2: Draw a bar to represent the frequency of each interval.

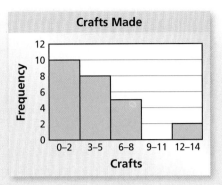

Exercises

Display the data in a histogram.

5.

Heights of Gymnasts	
Heights (in.)	Frequency
50–54	1
55–59	8
60–64	5
65–69	2

6.

Minutes Studied	
Minutes	Frequency
0–19	5
20–39	9
40–59	12
60–79	3

10.3 Shapes of Distributions (pp. 450–457)

Describe the shape of each distribution.

a.

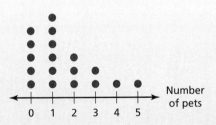

b.

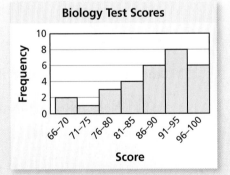

Most of the data are on the left, and the tail extends to the right.

∴ So, the distribution is skewed right.

Most of the data are on the right, and the tail extends to the left.

∴ So, the distribution is skewed left.

Exercises

7. Describe the shape of the distribution.

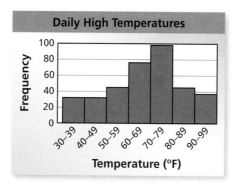

8. Choose the most appropriate measures to describe the center and the variation. Find the measures you chose.

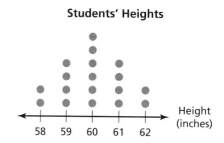

10.4 Box-and-Whisker Plots (pp. 458–465)

Make a box-and-whisker plot for the weights (in pounds) of pumpkins sold at a market.

16, 20, 14, 15, 12, 8, 8, 19, 14, 10, 8, 16

Step 1: Order the data. Find the median and the quartiles.

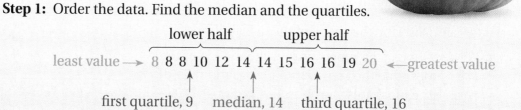

Step 2: Draw a number line that includes the least and the greatest values. Graph points above the number line that represent the five-number summary.

Step 3: Draw a box using the quartiles. Draw a line through the median. Draw whiskers from the box to the least and the greatest values.

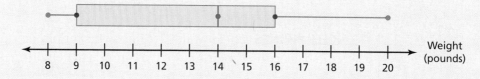

Exercises

Make a box-and-whisker plot for the data.

9. Ages of volunteers at a hospital:
14, 17, 20, 16, 17, 14, 21, 18

10. Masses (in kilograms) of lions:
120, 200, 180, 150, 200, 200, 230, 160

10 Chapter Test

Make a stem-and-leaf plot of the data.

1.
Quiz Scores (%)			
96	88	80	72
80	94	92	100
76	80	68	90

2.
CDs Sold Each Day				
45	31	29	38	38
67	40	62	45	60
40	39	60	43	48

3. Find the mean, median, mode, range, and interquartile range of the data.

Cooking Time (minutes)

Stem	Leaf
3	5 8
4	0 1 8
5	0 4 4 4 5 9
6	0

Key: 4 | 1 = 41 minutes

4. Display the data in a histogram.

Television Watched Per Week	
Hours	Frequency
0–9	14
10–19	16
20–29	10
30–39	8

5. **WATER** The dot plot shows the number of glasses of water that the students in a class drink in one day.

   a. Describe the shape of the distribution.

   b. Choose the most appropriate measures to describe the center and the variation. Find the measures you chose.

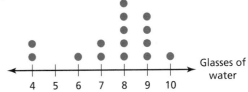

Make a box-and-whisker plot for the data.

6. Ages (in years) of dogs at a vet's office: 1, 3, 5, 11, 5, 7, 5, 9

7. Lengths (in inches) of fish in a pond: 12, 13, 7, 8, 14, 6, 13, 10

8. Hours practiced each week: 7, 6, 5, 4.5, 3.5, 7, 7.5, 2, 8, 7, 7.5, 6.5

9. **CELL PHONES** The double box-and-whisker plot compares the battery life (in hours) of two brands of cell phones.

   a. What is the range of the upper 75% of each brand?

   b. Which battery has a longer battery life? Explain.

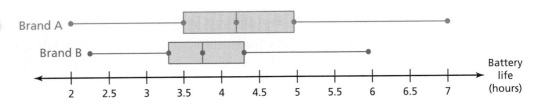

10 Cumulative Assessment

1. Research scientists are measuring the number of days lettuce seeds take to germinate. In a study, 500 seeds were planted. Of these, 473 seeds germinated. The box-and-whisker plot summarizes the number of days it took the seeds to germinate. What can you conclude from the box-and-whisker plot?

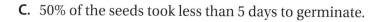

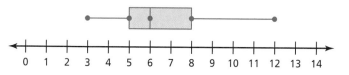

   A. The median number of days for the seeds to germinate is 12.

   B. 50% of the seeds took more than 8 days to germinate.

   C. 50% of the seeds took less than 5 days to germinate.

   D. The median number of days for the seeds to germinate was 6.

2. You are comparing the costs of buying bottles of water at the supermarket. Which of the following has the least cost per liter?

   F. six 1-liter bottles for $1.80

   G. one 2-liter bottle for $0.65

   H. eight $\frac{1}{2}$-liter bottles for $1.50

   I. twelve $\frac{1}{2}$-liter bottles for $1.98

3. What number belongs in the box to make the equation true?

$$3\frac{1}{2} \div 5\frac{2}{3} = \frac{7}{2} \times \boxed{}$$

   A. $\frac{17}{3}$

   B. $\frac{13}{2}$

   C. $\frac{3}{17}$

   D. $\frac{3}{2}$

4. What is the mean number of seats?

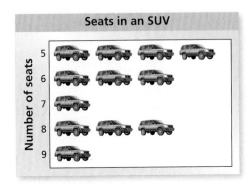

F. 2.4 seats

G. 5 seats

H. 6.5 seats

I. 7 seats

5. On Wednesday, the town of Mims received 17 millimeters of rain. This was x millimeters more rain than the town received on Tuesday. Which expression represents the amount of rain, in millimeters, the town received on Tuesday?

A. $17x$

B. $17 - x$

C. $x + 17$

D. $x - 17$

6. One of the leaves is missing in the stem-and-leaf plot.

The median of the data set represented by the stem-and-leaf plot is 38. What is the value of the missing leaf?

Stem	Leaf
1	3 4
2	
3	4 5 7 7 ? 9
4	0 1 1 4
5	0 2 3

Key: $1 | 4 = 14$

7. Which property is demonstrated by the equation below?

$$723 + (884 + 277) = 723 + (277 + 884)$$

F. Associative Property of Addition

G. Commutative Property of Addition

H. Distributive Property

I. Identity Property of Addition

8. A student took 5 tests this marking period and had a mean score of 92. Her scores on the first 4 tests were 90, 96, 86, and 92. What was her score on the fifth test?

   A. 92

   B. 93

   C. 96

   D. 98

9. At the end of the school year, your teacher counted up the number of absences for each student. The results are shown in the histogram below.

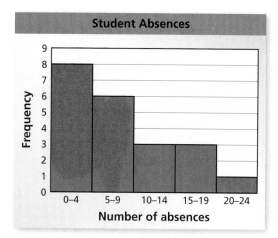

   Based on the histogram, how many students had fewer than 10 absences?

10. The 16 members of a camera club have the ages listed below.

   40, 22, 24, 58, 30, 31, 37, 25, 62, 40, 39, 37, 28, 28, 51, 44

   Part A Order the ages from least to greatest.

   Part B Find the median of the ages.

   Part C Make a box-and-whisker plot for the ages of the camera club members.

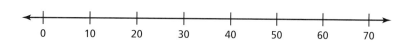

Cumulative Assessment   473

Appendix A
My Big Ideas Projects

A.1 **Literature Project**
Alice's Adventures in Wonderland

A.2 **History Project**
Mathematics in Ancient Egypt

My Big Ideas Projects

A.3 Art Project
Polyhedra in Art

A.4 Science Project
Why Does Ice Float?

A.1 Literature Project

Alice's Adventures in Wonderland

1 Getting Started

Lewis Carroll was a mathematician who also wrote literature books for children.

Essential Question How does mathematical problem solving influence a story plot?

Read *Alice's Adventures in Wonderland* by Lewis Carroll. In each chapter, rewrite one of the episodes so that it contains some of the math that you have studied this year.

Sample: You could rewrite Alice's shrinking adventure as follows.

Alice's normal height was 4′7″ or 55 inches. After drinking the bottle marked "Drink Me," Alice could feel herself shrinking more and more. After several minutes, Alice was only two-elevenths of her original height. Would Alice have to bend over to walk through a door that is only 1 foot high?

The illustrations on these two pages are from the 1907 edition of *Alice's Adventures in Wonderland*. They were drawn by Arthur Rackham (1867–1939), a book illustrator from England.

2 Things to Remember

- You can download each chapter of the book at *BigIdeasMath.com*.
- Add your own illustrations to your project.
- Try to include as many different math concepts as possible. Your goal is to include at least one concept from each of the chapters you studied this year.
- Organize your math stories in a folder, and think of a title for your report.

A.2 History Project

Mathematics in Ancient Egypt

1 Getting Started

The ancient Egyptian civilization in North Africa began around 3150 B.C. with the union of Upper and Lower Egypt under the first pharaoh. The rule of the pharaohs ended in 31 B.C. when the Romans conquered Egypt and made it a province.

The ancient Egyptians' achievements included a system of mathematics, glass technology, medicine, literature, irrigation and agricultural techniques, and art. Their mastery of surveying, quarrying, and construction techniques allowed them to build pyramids, temples, and obelisks.

Essential Question How is mathematical knowledge that was originally discovered by the ancient Egyptians used today?

Sample: To multiply two whole numbers, the ancient Egyptians repeatedly multiplied by 2. The "doubling lists" used to multiply 53 by 85 are shown. The lists stop when the last number in the first list is greater than the first number in the second list.

- 1   53
- 2   106
- 4   212
- 8   424
- 16   848
- 32   1696
- 64   3392

Because $85 = 1 + 4 + 16 + 64$,
$85 \cdot 53 = 53 + 212 + 848 + 3392 = 4505$.

Ancient Papyrus

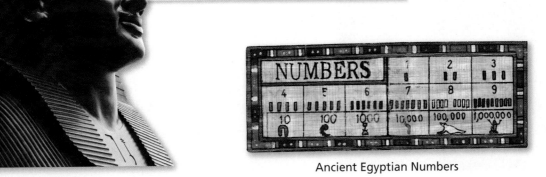

Ancient Egyptian Numbers

The Blue Sphinx

2 Things to Include

- Explain how the ancient Egyptians multiplied two whole numbers. Give an example.
- Explain how the ancient Egyptians divided two whole numbers. Give an example.
- How did the ancient Egyptians write whole numbers? Give some examples.
- How did the ancient Egyptians write fractions? Give some examples.
- How did the ancient Egyptians use ratios to measure the height of a pyramid? Give an example.
- Write your name using ancient Egyptian hieroglyphics.
- Describe how ancient Egyptians used mathematics. How does this compare with the ways in which mathematics is used today?

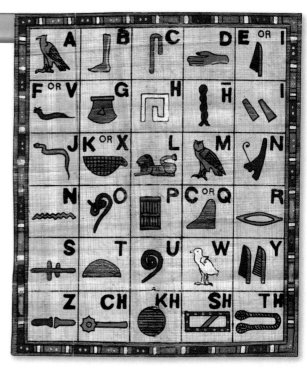

Ancient Egyptian Alphabet

3 Things to Remember

- Add your own illustrations to your project.
- Try to include as many different math concepts as possible. Your goal is to include at least one concept from each of the chapters you studied this year.
- Organize your report in a folder, and think of a title for your report.

What does this spell?

A.3 Art Project

Polyhedra in Art

1 Getting Started

Polyhedra is the plural of *polyhedron*. Polyhedra have been used in art for many centuries, in cultures all over the world.

Essential Question Do polyhedra influence the design of games and architecture?

Some of the most famous polyhedra are the five Platonic solids shown at the right.

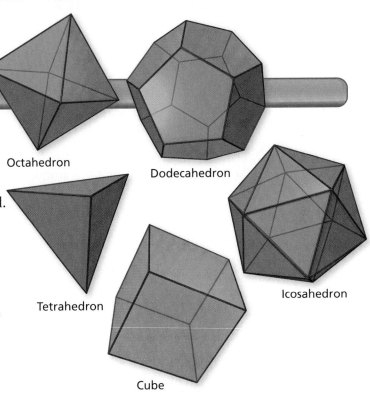

Mosaic by Paolo Uccello, 1430 A.D.

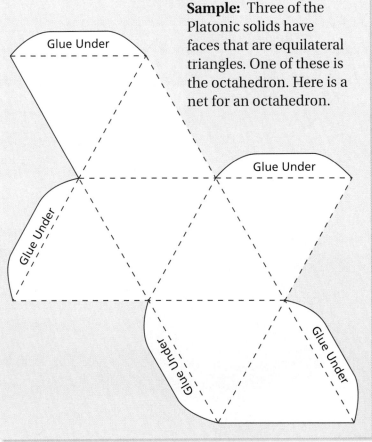

Sample: Three of the Platonic solids have faces that are equilateral triangles. One of these is the octahedron. Here is a net for an octahedron.

A6   Appendix A   My Big Ideas Projects

2 Things to Include

- Explain why the platonic solids are sometimes referred to as the cosmic figures.
- Draw a net for an icosahedron or a dodecahedron. Cut out the net and fold it to form the polyhedron.
- Describe the 13 polyhedra that are called Archimedean solids. What is the definition of this category of polyhedra? Draw a net for one of them. Then cut out the net and fold it to form the polyhedron.
- Find examples of polyhedra in games and architecture.

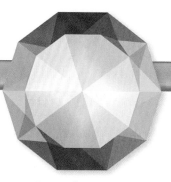

Faceted Cut Gem

Origami Polyhedron

3 Things to Remember

- Add your own illustrations or paper creations to your project.
- Organize your report in a folder, and think of a title for your report.

Concrete Tetrahedrons by Ocean

Bulatov Sculpture

A.4 Science Project

Why Does Ice Float?

1 Getting Started

Among the planets in our solar system, Earth is clearly the "water planet." Water occurs on Earth's surface as a liquid, solid, and gas. Ocean waters cover about 70% of Earth's surface. Fresh water in lakes and rivers covers less than 1%. Thick sheets of ice permanently cover Earth's polar regions, and glaciers occur in its higher mountains. Water in the form of clouds covers about half of Earth's surface at any time.

Essential Question How does floating ice affect you and the world in which you live?

There are three parts to this question. First, you have to understand why things float in water. Make lists of things that float in water and things that do not float in water. How can you describe, scientifically, the difference between these two types of objects?

Second, you have to discover what is special about water. Almost all other elements and compounds have the property that their solid forms sink in their liquid forms.

Third, you have to research how water in solid form influences the environment and living organisms. You may want to explore ocean currents, weather, and various cycles including the water cycle.

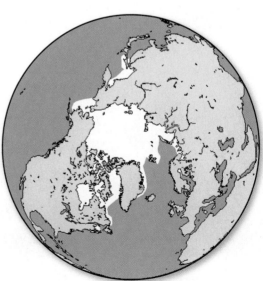

Earth's North Polar Ice Cap

2 Things to Include

- Does vegetable cooking oil float in water? Does ice float in vegetable oil?

- Compare the density of water and the density of ice. How do these densities relate to the fact that ice floats in water?

- What percent of an iceberg is above the water? What percent is below?

- What environmental and biological systems would not work if ice sunk in the oceans instead of floating?

- Explain how the floating of ice helps keep the oceans warm and therefore helps aquatic life exist in Earth's polar oceans.

3 Things to Remember

- Add your own illustrations to your project.

- Try to include as many different math concepts as possible. Your goal is to include at least one concept from each of the chapters you studied this year.

- Organize your report in a folder, and think of a title for your report.

Floating Iceberg

Section A.4   Science Project   **A9**

Selected Answers

Section 1.1 — Whole Number Operations
(pages 7–9)

1. addition
3. division
5. addition
7. **a.** dividend **b.** quotient **c.** divisor
9. $4785 - 3391$; 1394 people
11. 4785×2; 9570 people
13. 7081
15. 2462
17. 433
19. 6944
21. 31
23. 60
25. $47\frac{110}{173}$
27. The partial product 39 should be moved to the left so that the 3 is under the 2 and the 9 is under the 7. The answer should be 663.
29. multiplication
31. division
33. addition
35. 24 in.; 35 in.^{2}
37. 36 m; 80 m^2
39. You can use addition to check your answer by adding 93 to itself 6 times. You can use division to check your answer by dividing 558 by either 93 or 6.
41. no; If the remainder is greater than the divisor, then the quotient should be increased until the remainder is less than the divisor or equal to zero.
43. 46 tokens
45. **a.** $424 **b.** $\frac{3}{4}$ qt, or $\frac{3}{16}$ gal
47 and 49.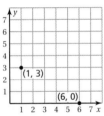
51. A

Section 1.2 — Powers and Exponents
(pages 14 and 15)

1. An exponent indicates the number of times the base is used as a factor. A power is the entire expression (base and exponent). A power is a product of repeated factors.
3. $3 + 3 + 3 + 3 = 3(4)$ does not belong because it shows a product as a sum of repeated addends, whereas the other three show powers as products of repeated factors.
5. 13^2
7. 2^5
9. 8^4
11. 7^6
13. The base is written as the exponent and the exponent is written as the base. $4 \cdot 4 \cdot 4 = 4^3$
15. 64
17. 196
19. 65,536
21. 1,419,857
23. 8,000,000 people
25. not a perfect square
27. perfect square
29. perfect square
31. not a perfect square

33. 40,000 cm^2 **35.** 8 squares

37.

Power	4^6	4^5	4^4	4^3	4^2	4^1
Value	4096	1024	256	64	16	4

As the exponent decreases, the value of the power is divided by 4. $4^0 = 1$

39. 13 blocks; add $7^2 - 6^2$ blocks; 19 blocks; add $10^2 - 9^2$ blocks; 39 blocks; add $20^2 - 19^2$ blocks

41. 165   **43.** 7

Section 1.3 Order of Operations
(pages 20 and 21)

1. Using the order of operations for $12 - 8 \div 2$, you divide 8 by 2 and then subtract the result from 12. Using the order of operations for $(12 - 8) \div 2$, you subtract 8 from 12 and then divide by 2.

3. 57   **5.** 2   **7.** 5

9. 24   **11.** 88   **13.** 2

15. Addition was performed before multiplication. $9 + 2 \times 3 = 9 + 6 = 15$

17. 8 pages   **19.** 25   **21.** 47

23. 8   **25.** 22 people   **27.** 1

29. $34

31. $23; Add the prices of the items you buy. Then subtract the amount of the gift card from the total.

33. **a.** $27 \div 3 + 5 \times 2 = 19$ **b.** *Sample answer:* $9^2 + 11 - 8 \times 4 \div 1 = 60$
  c. $5 \times 6 - 15 + 9 = 24$ **d.** $14 \times 2 \div 7 - 3 + 9 = 10$

35. 6.1   **37.** 0.9

Section 1.4 Prime Factorization
(pages 28 and 29)

1. The prime factorization of a composite number is the number written as a product of its prime factors.

3. 6, 9 does not belong because it is a factor pair of 54 and the others are factor pairs of 56.

5. 3, 5, 9   **7.** None, 1709 is a prime number.

9. 1, 22; 2, 11   **11.** 1, 39; 3, 13   **13.** 1, 54; 2, 27; 3, 18; 6, 9

15. 1, 61   **17.** 5 • 5 or 5^2 **19.** 2 • 13

21. 2 • 3 • 3 • 3 or 2 • 3^3 **23.** 7 • 11

Section 1.4 Prime Factorization (continued)
(pages 28 and 29)

25.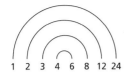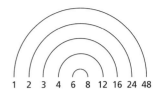

27. 1575   **29.** 4   **31.** 36

33. yes; 2 is a prime number because it only has 1 and itself as factors. The rest of the even whole numbers have 2 as a factor.

35. Use 36 objects to help you determine the possible group sizes.

37. cupcake table; Because 60 has more factors than 75, there are more rectangular arrangements.

39. 6 prisms; There are 6 unique arrangements of length, width, and height using the factors of 40.
(Note that $1 \times 1 \times 40$ names the same prism as $40 \times 1 \times 1$.);
$1 \times 1 \times 40, 1 \times 2 \times 20, 1 \times 4 \times 10, 1 \times 5 \times 8, 2 \times 2 \times 10, 2 \times 4 \times 5$

41. 357   **43.** 1248

Section 1.5 Greatest Common Factor
(pages 34 and 35)

1. The GCF is the greatest factor that is shared by the two numbers.

3. What is the greatest prime factor of 24 and 32?; 2; 8

5. 2   **7.** 3   **9.** 1   **11.** 17

13. 15   **15.** 9   **17.** 1

19. 7 is the greatest common *prime* factor. The GCF is $2 \cdot 7 = 14$.

21. 23 packets   **23.** 7   **25.** 14

27. *Sample answer:* Prime factorization because it is tedious to find all the factors of large numbers.

29. always   **31.** 12; 6 red, 5 pink, and 4 yellow

33. a. Because 73 is a prime number and the GCF of the three numbers is 1.

    b. 18; The GCF of 54 and 36 is 18. 18 divides evenly into 72 leaving one banana left over.

35. Commutative Property of Addition

37. Commutative Property of Multiplication

39. B

Section 1.6 — Least Common Multiple
(pages 40 and 41)

1. The LCM of two numbers is the least of the multiples shared by the two numbers.
3. 21
5. 60
7. 12
9. 40
11. 36
13. 108
15. 66
17. 350
19. 15 days
21. D; This model represents multiples of 4 and 6 which have an LCM of 12. The other models represent multiples of 3 and 8, 8 and 12, and 6 and 8, which have an LCM of 24.
23. 165
25. 120
27. 1260
29. always
31. never
33. 300th caller
35. a. 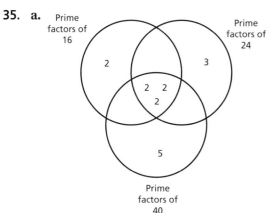   b. 240   c. 80; 120
37. 3^2
39. 17^5

Extension 1.6 — Adding and Subtracting Fractions
(page 43)

1. $\dfrac{4}{24}, \dfrac{9}{24}$
3. $\dfrac{15}{36}, \dfrac{8}{36}$
5. <
7. =
9. $1\dfrac{5}{12}$
11. $\dfrac{17}{60}$
13. $5\dfrac{11}{18}$
15. $1\dfrac{1}{12}$
17. *Sample answer:* The LCD method uses numbers that are easier to work with, but there is extra work in finding the LCD. Using the other method, there are no preliminary steps for finding the LCD, but there may be more simplifying in the solution.

Section 2.1 Multiplying Fractions
(pages 59–61)

1. Multiply numerators and multiply denominators, then simplify the fraction.

3. *Sample answer:* $3\frac{1}{2} \times 3\frac{1}{7} = 11$

5. $\frac{5}{16}$ 7. $\frac{3}{28}$ 9. $\frac{5}{8}$ 11. $\frac{1}{3}$ 13. $5\frac{1}{4}$

15. $\frac{16}{45}$ 17. $\frac{5}{27}$ 19. $1\frac{19}{30}$ 21. $\frac{3}{10}$

23. $\frac{4}{7} > \left(\frac{9}{10} \times \frac{4}{7}\right)$; Because $\frac{9}{10} < 1$, the product will be less than $\frac{4}{7}$.

25. $\frac{5}{6} = \left(\frac{5}{6} \times \frac{7}{7}\right)$; Because $\frac{7}{7} = 1$, by the Multiplication Property of One, the two expressions are equal.

27. 2 29. 2 31. 2 33. $1\frac{1}{2}$

35. $1\frac{3}{14}$ 37. $36\frac{2}{3}$ 39. $6\frac{4}{9}$ 41. $11\frac{3}{8}$

43. You must first rewrite the mixed number as improper fractions and then multiply.

$2\frac{1}{2} \times 7\frac{4}{5} = \frac{5}{2} \times \frac{39}{5}$

$= \frac{\overset{1}{\cancel{5}} \times 39}{2 \times \underset{1}{\cancel{5}}}$

$= \frac{39}{2}$, or $19\frac{1}{2}$

45. **a.** 7 ft^2 **b.** $10\frac{1}{3}$ ft^2

47. $\frac{2}{15}$ 49. $26\frac{2}{5}$ 51. $\frac{9}{25}$ 53. 462 in.^{2}

55. 4 mi

57. Which units of measure would make the calculations easier?

59. **a.** $\frac{3}{50}$ **b.** 45 people

61. $3^2 \cdot 5$ 63. $2^2 \cdot 3 \cdot 5$

Section 2.2 Dividing Fractions
(pages 67–69)

1. *Sample answer:* $\frac{2}{5}, \frac{5}{2}$ 3. B 5. A

7. $\frac{1}{8}$ 9. $\frac{5}{2}$ 11. $\frac{1}{2}$ 13. 16 15. $\frac{1}{14}$

17. $\frac{1}{3}$ 19. 3 21. $\frac{2}{27}$ 23. $\frac{27}{28}$ 25. $20\frac{1}{4}$

27. You need to invert the second fraction before you multiply.

$$\frac{4}{7} \div \frac{13}{28} = \frac{4}{7} \times \frac{28}{13} = \frac{4 \times \overset{4}{\cancel{28}}}{\underset{1}{\cancel{7}} \times 13} = \frac{16}{13}, \text{ or } 1\frac{3}{13}$$

29. Round $\frac{2}{5}$ to $\frac{1}{2}$ and $\frac{8}{9}$ to 1. $\frac{1}{2} \div 1 = \frac{1}{2}$, which is not close to the incorrect answer of $\frac{20}{9}$.

31. $5\frac{5}{8}$ times **33.** yes **35.** yes **37.** $\frac{1}{3}$

39. >; When you divide a number by a fraction less than 1, the quotient is greater than the number.

41. >; When you divide a number by a fraction less than 1, the quotient is greater than the number.

43. $\frac{1}{216}$ **45.** $1\frac{1}{6}$ **47.** 2 **49.** $\frac{3}{26}$ **51.** $\frac{2}{3}$

53. $2\frac{2}{5}$ hours, or 2 hours 24 minutes

55. a. $3\frac{3}{4}$ times **b.** $3\frac{1}{3}$ times **c.** $\frac{1}{4}$

57. a. $67\frac{1}{5}$ gal **b.** $8\frac{2}{5}$ gal **c.** $33\frac{3}{5}$ gal

59. 6 **61.** 5

Section 2.3 Dividing Mixed Numbers
(pages 74 and 75)

1. $\frac{3}{22}$ **3.** sometimes; The reciprocal of $\frac{2}{2}$ is $\frac{2}{2}$, which is improper.

5. 3 **7.** $9\frac{3}{4}$ **9.** $3\frac{18}{19}$ **11.** $\frac{9}{10}$

13. $12\frac{1}{2}$ **15.** $1\frac{1}{5}$ **17.** $\frac{2}{7}$ **19.** $1\frac{5}{18}$

21. The mixed number $1\frac{2}{3}$ was not written as an improper fraction before inverting.

$$3\frac{1}{2} \div 1\frac{2}{3} = \frac{7}{2} \div \frac{5}{3}$$
$$= \frac{7}{2} \times \frac{3}{5}$$
$$= \frac{7 \times 3}{2 \times 5}$$
$$= \frac{21}{10}, \text{ or } 2\frac{1}{10}$$

23. 14 hamburgers

25. no; The model shows $2\frac{1}{2} \div 1\frac{1}{6} = 2\frac{1}{7}$. There are 2 full groups of $1\frac{1}{6}$ plus one piece remaining, which represents $\frac{1}{7}$ of $1\frac{1}{6}$.

27. 4 **29.** $1\frac{1}{3}$ **31.** $5\frac{1}{6}$ **33.** $\frac{7}{54}$ **35.** $12\frac{1}{2}$ **37.** $\frac{22}{35}$

39. a. 6 ramps; *Sample answer:* The estimate is reasonable because $12\frac{1}{2}$ was rounded down.
b. 6 ramps; $1\frac{1}{4}$ feet left over

41. 0.43 **43.** 3.8 **45.** C

Section 2.4 Adding and Subtracting Decimals
(pages 82 and 83)

1. Estimating allows you to check that your answer is reasonable.
3. $1.15 + 0.43 = 1.58$
5. 11.029
7. 22.899
9. 29.937
11. 1.46
13. 4.366
15. 2.644
17. Line up the decimal points before adding. Insert a 0 at the end of the second number so that both numbers have the same number of decimal places. $6.058 + 3.95 = 10.008$.
19. $8.30
21. 19.58
23. 10
25. 15.606
27. the decimal parts in the sum total 1; the decimal parts in the difference are exactly the same
29. 34.995 m
31. 4.816 AU
33. 20.189 AU
35. 6.85 units
37. $\frac{1}{4}$
39. $\frac{1}{20}$

Section 2.5 Multiplying Decimals
(pages 89–91)

1. Place the decimal point so that there are two decimal places. $1.2 \times 2.4 = 2.88$
3. 8.722
5. 19.5750
7. 4
9. 3.15
11. 0.21
13. 33.6
15. 115.04
17. 21.45
19. 13.888
21. 2.4
23. 0.0342
25. The decimal is in the wrong place. $0.0045 \times 9 = 0.0405$
27. 30.06 lb
29. 3 mm; 9 mm
31. 0.024
33. 0.000072
35. 0.03
37. 0.000012
39. 109.74
41. 3.886
43. 13.7104
45. 51.3156
47. $3.24
49. $1284.78
51. $7.12 \times 8.22 \times 100 = 7.12 \times 822 = 5852.64$
53. 137
55. 23.112
57. 71.984
59. 36.225
61. 2; 3; 4
63. Each number is 0.1 times the previous number; 0.0015, 0.00015, 0.000015
65. Each number is 1.5 times the previous number; 25.3125, 37.96875, 56.953125
67. a. 190.06 mi
    b. 91.29 mi
69. Which framing is thicker?
71. 5
73. 7

A16 Selected Answers

Section 2.6 Dividing Decimals
(pages 97–99)

1. $4\overline{)2.44}$ = 0.61
3. $6.38 \div 11 = 0.58$
5. $47\overline{)136}$
7. $216\overline{)1850}$
9. 13
11. 9
13. 6.7
15. 1.3
17. 9.4
19. 3.2
21. 0.098
23. 3.57
25. A zero should be placed before the 8 in the quotient.
$$6\overline{)0.516} = 0.086$$
$$\underline{48}$$
$$36$$
$$\underline{36}$$
$$0$$
27. the 5-ounce bottle; The price per ounce is $2.06 for the 5-ounce bottle and $2.12 for the 4-ounce bottle.
29. 1.62
31. 10.12
33. 8.046
35. $26.96
37. 9
39. 0.08
41. 400
43. 460
45. about 1.33
47. about 12.21
49. 113 tickets
51. 850 songs
53. =
55. <
57. about 5357 bees
59. When dividing, make sure your units cancel.
61. $1\frac{1}{6}$
63. $\frac{1}{20}$
65. B

Section 3.1 Algebraic Expressions
(pages 115–117)

1. 3(4) + 5 does not belong because it is a numerical expression and the other three are algebraic expressions.
3. decrease; When you subtract greater and greater values from 20, you will have less and less left.
5. 20 × 6; $120
7. 20 − 12; $8
9. Terms: g, 12, $9g$
   Coefficients: 1, 9
   Constant: 12
11. Terms: $2m^2$, 15, $2p^2$
    Coefficients: 2, 2
    Constant: 15
13. Terms: $8x$, $\frac{x^2}{3}$
    Coefficients: 8, $\frac{1}{3}$
    Constant: none
15. a. Terms: 2ℓ, $2w$; Coefficients: 2, 2: Constant: none
    b. The coefficient 2 of ℓ represents that there are 2 lengths on the rectangle. The coefficient 2 of w represents that there are 2 widths on the rectangle.
17. g^5
19. $5.2y^3$
21. $2.1xz^4$
23. $25d^2$
25. 9
27. 11
29. 10
31. 6
33. 5
35. 4

Section 3.1
Algebraic Expressions (continued)
(pages 115–117)

37. Multiplication should be done first, then addition.
$5m + 3 = 5 \cdot 8 + 3$
     $= 40 + 3$
     $= 43$

39. 34 mm; 118 mm

41.

x	2	4	8
64 ÷ x	32	16	8

43. 23   **45.** $2\frac{5}{6}$ **47.** 22

49. 46   **51.** 24

53. Start by drawing a visual image of moving 2000 feet in exactly 10 minutes.

55. 64 in.^{3} **57.** 512   **59.** 256

Section 3.2
Writing Expressions
(pages 122 and 123)

1. x take away 12; $x - 12$; $x + 12$ **3.** $8 - 5$ **5.** $28 \div 7$

7. $18 - 3$ **9.** $x - 13$ **11.** $18 \div a$

13. $7 + w$ or $w + 7$ **15.** $y + 4$ or $4 + y$ **17.** $2 \cdot z$ or $z \cdot 2$

19. The expression is not written in the correct order; $\frac{8}{y}$

21. a. $x \div 5$

   b. *Sample answer:* If the total cost is $30, then the cost per person is $x \div 5 = 30 \div 5 = \$6$. The result is reasonable.

23. *Sample answer:* The sum of n and 6; 6 more than a number n

25. *Sample answer:* A number b less than 15; 15 take away a number b

27. $\frac{y}{4} - 3$; 2   **29.** $8x + 6$; 46

31. a.

Game	1	2	3	4	5
Cost	$5	$8	$11	$14	$17

   b. $2 + 3g$
   c. $26

33. It might help to see the pattern if you make a table of the data in the bar graph.

35. $\frac{x}{4}$ **37.** 59   **39.** 140

A18   Selected Answers

Section 3.3 — Properties of Addition and Multiplication
(pages 130 and 131)

1. *Sample answer:* $\dfrac{1}{5} + \dfrac{3}{5} = \dfrac{3}{5} + \dfrac{1}{5}$
   $\dfrac{4}{5} = \dfrac{4}{5}$

3. *Sample answer:* $(5 \cdot x) \cdot 1 = 5 \cdot (x \cdot 1)$
   $= 5x$

5. Comm. Prop. of Mult.   7. Assoc. Prop. of Mult.   9. Add. Prop. of Zero

11. The grouping of the numbers did not change. The statement illustrates the Commutative Property of Addition because the order of the addends changed.

13. $(14 + y) + 3 = (y + 14) + 3$ Comm. Prop. of Add.
    $= y + (14 + 3)$ Assoc. Prop. of Add.
    $= y + 17$ Add 14 and 3.

15. $7(9w) = (7 \cdot 9)w$ Assoc. Prop. of Mult.
    $= 63w$ Multiply 7 and 9.

17. $(0 + a) + 8 = a + 8$ Add. Prop. of Zero

19. $(18.6 \cdot d) \cdot 1 = 18.6 \cdot (d \cdot 1)$ Assoc. Prop. of Mult.
    $= 18.6d$ Mult. Prop. of One

21. $(2.4 + 4n) + 9 = (4n + 2.4) + 9$ Comm. Prop. of Add.
    $= 4n + (2.4 + 9)$ Assoc. Prop. of Add.
    $= 4n + 11.4$ Add 2.4 and 9.

23. $z \cdot 0 \cdot 12 = (z \cdot 0) \cdot 12$ Assoc. Prop. of Mult.
    $= 0 \cdot 12$ Mult. Prop. of Zero
    $= 0$ Mult. Prop. of Zero

25. **a.** x represents the cost of a box of cookies.   **b.** $120x$

27. $7 + (x + 5) = x + 12$ 29. $(7 \cdot 2) \cdot y$ 31. $(17 + 6) + 2x$

33. $w \cdot 16$ 35. 98   37. 90

39. 37 is already prime.   41. 3×7^2 43. B

Section 3.4 — The Distributive Property
(pages 137–139)

1. *Sample answer:* You must distribute or give the number outside the parentheses to the numbers inside the parentheses.

3. $4 + (x \cdot 4)$ does not belong because it does not represent the Distributive Property.

5. 63   7. 516   9. 936   11. 504   13. $\dfrac{4}{7}$

15. $2\dfrac{1}{2}$ 17. $3x + 12$ 19. $6s - 54$ 21. $96 + 8a$ 23. $72 - 12k$

25. $63 + 9c$ 27. $40g + 24$ 29. $4x + 4y$ 31. $7p + 7q + 63$

33. The 6 was not distributed to the 8 inside the parentheses; $6(y + 8) = 6y + 48$

35. $5(r + 15)$ and $5r + 5 \cdot 15$, because they are equivalent expressions.

Section 3.4 — The Distributive Property (continued)
(pages 137–139)

37. $16(10 + x) = 160 + 16x$ **39.** $6x + 25$

41. $68 + 28k$ **43.** $19y + 5$ **45.** $3d + 1$ **47.** $5v$

49. $2.7w - 14.04$ **51.** $\dfrac{11}{4}z + \dfrac{3}{10}$ **53.** $7x + 12y$ **55.** $x = 8$ **57.** $x = 3$

59. Area: $8x + 64$
    Perimeter: $2x + 32$

61. Area: $9x + 108$
    Perimeter: $2x + 42$

63. **a.** 6.2   **b.** 14
Sample answer: The preferred method is not the same for both expressions.
For part (a), evaluating inside the parentheses first requires easier and less calculations.
For part (b), using the Distributive Property will eliminate the fractions.

65. $7(x + 3) + 8 \cdot x + 3 \cdot x + 8 - 9 = 2(9x + 10)$

67. 34.006   **69.** 0.387

Extension 3.4 — Factoring Expressions
(page 141)

1. $7(1 + 2)$ **3.** $6(3 - 2)$ **5.** $12(5 - 3)$ **7.** $28(3 + 1)$

9. $2(x + 5)$ **11.** $13(2x - 1)$ **13.** $9(4x + 1)$ **15.** $5(2x - 5y)$

17. yes; yes; Because a and b are divisible by c, you can factor c out of each expression. Because c is a factor of the sum and the difference, each expression is divisible by c.

19. $(x + 4)$ feet

Section 4.1 — Areas of Parallelograms
(pages 156 and 157)

1. The area of a polygon is the amount of surface it covers. The perimeter of a polygon is the distance around the polygon.

3. 18 ft^2 **5.** 187 km^2 **7.** 243 in.^{2}

9. 15 m was used for the height instead of 13 m.
    $A = 8(13) = 104$ m^2

11. 12 units^2 **13.** 24 units^2 **15.** 72 m^2

17. What shape could have an area of 128 square feet? What shape could have an area of s^2 square feet?

19. 287 in.^{2}

21. n^2bh where b represents the base and h represents the height of the original parallelogram, or n^2A where A represents the area of the original parallelogram.

23. 1640   **25.** 118

Section 4.2 — Areas of Triangles
(pages 162 and 163)

1. yes; To find the area of the triangle, you must also know the height of the triangle. That is, the perpendicular distance from the base to the opposite vertex.

3. 6 cm^2

5. 1620 in.^2

7. 1125 cm^2

9. The side length of 13 meters was used instead of the height.
$A = \dfrac{1}{2}(10)(12) = 60 \text{ m}^2$

11. 324 cm^2

13. 90 mi^2

15. Sample answer:

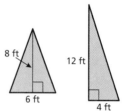

17. x^2 times greater

19. 4 times greater

21. Mult. Prop. of One

23. Assoc. Prop. of Add.

Section 4.3 — Areas of Trapezoids
(pages 170 and 171)

1. bases: 4 ft and 7 ft; height: 15 ft

3. $2\ell + 2w$; This is an expression for the perimeter of a rectangle. The other three are expressions for area (triangle, rectangle, and trapezoid).

5. 24 units^2

7. 28 in.^2

9. 105 ft^2

11. 8 units^2

13. 12 units^2

15. 60 in.^2

17. 78 mi^2

19. 18 ft

21. Use the strategy *Solve a Simpler Problem* by assigning values to the variables.

23 and 25.

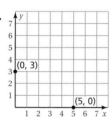

27. C

Extension 4.3 — Areas of Composite Figures
(page 173)

1. 36 units^2

3. 20 units^2

5. $126\dfrac{1}{2} \text{ cm}^2$

7. 3400 yd^2
Sample answer: Separate the figure into a trapezoid and a square.

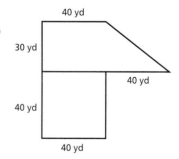

Section 4.4 — Polygons in the Coordinate Plane
(pages 178 and 179)

1. Plot the points that represent the vertices of the polygon and connect the points in order.

3.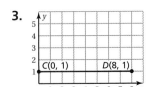
Length of CD is 8 units.

5.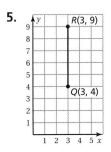
Length of QR is 5 units.

7.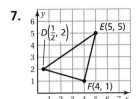

9.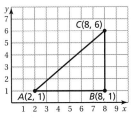

11.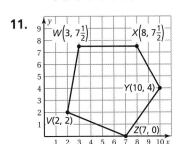

13. 24 units; 36 units^2

15. 28 units; 45 units^2

17. a. square
    b. 28 ft; 49 ft^2

19. Sample answer:
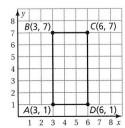

21. Sample answer:

23. 27 miles; There are only two ways to go from station P to station L. Traveling from station P to N to M to L is 27 miles. Traveling from station P to J to K to L is 33 miles.

25. 2.5 times larger   27. 2   29. $\frac{5}{16}$ 31. D

Section 5.1 — Ratios
(pages 194 and 195)

1. consonants; For every 5 vowels there are 7 consonants, so consonants outnumber the vowels.

3. 2 out of every 5; This ratio is 2 : 3, all other ratios are 2 : 5.

5.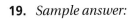

7. 6 to 4, or 6 : 4; For every 6 basketballs, there are 4 soccer balls.

9. 3 to 7, or 3 : 7; For every 3 shirts, there are 7 pants

11. 8 to 15, or 8 : 15; 8 out of 15 movies are comedies.

13. 15 to 3, or 15 : 3; Out of 15 movies, 3 are dramas.

15. 9 h

17. 12 : 16

19. 6 black pieces; The ratio of black to red is 3 : 5, so each part is 16 ÷ 8 = 2. So, there are 3 • 2 = 6 black pieces and 5 • 2 = 10 red pieces.

21. It may be helpful to organize your results in a table.

23. 4 pints of soda water, 8 pints of fruit punch concentrate, 20 pints of ginger ale; Yes; *Sample answer:* There is twice as much fruit punch as soda water (as in the original ratio). There is 5 times as much ginger ale as soda water (as in the original ratio).

25. 4.6

27. 2.53

29. B

Section 5.2 Ratio Tables
(pages 201–203)

1. Two ratios are equivalent if they can be written as the same ratio.

3. 12 : 15 does not belong because all other ratios are equivalent.

5. The ratio of ladybugs to bees can be described by 12 : 4, 6 : 2, or 3 : 1.

7.

Violins	8	24
Cellos	3	9

8 : 3 and 24 : 9

9.

Burgers	3	6	9
Hot Dogs	5	10	15

3 : 5, 6 : 10, and 9 : 15

11.

Forks	16	8	48
Spoons	10	5	30

16 : 10, 8 : 5, and 48 : 30

13.

You	3	6	9	12
Friend	4	8	12	16

16 tickets

15.

First	100	10	60
Second	60	6	36

$60

17. Adding the same number, 5 in this case, to each part of the ratio does not create equivalent ratios. You can add corresponding parts of equivalent ratios to create new equivalent ratios.

Sample answer:

A	3	6	9
B	7	14	21

19. 28 basketballs

21. Add the corresponding quantities of Recipes B and D to create Recipe E.

23. Subtract the corresponding quantities of Recipe B from Recipe C to create Recipe A.

25. *Sample answer:* Add the corresponding quantities of Recipes B and F to create a batch with 11 servings.

Section 5.5

Percents (continued)
(pages 222 and 223)

33. 81.25% **35.** 82.5%

37. Organize the percents and fractions in a table. What operation should you use to compare Illinois to Hawaii?

39. You can shade half of one of the squares.

41. $\frac{1}{2}$ **43.** 16 **45.** D

Section 5.6

Solving Percent Problems
(pages 229–231)

1. Twenty percent of what number is 30?; 6; 150

3–21. Explanations will vary.

3. 12 **5.** 35 **7.** 9 **9.** 12.5 **11.** 21

13. 20.25 **15.** 24 **17.** 14 **19.** 84 **21.** 94.5

23. The percent was not written as a fraction before multiplying; $40\% \times 75 = \frac{2}{5} \times 75 = 30$

25. 35.2 in.

27–35. Explanations will vary.

27. 140 **29.** 84 **31.** 80 **33.** 25 **35.** 20

37. The percent should be written as a fraction before dividing; $5 \div 20\% = 5 \div \frac{1}{5} = 25$

39. a. 50 students **b.** 18 students **41.** 21 cars

43. = **45.** > **47.** 48 min

49. a. 432 in.^{2}

  b. 37.5%; Because the length is doubled, the width of the rectangle is now half of 75% of its length, or 37.5%.

51. *Sample answer:* Because 30% of *n* is equal to 2 times 15% of *n* and 45% of *n* is equal to 3 times 15% of *n*, you can write 30% of $n = 2 \times 12 = 24$ and 45% of $n = 3 \times 12 = 36$.

53. 97.2% **55.** 16.5 **57.** 26.28

Section 5.7

Converting Measures
(pages 236 and 237)

1. yes; Because 1 centimeter is equal to 10 millimeters, the conversion factor equals 1.

3. Find the number of inches in 5 cm; 5 cm ≈ 1.97 in.; 5 in. = 12.7 cm

5. person weighing 75 kg; 75 kg ≈ 166.67 lb and 166.67 lb > 110 lb

7. 1.5 **9.** 12.63 **11.** 1.22 **13.** 0.19 **15.** 37.78 **17.** 14.49

19. a. about 60.67   **b.** about 8.04 km

21. <   **23.** >   **25.** >   **27.** 1320   **29.** 111.8   **31.** 0.001

33. When using conversion factors, make sure your units cancel.

35. about 669,600,000 mi/h

37. 30   **39.** 18   **41.** C

Section 6.1 Integers
(pages 252 and 253)

1. 8, −9, 22

3. *Sample answer:* below, under, lose

5.

7. number line with point at 15 (from 11 to 19)

9. 37,500   **11.** −56   **13.** 5   **15.** −318

17. number line with points at −8 and 8 (from −8 to 8)

19. number line with points at −9 and 9 (from −12 to 12)

21. number line with points at −150 and 100

23. number line with points at −400 and 400

25.

27. −8   **29.** 18

31. a. *Sample answer:* Choosing 8, the opposite is −8.
  b. *Sample answer:* 8
  c. The opposite of the opposite of an integer is the integer.; Yes; *Sample answer:*

  Case 1: 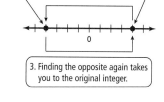   Case 2: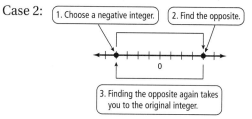

  Case 3: Choose 0. The opposite of 0 is 0, so the opposite of the opposite of 0 is 0.
  d. −(−(−6)) is the opposite of the opposite of −6; −(−(−6)) = −6

33. $\dfrac{3}{8}, \dfrac{1}{2}, \dfrac{3}{4}, \dfrac{7}{8}$ **35.** B

Section 6.2 Comparing and Ordering Integers
(pages 258 and 259)

1. On a number line, numbers to the left are less than numbers to the right. Numbers to the right are greater than numbers to the left.

3. The value of *a* is less than the value of *b* because *a* is to the left of *b*.

5. <   **7.** >   **9.** >   **11.** >

Section 6.2

Comparing and Ordering Integers *(continued)*
(pages 258 and 259)

13. The explanation about where the integers are located on a number line is incorrect; $-7 < -3$; So, -7 is to the left of -3 on a number line.

15. $-4, -3, -2, 1, 2$ **17.** $-7, -4, 2, 3, 6$ **19.** $-20, -10, -5, 15, 25$ **21.** oxygen

23. always; The opposite of a positive integer is a negative integer. Positive integers are greater than negative integers.

25. a. Florida, Louisiana, Arkansas, Tennessee, California
   b. California, Louisiana, Florida, Arkansas, Tennessee
   c. An elevation of 0 feet represents sea level.

27. no; In order for the median to be below 0°F, at least 6 of the temperatures must be below 0°F.

29.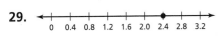

31.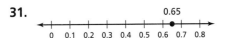

33. B

Section 6.3

Fractions and Decimals on the Number Line
(pages 264 and 265)

1. a **3.** -2.6 **5.** *Sample answer:* $-2\frac{1}{4}$

7.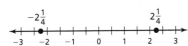

9.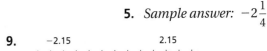

11. $<$ **13.** $<$ **15.** $>$ **17.** $>$

19. the larger sand dollar

21. $-1, -\frac{3}{4}, -\frac{5}{8}, -\frac{1}{20}, 0$

23. $-5, -4.9, -4.35, -4.3, -4$

25. Write the numbers as decimals instead of finding a common denominator.

27. 1, 2, and any integer less than -3

29.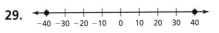

31. (number line from -20 to 20 with points at -15 and 10)

Section 6.4

Absolute Value
(pages 272 and 273)

1. Find the distance between the number and 0 on a number line.

3. What integer is 3 units to the left of 0?; -3; 3

7. 2 **9.** 8.35

11. $3\frac{2}{5}$ **13.** 14.06

5.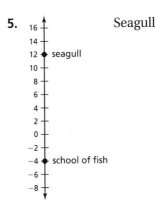

15. −10, 10   **17.** <   **19.** >   **21.** <

23. Scientist B   **25.** A; You owe more than $25, so debt > $25.

27. −2, 0, $|-1|$, $|4|$, 5   **29.** −11, 0, $|3|$, $|-6|$, 9, 10

31. 0   **33.** −1

35. sometimes; If the number is negative then its absolute value is greater, but if the number is positive or zero then it is equal to its absolute value.

37. never; The absolute value of a positive number is the number itself.

39. *Sample answer:* $x = -2$, $y = -3$

41. 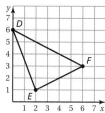   **43.**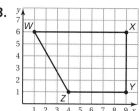

Section 6.5 — The Coordinate Plane
(pages 279–281)

1. 4

3. (2, −3); (2, −3) is in Quadrant IV. The other three points are in Quadrant II.

5. (3, 1)   **7.** (−2, 4)   **9.** (2, −2)   **11.** (−4, 2)   **13.** (4, 0)

15–21. See graph below.   **15.** Quadrant I

17. *y*-axis

19. Quadrant IV

21. *x*-axis

23. The numbers are reversed. To plot (4, 5), start at (0, 0) and move 4 units right and 5 units up.

25. 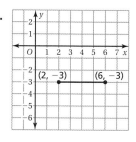 4   **27.** 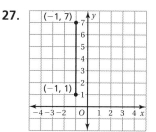 6   **29.** 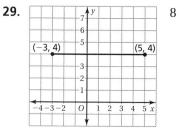 8

31. (−2, 1)

Section 6.5 The Coordinate Plane (continued)
(pages 279–281)

33.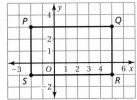

22 units; 28 units^2

35. a. about 142,000

b. 2011 and 2012

c. about 10,000

37. a.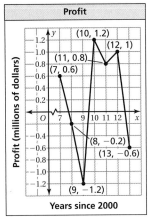

b. *Sample answer:* There are four years where the profit is positive and three years where the profit is negative. The profit decreased from 2007 to 2009. The profit increased the most from 2009 to 2010.

c. $1.6 million

d. Because the *x*-axis represents the number of years since 2000, then 0 represents 2000. So, you could graph the profits for 1990 to 2006 by using *x*-values from −10 to 6.

39. Quadrant III

41. Quadrant I or Quadrant IV

43. origin

45. never; All points in Quadrant III have negative *y*-coordinates.

47. Reptiles

49. no; Quadrants III or IV

51. Because the rain forest is in Quadrant IV, the *x*-coordinate of the point will be positive and the *y*-coordinate of the point will be negative.

53. *Sample answer:* (−6, 3), (−2, 3), (−2, −9), (2, −9)

55. $y − 4$ **57.** $x + 9$ **59.** C

Extension 6.5 Reflecting Points in the Coordinate Plane
(page 283)

1. a. (3, −2)   **3. a.** (−5, 6)   **5. a.** (0, 1)   **7. a.** (2.5, −4.5)

  b. (−3, 2)   **b.** (5, −6)   **b.** (0, −1)   **b.** (−2.5, 4.5)

9. (−4, −5)   **11.** (2, 2)   **13.** (3, 9); 18 units

15. a. (−4, −5); (1, −7); (2, 2); (−6.5, 10.5); yes; *Sample answer:* The order of the reflections does not matter. You are still reflecting the points in both axes.

  b. *Sample answer:* Use the opposite of each coordinate.

Section 7.1 Writing Equations in One Variable (pages 298 and 299)

1. An equation has an equal sign and an expression does not.
3. *Sample answer:* A number *n* subtracted from 28 is 5.
5. What was the high temperature if it was 4° less than 62°F? 58°F
7. $y - 9 = 8$
9. $w \div 5 = 6$
11. $5 = \frac{1}{4}c$
13. $n - 9 = 27$
15. $6042 = 1780 + a$
17. $16 = 3x$
19. $326 = 12(14) + 6(5) + 16x$
21. It might be helpful to organize the given information visually.
23. 13
25. 28
27. B

Section 7.2 Solving Equations Using Addition or Subtraction (pages 305–307)

1. Substitute your solution back into the original equation and see if you obtain a true statement.
3. subtraction
5. so that *x* is by itself; so that the two sides remain equal
7. yes
9. no
11. yes
13. $t = 1$
15. What number plus 5 equals 12?; $a = 7$
17. 20 is what number minus 6?; $d = 26$
19. $z = 16$
21. $p = 3$
23. $h = 34$
25. $q = 11$
27. $x = \frac{7}{30}$
29. $a = 11.8$
31. They must apply the same operations to both sides.

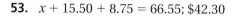

33. $x - 8 = 16$; 24th floor
35. Subtract 3 from each side. (Subtraction Property of Equality); Subtract 3 from 3. (Subtract.); Add *x* and 0. (Addition Property of Zero)
37. $k + 7 = 34$; $k = 27$
39. $93 = g + 58$; $g = 35$
41. $y = 15$
43. $v = 28$
45. $d = 54$
47. $x + 34 + 34 + 16 = 132$; 48 in.
49. Addition is commutative.
51. Begin by writing the characteristics of each problem.
53. $x + 15.50 + 8.75 = 66.55$; $42.30
55. **a.** $5.25
    b. no; You have $5.25 left and it costs $9.75 to ride each ride once.

57. 96
59. 5
61. C

Selected Answers **A31**

Section 7.3

Solving Equations Using Multiplication or Division (pages 312 and 313)

1. 12
3. $\dfrac{4x}{4} = \dfrac{24}{4}$
5. $8 \cdot 3 = (n \div 3) \cdot 3$
7. $s = 70$
9. $x = 24$
11. $a = 4$
13. $y = 10$
15. $x = 15$
17. $d = 78$
19. $c = 66$
21. $n = 2.56$
23. They should have multiplied by 4.
    $x \div 4 = 28$
    $(x \div 4) \cdot 4 = 28 \cdot 4$
    $x = 112$
25. $3x = 45$; 15 items
27. 9 units
29. 8 units
31. 20 cards
33. $x = 6$; Because $5x$ is on both sides of the equation, $3x$ must be equal to 18 so that the equation is true.
35. length: 20 in.; width: 5 in.
37. $\dfrac{t}{3} = 7$

Section 7.4

Writing Equations in Two Variables (pages 319–321)

1. *Sample answer:* An independent variable can change freely. A dependent variable depends on the independent variable.
3. $n = 4n - 6$; This one is not an equation in two variables.
5. $A = 9h$, where A is the area in square feet and h is the height in feet; A depends on h.
7. yes
9. no
11. yes
13. w is independent and A is dependent.
15. p is independent and t is dependent.
17. $270

19 and 21. Sample answers are given.

	Independent Variable	Dependent Variable
19.	The speed you are pedaling a bike	Time it takes to stop your bike
21.	The number of years of education	The amount of money you earn

23. *Sample answer:* $c = 25m + 35$ where m is the number of months and c is the total cost of the gym membership.

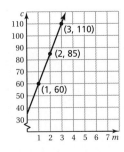

25. Methods to create the graph will vary.

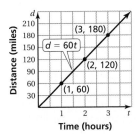

A32 Selected Answers

27. $d = \dfrac{5}{6}t$;

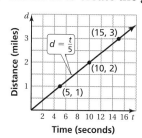

29. $d = 240t$;

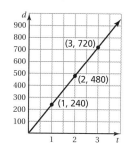

31. 1

33. no; By definition, the independent variable can change freely.

35. 50 city blocks

37. Methods to create the graph will vary.

39. a. no; It does not make sense to draw a line between the points to show the solutions because you cannot sell part of a ticket.

b. $c = 10n$

41. 80%

43. 68%

Section 7.5 Writing and Graphing Inequalities
(pages 329–331)

1. Both phrases refer to numbers that are greater than a given number. The difference is that "greater than or equal to" includes the number itself, whereas "greater than" does not.

3. The graph of $x \leq 6$ has a closed circle at 6. The graph of $x < 6$ has an open circle at 6.

5. $k < 10$

7. $z < \dfrac{3}{4}$

9. $1 + y \leq -13$

11. yes   **13.** yes   **15.** no   **17.** B   **19.** D

21. $x < 1$; A number x is less than 1.

23. $x \geq -4$; A number x is at least -4.

25.–**35.** *(number line graphs)*

37. $x \geq 1$ means that 1 is also a solution, so a closed circle should be used.

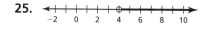

39. a. $b \leq 3$;

b. $\ell \geq 18$;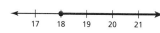

41. The cost of the necklace and another item should be less than or equal to $33.

43. sometimes; The only time this is not true is if $x = 5$.

45. $p \leq 375$

47. $x = 9$

49. $x = 28$

51. D

Section 7.6 Solving Inequalities Using Addition or Subtraction (pages 336 and 337)

1. Sample answer: $x + 7 \geq 143$

3. By solving the inequality to obtain $x \leq 1$, the graph has a closed circle at 1 and an arrow pointing in the negative direction.

5. $x < 9$;

7. $5 \geq y$;

9. $6 > x$;

11. $y < 106$;

13. $3 < x$;

15. $\frac{1}{4} \leq n$;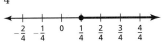

17. To solve the inequality, 9 should be added to both sides, not subtracted.

    $28 \geq t - 9$
    $\underline{+9 \quad +9}$
    $37 \geq t$

19. $x + 18.99 \leq 24$; $x \leq \$5.01$

21. $x - 3 > 15$; $x > 18$

23. $11 > s$;

25. $34{,}280 + d + 1000 > 36{,}480$; $d > 1200$ dragonflies

27. The estimate for running a mile should be greater than 4 minutes, because the world record is under 4 minutes.

29. $t = 48$

31. $x = 9$

33. C

Section 7.7 Solving Inequalities Using Multiplication or Division (pages 342 and 343)

1. The solution of $2x \geq 10$ includes the solution of $2x = 10$, $x = 5$, and all other x values that are greater than 5.

3. Div. Prop. of Ineq.

5. Sample answer: $\frac{x}{2} \geq 4$, $2x \geq 16$

7. $n > 12$;

9. $c \geq 99$;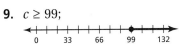

11. $x \geq 5$;

13. $p \leq 6$;

15. $x < 15$;

17. $v \leq 81$;

19. $w \leq 32$;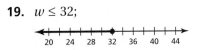

21. $x \geq 48$;

23. $8x < 168$; $x < 21$ ft

25. $8n < 72$; $n < 9$

27. $225 \geq 12w$; $18.75 \geq w$

29.

31. $80x > 2 \cdot 272$; $x > 6.8$ yards per play

33. Sample answer: the number of gallons of milk you can buy with $20; the length of a park that has an area of at least 500 square feet

A34 Selected Answers

35. yes; $a > b$ and $x > y$

37. yes; $a > b$ and $x > y$

39. rectangle

41. parallelogram

Section 8.1 Three-Dimensional Figures
(pages 358 and 359)

1. false; It has two triangular faces.

3. true

5. false; Some are perpendicular and some are neither (skew).

7. front: 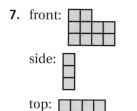 10 cubes
side:
top:

9. front: 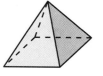 9 cubes
side:
top:

11. 10 faces, 24 edges, and 16 vertices

13.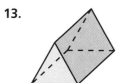

15.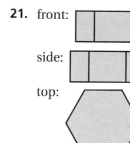

17. front: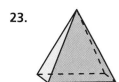
side:
top:

19. front: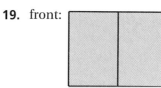
side:
top:

21. front:
side:
top:

23.

25.

27. *Answer should include, but is not limited to:* an original drawing of a house; a description of any solids that make up any part of the house

29. *Sample answer:*

a.

Triangular prism
6 vertices
9 edges

Square pyramid
5 vertices
8 edges

b. More than one solid can have the same number of faces, so knowing the number of edges and vertices can help you draw the intended solid.

31. 12 cm^2

33. D

Section 8.2 Surface Areas of Prisms
(pages 364 and 365)

1. Find the sum of the areas of the faces.

3. 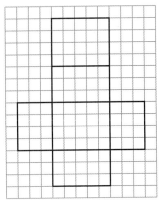 94 units²

5. 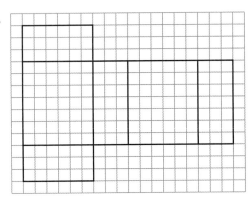 162 units²

7. 198 cm²
9. 17.6 ft²
11. 57.1 mm²
13. 136 ft²
15. Draw diagrams of the given information.
17. 364 ft²
19. 165 ft²
21. C

Section 8.3 Surface Areas of Pyramids
(pages 372 and 373)

1. Find the sum of the areas of the faces.

3.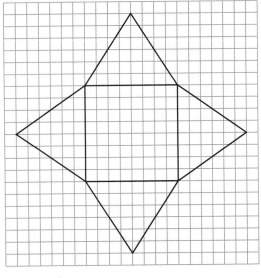
160 units²

5.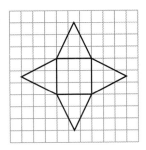
27 units²

7. 172.8 yd²
9. 224.4 ft²
11. 55 m²
13. 21,274.4 ft²
15. 4
17. no; You can place the four triangles on top of the square and it covers the entire square. But when you lift up the triangles, they do not touch. So, they do not form a pyramid.

19.

Apples	10	5	30
Oranges	4	2	12

10 : 4, 5 : 2, 30 : 12

A36 Selected Answers

Section 8.4 Volumes of Rectangular Prisms
(pages 378 and 379)

1. The volume of an object is the amount of space it occupies. The surface area of an object is the sum of the areas of all of its faces.

3. How much does it take to cover the rectangular prism?; 310 cm^2; 350 cm^3

5. $1\frac{5}{16}$ cm^3

7. $\frac{15}{16}$ m^3

9. $12\frac{1}{2}$ m^3

11. $220.5 = 7 \cdot w \cdot 7$; 4.5 cm

13. Use unit cubes to visualize filling the fish tank.

15. 1728 1-inch cubes; There are 1728 1-inch cubes in a cube with a side length of 1 foot. The area of the cube with a side length of 1 foot is 1 cubic foot, or 1728 cubic inches. So, 1 cubic foot is equal to 1728 cubic inches. You can use the conversion factors $\frac{1728 \text{ in.}^3}{1 \text{ ft}^3}$ and $\frac{1 \text{ ft}^3}{1728 \text{ in.}^3}$ to convert between cubic inches and cubic feet.

17. 1152 cm^3

19. yes

21. no

Section 9.1 Introduction to Statistics
(pages 394 and 395)

1. A statistical question is one for which you do not expect to get a single answer. Instead, you expect a variety of answers, and you are interested in the distribution and tendency of those answers. *Sample answer:* How old are the teachers in middle school?

3. yes; There are many different answers.

5. *Sample answer:* 2 pets; no

7. 100 senators; yes

9. not statistical; There is only one answer.

11. statistical; There are many different answers.

13.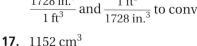

Most of the registrations are in a cluster from 21 to 26. The peak is 25. There is a gap between 16 and 21.

15.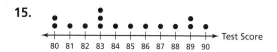

The test scores are spread out pretty evenly with no clusters or gaps. The peak is 83.

17. **a.** 21 earthworms

   b. *Sample answer:* Use a centimeter ruler. The units are centimeters.

   c. *Sample answer:* "What is the length of an earthworm?"; The lengths are spread out pretty evenly from 15 centimeters to 28 centimeters.

19. *Sample answer:*
    Anemometer; miles per hour

21. *Sample answer:*
    Richter scale; magnitude

23. *Sample answer:* 65 mi/h; Most of the data cluster around 65, and 65 miles per hour is a common speed limit.

25. Does changing the order of the bars in the bar graph affect the distribution?

27. no

29. D

Section 9.2 — Mean
(pages 400 and 401)

1. Add the data values then divide by the number of data values.

3. yes; Because of the variability of the answers to a statistical question, the mean gives an average of the answers. That way, you can use only one value, the mean, to answer the statistical question.

5. 1 movie seen this week; Find the total number of movies and divide by the number of people.

7. 3 brothers and sisters

9. 16 visits

11. **a.** yes; There will be variability in the lengths of the commercial breaks.

    b. 3.45 minutes

13. *Sample answer:* 20, 21, 21, 21, 21, 22; 20, 20.5, 20.5, 21.5, 21.5, 22

15. 3.9 inches; No, neither team has a height that is much shorter or taller than the other heights. So, you can say that the Tigers are taller than the Dolphins on average.

17. There are 5 different allowance amounts and 24 students. Each amount is used more than once.

19. 9

21. 18.5

23. B

Section 9.3 — Measures of Center
(pages 407–409)

1. *Sample answer:* 1, 2, 3, 4, 5, 6

3. outlier; The other three are measures of center.

5. 5.5

7. median: 7; mode: 3

9. median: 92.5; mode: 94

11. median: 17; mode: 12

13. The data were not ordered from least to greatest; The median is 55.
    49, 50, 51, 55, 58, 59, 63

15. singing

17. mean: 35.875; median: 44; mode: 48
    Sample answer: The median is probably best, because it is close to most of the data. The mean is less than most of the data and the mode is the greatest value.

19. mean: 12; median: 8; mode: 2
    Sample answer: The median is the best measure, because the mean is greater than most of the data and the mode is the least value.

21. With Outlier Without Outlier
    mean: 48.5   mean: 53
    median: 53   median: 54
    mode: none   mode: none
    The outlier reduces the median slightly, but reduces the mean more. There is no mode with or without the outlier.

23. mean: 7.61; median: 7.42; no mode

25. a. mean: 94°F; median: 91°F; mode: 91°F
 Sample answer: Both the median and mode are the best measures for the data, because both are very close to most of the values.

b. mean: 77°F; median: 77°F; modes: 77°F and 78°F
 Sample answer: Both the mean and median are the best measures for the data, because there are two modes.

27. 10 hours; Using the mean as the average, you would need to work 12 hours. Using the median as the average, you would need to work 10 or more hours. Using the mode as the average, you would need to work 10 hours. So, the minimum number of hours is 10 and you can use the median or mode to justify your answer.

29. Ordering the data makes it easier to find the median and the mode.

31. a. mean: $1794; median: $1790; mode: $1940

b. mean: $1883.70; median: $1879.50; mode: $2037
 The mean, median, and mode all increased by 5%.

c. annual salaries: $23,280, $19,920, $22,320, $25,200, $20,640, $18,480, $21,120, $23,280, $21,840, $19,200; mean: $21,528; median: $21,480; mode: $23,280; They are 12 times the mean, median, and mode of the monthly salary.

33. 13   **35.** 119   **37.** D

Section 9.4 — Measures of Variation
(pages 416 and 417)

1. A measure of center represents the center of a data set, but a measure of variation describes the distribution of a data set.

3. What is the range of the data?; 20; 12

5. median = 81.5; median of lower half = 67; median of upper half = 92; The data is spread out.

7. 23   **9.** 7.3

11. median = 37; Q_1 = 33.5; Q_3 = 40.5; IQR = 7   **13.** median = 133.5; Q_1 = 128; Q_3 = 139; IQR = 11

15. range = $21\frac{3}{4}$ ft; The distances traveled by the paper airplane vary by no more than $21\frac{3}{4}$ feet; IQR = 11 ft; The middle half of the distances traveled by the paper airplane vary by no more than 11 feet.

17. Exercise 11: 54
 Exercise 12: none
 Exercise 13: 106 and 158
 Exercise 14: 38

19. a. range = 172 points; IQR = 42 points

b. The outlier is 193 points; range = 101; IQR = 34; range

Section 10.2 Histograms
(pages 445–447)

1. The *Test Scores* graph is a histogram because the number of students (frequency) achieving the test scores are shown in intervals of the same size (20).

3. No bar is shown on that interval.

5. *Sample answer:*

Interval	Tally	Total
20–29	I	1
30–39	III	3
40–49	JHT II	7
50–59	IIII	4

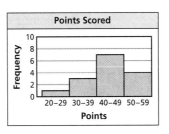

7.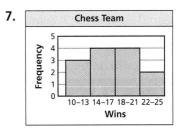

9. There should not be space between the bars of the histogram.

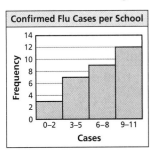

11. The frequency is the number of songs not the percent of songs. The statement should be "12 of the songs took 5–8 seconds to download."

13. Pennsylvania; You can see from the intervals and frequencies that Pennsylvania counties are greater in area, which makes up for it having fewer counties.

15. **a.** yes; The stem-and-leaf plot shows that 10 pounds is a data value.

    b. no; Both displays show that 11 residents produced between 20 and 29 pounds of garbage.

17. Begin by ordering the data.

19. 45

21. 22.4

23. D

Section 10.3 Shapes of Distributions
(pages 454 and 455)

1. The shape of a skewed distribution will have a tail on one side. The shape of a symmetric distribution is when the data on the left are a mirror image of the data on the right.

3.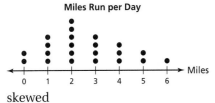
skewed

5. skewed left

7. skewed right

A42 Selected Answers

9.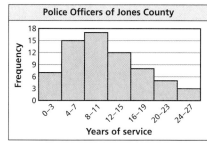

skewed right   symmetric

Jones County; The distribution of Jones County is skewed right, so most of the data values are on the left.

11. no; Distributions can have any shape.

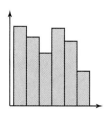

13. a. 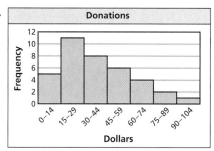 skewed right

b.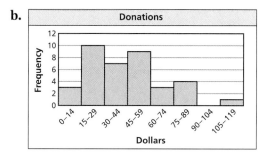

Both distributions are skewed right. The original donation distribution is more skewed right than the distribution when the increases are added to the donations. Some of the data values moved into different intervals when $5 is added to each donation, which is why the distributions are not exactly the same.

15. median = 70; Q_1 = 65.5; Q_3 = 75; IQR = 9.5   **17.** A

Extension 10.3 Choosing Appropriate Measures
(page 457)

1. median and interquartile range; median: $32; IQR: $6

3. no; You do not know the actual values in the data set. You can approximate the mean and MAD but your answers will not be exact.

5. *Sample answer:*

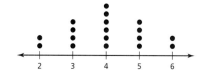

Section 10.4
Box-and-Whisker Plots
(pages 463–465)

1. Order the data. The first number is the *least value* and the last number is the *greatest value*. The middle value is the *median*. The middle value of the lower half of the data is the *first quartile*. The middle value of the upper half of the data is the *third quartile*.

3. Is the distribution skewed right?; yes; no

5.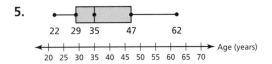

7.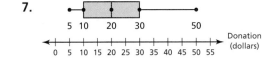

9. The data should be ordered before finding the five-number summary.

11. **a.** about $\frac{1}{2}$

    b. The right whisker is longer than the left whisker. So the data are more spread out above the third quartile than below the first quartile.

    c. 150; The middle half of the data varies by no more than 150 gallons.

13. skewed left; The left whisker is longer than the right whisker, and most of the data are on the right.

15. symmetric; The whiskers are about the same length, and the median is in the middle of the box.

17. **a.** School 1 is skewed left and School 2 is skewed right.

    b. School 2; The range for School 2 is a half hour greater than the range for School 1. Also, the IQR of School 2 is greater than the IQR of School 1.

    c. School 1; School 1 has more data on the left than School 2. So, School 1 is more likely to have recess before lunch.

19.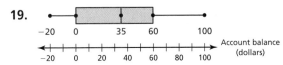

21. Use the median to describe the center and the interquartile range to describe the variation.

23. When the least value and the first quartile are equal, there is no whisker on the left. When the greatest value and the third quartile are equal, there is no whisker on the right.

25. **a.** Team 1; There is less variability in the data.

    b. 24 games

    c. Team 1; In 75% of the games, Team 1 scored 6 runs or more. However, Team 2 scored 6 runs or less in 75% of the games.

    d. Team 1; *Sample answer:* By looking at the shapes of the distributions, you can see that the majority of the data for Team 1 is greater than the majority of the data for Team 2.

27. >

29. >

Key Vocabulary Index

Mathematical terms are best understood when you see them used and defined *in context*. This index lists where you will find key vocabulary. A full glossary is available in your Record and Practice Journal and at *BigIdeasMath.com*.

absolute value, 270
algebraic expression, 112
base, 12
box-and-whisker plot, 460
coefficient, 112
common factors, 32
common multiples, 38
composite figure, 172
constant, 112
conversion factor, 234
coordinate plane, 276
dependent variable, 316
edge, 356
equation, 296
equation in two variables, 316
equivalent expressions, 128
equivalent rates, 206
equivalent ratios, 198
evaluate, 18
exponent, 12
face, 356
factor pair, 26
factor tree, 26
factoring an expression, 140
first quartile, 414
five-number summary, 460
frequency, 441
frequency table, 441
graph of an inequality, 328
greatest common factor, 32
histogram, 442
independent variable, 316
inequality, 326
integers, 250
interquartile range, 414
inverse operations, 303
leaf, 436
least common denominator, 42
least common multiple, 38
like terms, 136
mean, 398
mean absolute deviation, 420
measure of center, 404
measure of variation, 414
median, 404

metric system, 234
mode, 404
negative numbers, 250
net, 362
numerical expression, 18
opposites, 250
order of operations, 18
origin, 276
outlier, 399
percent, 220
perfect square, 13
polygon, 152
polyhedron, 356
positive numbers, 250
power, 12
prime factorization, 26
prism, 356
pyramid, 356
quadrants, 276
quartiles, 414
range, 414
rate, 206
ratio, 192
ratio table, 198
reciprocals, 64
solid, 356
solution, 302
solution of an equation in two variables, 316
solution of an inequality, 327
solution set, 327
statistical question, 392
statistics, 392
stem, 436
stem-and-leaf plot, 436
surface area, 362
terms, 112
third quartile, 414
unit analysis, 234
unit rate, 206
U.S. customary system, 234
variable, 112
Venn diagram, 30
vertex, 356
volume, 374

Student Index

This student-friendly index will help you find vocabulary, key ideas, and concepts. It is easily accessible and designed to be a reference for you whether you are looking for a definition, real-life application, or help with avoiding common errors.

A

Absolute value, 268–273
   defined, 270
   error analysis, 272
   real-life application, 271
Addition
   Associative Property of, 127–131
   Commutative Property of, 126–131
   of decimals, 78–83
      error analysis, 82
      real-life application, 81
      writing, 82
   equations
      error analysis, 305
      real-life application, 304
      solving by, 300–307
   of fractions, 42–43
   inequalities
      solving by, 332–337
   Property
      of Equality, 303
      error analysis, 130
      of Inequality, 334
      real-life application, 129
      of Zero, 129
   whole numbers, 4
Addition Property of Equality, 303
   real-life application, 304
Addition Property of Inequality, 334
Addition Property of Zero, 129
Algebra
   equations
      in one variable, 294–299
      solving, 300–313
      in two variables, 314–321
   expressions
      simplifying, 135–139
      writing, 110–123
   inequalities
      graphing, 324–331
      solving, 332–343
      writing, 326
Algebraic expression(s), 110–117
   defined, 112
   error analysis, 115, 116

evaluating
   with two operations, 114
   with two variables, 113
factoring, 141
like terms
   combining, 136
   defined, 136
real-life application, 114
simplifying, 135
terms
   defined, 112
   like, 136
writing, 112–113
Area
   of a composite figure, 172–173
      real-life application, 173
   of a parallelogram, 152–157
      error analysis, 156
      formula, 154
      real-life application, 155
   of a polygon
      real-life application, 177
   of a trapezoid, 166–171
      error analysis, 170
      formula, 168
      real-life application, 169
      writing, 167
   of a triangle, 158–163
      error analysis, 162
      formula, 160
      real-life application, 161
      research, 163
      writing, 163
Associative Property of Addition, 126–131
   error analysis, 130
Associative Property of Multiplication, 126–131

B

Base, defined, 12
Base ten blocks for modeling
   decimal addition, 78
   decimal division, 92–93
   decimal multiplication, 84
   decimal subtraction, 78

Box-and-whisker plot(s), 458–465
   defined, 460
   error analysis, 463
   five-number summary, 460

C

Choose Tools, *Throughout. For example, see:*
   data displays
      choosing, 465
      measures of center, 465
      variation of distribution, 465
   decimals
      adding, 82
      ordering, 261
      subtracting, 82
   equations in two variables, 315, 320
   parallelograms, area of, 156
   ratios, graphing, 215
   ratio tables, 215
Coefficient, defined, 112
Common Error
   polygons in the coordinate plane, 177
   reflecting points in the coordinate plane, 283
   writing expressions, 120, 129
Common factors, defined, 32
Common multiple(s), 36–41
   defined, 38
Commutative Property of Addition, 126–131
   real-life application, 129
Commutative Property of Multiplication, 126–131
Comparing
   integers, 254–259
   ratios, 210–215
Composite figure(s)
   area of, 172–173
      real-life application, 173
   defined, 172
Concept circle, 410
Connections to math strands
   Algebra, 116, 117, 138

Geometry, 91, 99, 130, 139, 141, 195, 231, 283, 299, 306, 342, 373
Constant, defined, 112
Conversion factor, defined, 234
Coordinate plane
  defined, 276
  distances in, 277
  finding distances, 176
  graphing in, 274–283
    error analysis, 279
  ordered pairs, 276–277
  origin, 276
  polygons in the, 174–179
    error analysis, 178
    real-life application, 177
    writing, 178
  quadrant, 276
  real-life application, 278
  reflecting points in, 282–283
Critical Thinking, *Throughout. For example, see:*
  absolute value, 273
  area of a polygon, 163, 171
  data displays
    box-and-whisker plots, 465
    histograms, 445, 447
    shapes of distribution, 455
    stem-and-leaf plots, 439
  decimals, 83
  division, 9
  equations, solving, 307
  greatest common factor, 35
  inequalities
    graphing, 331
    solving, 343
  least common multiples, 41
  mean absolute deviation, 423
  measures of center, 409
  percents, 223
    modeling, 223
  prime numbers, 29
  rectangular prism, volume and surface area, 378
  statistical questions, 394
  surface area
    of a prism, 365
    of a pyramid, 373
  trapezoids, 171
  triangles, 162, 163, 171
  unit conversion, 237
Customary system, *See* U.S. customary system

D

Data analysis, *See* Data displays; Statistics
Data displays, *See also* Graphs and graphing
  box-and-whisker plots, 458–465
    defined, 460
    error analysis, 463
    five-number summary, 460
    shapes of, 462
  comparing, 444
  dot plots, 392–395
  frequency table, 441
  histograms, 440–447
    defined, 442
    error analysis, 445, 446
    modeling, 441
  shapes of distribution, 450–457
    appropriate measures of center and variation, 456–457
    box-and-whisker plots, 462
    skewed left, 452
    skewed right, 452
    symmetric, 452
  stem-and-leaf plots, 434–439
    defined, 436
    error analysis, 438
    leaf, 436
    stem, 436
    writing, 438
Decimal(s)
  adding, 78–83
    error analysis, 82
    modeling, 78
    real-life application, 81
    writing, 82
  dividing, 92–99
    by decimals, 95
    error analysis, 97, 98
    modeling, 92–93
    real-life application, 96
    by whole numbers, 94
    writing, 93
  graphing, 262–263
  multiplying, 84–91
    by decimals, 87
    error analysis, 89, 90
    modeling, 84–85
    real-life application, 88
    by whole numbers, 86

  ordering, 260–265
    negative, 262
  subtracting, 78–73
    error analysis, 82
    modeling, 78–79
    real-life application, 81
    writing, 82
Definition and example chart, 216
Denominator(s), least common, 42
Dependent variable, defined, 316
Diagram(s)
  double number line, 205
  modeling, 191
  tape, 191, 193
  Venn, 30–31, 36, 37
    defined, 30
Different Words, Same Question, *Throughout. For example, see:*
  absolute value, 272
  box-and-whisker plots, 463
  converting units, 236
  fractions
    dividing, 74
    multiplying, 74
  greatest common factor, 34
  inequalities, 329
  measures of variation, 416
  percents, 229
  surface area of a prism, 364
  triangles, 162
  unit rates, 208
  volume of a prism, 378
  writing equations, 298
  writing expressions, 122
Distance formula, 318
Distributive Property, 132–141
  error analysis, 137, 138
  modeling, 132
  real-life application, 135
  writing, 137
Division
  of decimals, 92–99
    by decimals, 95
    error analysis, 97, 98
    modeling, 92–93
    real-life application, 96
    by whole numbers, 94
    writing, 93
  equations
    real-life application, 311
    solving by, 308–313

of fractions, 62–69
   error analysis, 67
   by a fraction, 65
   modeling, 63
   reciprocals, 64
   by a whole number, 66
inequalities
   solving by, 338–343
of mixed numbers, 70–75
   error analysis, 74
   modeling, 70–71
   real-life application, 73
Property
   of Equality, 311
   of Inequality, 340
whole numbers, 4–6
   error analysis, 8
   by a fraction, 65
   real-life application, 6
Division Property of Equality, 311
   error analysis, 312
   real-life application, 311
   writing, 312
Division Property of Inequality, 340
   real-life application, 341
Dot plot(s), 392–395

E

Edge, defined, 356
Equality
   Addition Property of, 303
   Multiplication Property of, 310
   Subtraction Property of, 303
Equation(s)
   defined, 296
   inverse operations, 303
   in one variable
     error analysis, 298
     real-life application, 297
     writing, 294–299
   solution, 302
   solving
     by addition, 300–307
     by division, 308–313
     error analysis, 305, 312
     by multiplication, 308–313
     real-life application, 304, 311
     by subtraction, 300–307
     writing, 305, 312
Equation in two variables, 314–321
   defined, 316
   dependent variable, 316
   error analysis, 319

   graphing, 317
   independent variable, 316
   real-life application, 318
   solution of, 316
   writing, 314–321
Equivalent expressions
   defined, 128
   identifying, 140
   using properties to write,
     128–129
Equivalent rate(s), 205–209
   defined, 206
   finding, 207
Equivalent ratios
   defined, 198
   error analysis, 202
Error Analysis, *Throughout. For
   example, see:*
   absolute value, 272
   area
     of a parallelogram, 156
     of a trapezoid, 170
     of a triangle, 162
   Associative Property of Addition,
     130
   converting units, 236
   coordinate plane, 279
   data displays
     box-and-whisker plots, 463
     histograms, 445, 446
     stem-and-leaf plots, 438
   decimals
     adding, 82
     dividing, 97, 98
     multiplying, 89, 90
     subtracting, 82
   Distributive Property, 137, 138
   equations
     in one variable, 298
     solving, 305, 312
     in two variables, 319
   exponents, 14
   expressions
     algebraic, 115, 116
     writing, 122
   fractions
     dividing, 67
     multiplying, 59, 60
     writing, 222
   greatest common factor, 34
   inequalities
     graphing, 330
     solving, 336, 342

   integers
     comparing, 258
     positive, 252
   mixed numbers
     dividing, 74
     multiplying, 60
   order of operations, 20
   percents, 222
     finding the whole, 230
     of a number, 229
   polygons in the coordinate
     plane, 178
   powers, 14
   prime factorization, 28
   ratio tables, 202
   statistics
     mean absolute deviation, 422
     measures of center, 407
     measures of variation, 416
   whole numbers
     dividing, 8
     multiplying, 8
Evaluate, defined, 18
Example and non-example chart,
   322
Exponent(s), 10–15
   base of, 12
   defined, 12
   error analysis, 14
   order of operations, 18
   powers and, 10–15
   writing, 11
Expression(s)
   algebraic, 110–117
     defined, 112
     error analysis, 115, 116
     evaluating, 113–114
     factoring, 141
     like terms, 136
     real-life application, 114
     simplifying, 135
     terms, 112
     writing, 112
   equivalent
     defined, 128
     identifying, 140
     using properties to write,
        128–129
   evaluating, 18
     error analysis, 20
   factoring, 140–141
     defined, 140

numerical
    defined, 18
    factoring, 140
    writing, 118–123
        error analysis, 122
        as powers, 12
        real-life application, 121, 129

F

Face, defined, 356
Factor(s)
    common, defined, 32
    greatest common, 30–35
        defined, 32
        error analysis, 34
        real-life application, 33
    pairs, defined, 26
    prime factorization, 24–29
        error analysis, 28
        writing, 27
    repeated, 11
Factor pairs, defined, 26
Factor tree, defined, 26
Factoring an expression, 140–141
    defined, 140
First quartile, defined, 414
Five-number summary, defined, 460
Formulas
    area
        of a parallelogram, 152–157
        of a trapezoid, 168
        of a triangle, 160
    distance, 318
    volume of a rectangular prism, 29, 374–376
Four square, 164
Fraction(s)
    adding, 42–43
    dividing, 62–69
        error analysis, 67
        by a fraction, 65
        mixed numbers by, 72
        modeling, 63
        reciprocals, 64
        a whole number by, 65
    graphing, 262–263
    least common denominator, 42
    multiplying, 54–61
        error analysis, 59, 60
        modeling, 61

real-life application, 57, 58
        writing, 59
    ordering, 260–265
        negative, 262
        real-life application, 263
    percents and, 218–223
    subtracting, 42–43
    writing
        error analysis, 222
        as percents, 218–223
Frequency, defined, 441
Frequency table, defined, 441

G

Geometry
    coordinate plane, 174–179, 274–283
    nets, 360–363, 368–371
    parallelograms, 152–157
    perimeter, 177–178
    polygons, 152–157, 174–179
    polyhedrons, 354–359
    prisms, 356, 360–365, 374–379
    project, 359, 379
    pyramids, 356, 368–373
    solids, 354–373
    trapezoids, 166–171
    triangles, 158–163
Graph(s) and graphing, *See also* Data displays
    in the coordinate plane, 274–283
        error analysis, 279
        ordered pairs, 276–277
        origin, 276
        quadrant, 276
        real-life application, 278
    of equations in two variables, 317
    of inequalities, 324–331
    of ratios, 210–215
Graph of an inequality
    defined, 328
    error analysis, 330
    modeling, 330
    real-life application, 328
    writing, 329
Graphic Organizers
    concept circle, 410
    definition and example chart, 216
    example and non-example chart, 322

four square, 164
    information frame, 22
    information wheel, 124
    notetaking organizer, 76
    process diagram, 366
    summary triangle, 266
    word magnet, 448
Greatest common factor, 30–35
    defined, 32
    error analysis, 34
    modeling, 30–31, 37
    real-life application, 33
    writing, 34

H

Histogram(s), 440–447
    defined, 442
    error analysis, 445, 446
    frequency, 441
    frequency tables, 441
    modeling, 441

I

Independent variable, defined, 316
Inequalities
    defined, 326
    graphing, 324–331
        defined, 328
        error analysis, 330
        modeling, 330
        real-life application, 328
        writing, 329
    solution of, 327
    solution set, 327
    solving
        by addition, 332–337
        by division, 338–343
        error analysis, 336, 342
        by multiplication, 338–343
        real-life application, 335, 341
        by subtraction, 332–337
        writing, 336, 338
    writing, 324–331
        error analysis, 330
        modeling, 330
        real-life application, 328
Information frame, 22
Information wheel, 124
Integer(s), 248–253
    comparing, 254–259
        error analysis, 258

real-life application, 257
writing, 258
graphing, 251
negative, defined, 250
opposites, 250
ordering, 254–259
positive
defined, 250
error analysis, 252
writing, 250
real-life application, 251
writing, 249
Interquartile range, 414–417
defined, 414
Interval(s), 440–442
Inverse operations, defined, 303

L

Leaf, defined, 436
Least common denominator,
defined, 42
Least common multiple, 36–41
defined, 38
modeling, 36–37, 40
real-life application, 39
writing, 40
Like terms
combining, 136
defined, 136
Line plot, *See* Dot plot(s)
Logic, *Throughout. For example, see:*
area of parallelograms, 157
dividing mixed numbers, 75
histograms, 447
mean absolute deviation, 419
median, 447
reflecting points in the coordinate plane, 283
relationship of operations, 8
shapes of distribution, 455
solids, 356
solving inequalities, 343
statistics, 319, 419
unit rates, 209

M

Mean, 396–401
defined, 398
modeling, 397
research, 408

Mean absolute deviation, 418–423
defined, 420
error analysis, 422
finding, 420
real-life application, 421
Meaning of a Word
associate, 127
commute, 126
deviate, 418
distribute, 132
invert, 64
opposite, 250
percent, 218
skewed, 451
Measurement
conversion factor
defined, 234
metric system
defined, 234
metric units
converting to customary units, 232–237
unit analysis
defined, 234
writing, 236
unit conversion, 232–237
error analysis, 236
U.S. customary system
converting to metric units, 232–237
defined, 234
Measures of center, 396–409
choosing, 405
by shape of distribution, 456–457
defined, 404
error analysis, 407
mean, 396–391
defined, 398
median, 402–409
defined, 404
mode, 402–409
defined, 404
research, 408
Measures of variation, 412–417
defined, 414
error analysis, 416
interquartile range, 414–417
defined, 414
quartiles, 414–417
defined, 414
first, 414
third, 414

range, 412–417
defined, 414
writing, 417
Median, 402–409
defined, 404
error analysis, 407
research, 408
Mental Math, *Throughout. For example, see:*
Distributive Property, 132, 134
equations, 301
multiplying decimals, 86
percents, 225
using properties, 127
Metric system
converting units, 232–237
error analysis, 236
defined, 234
Mixed number(s)
dividing, 70–75
error analysis, 74
by a fraction, 72
modeling, 70–71
real-life application, 73
multiplying, 57–58
error analysis, 60
subtracting, 43
Mode, 402–409
defined, 404
research, 408
Modeling, *Throughout. For example, see:*
decimals
adding, 78
subtracting, 78–79
diagrams, 191
Distributive Property, 132
expressions
writing, 123
frequency tables, 441
greatest common factor
Venn diagram, 30
inequalities
graphing, 330
writing, 330
least common multiples, 36–37, 40
line graph, 280
line plots, 390
mixed numbers, dividing, 70–71
rates, 205
ratios, 191, 205
statistics, 390
mean, 397
whole number operations, 9

Multiples
  least common, 36–41
    modeling, 36–37, 40
    real-life application, 39
    writing, 40
Multiplication
  Associative Property of, 126–131
  Commutative Property of, 126–131
  of decimals, 84–91
    by decimals, 87
    error analysis, 89, 90
    modeling, 84–85
    real-life application, 88
  of fractions, 54–61
    error analysis, 59, 60
    modeling, 61
    real-life application, 57, 58
    writing, 59
  inequalities
    solving by, 338–343
  of mixed numbers, 57–58
    error analysis, 60
  Property
    of Equality, 310
    of Inequality, 340
    of One, 129
    of Zero, 129
  solving equations by, 308–313
  whole numbers, 4
    by decimals, 86
    error analysis, 8
Multiplication Property of Equality, 310
Multiplication Property of Inequality, 340
  error analysis, 342
Multiplication Property of One, 129
Multiplication Property of Zero, 129
Multiplicative Inverse Property, 310

N

Negative number(s)
  defined, 250
Net
  defined, 362
  of a prism, 360–363
    rectangular, 362
    triangular, 363
  of a pyramid, 368–373
    square, 370
    triangular, 371
Notetaking organizer, 76

Number line
  for graphing
    decimals, 262–263
    fractions, 262–263
    integers, 251
  for ordering
    decimals, 260–267
    fractions, 260–267
    integers, 254–259
Number Sense, *Throughout. For example, see:*
  algebraic expressions, 115
  box-and-whisker plots, 463
  Commutative Property of Addition, 130
  decimals
    dividing, 97
    multiplying, 89
    ordering, 264
  division, 8
  expressions
    writing, 123
  fractions, 74
    multiplying, 60, 61
    ordering, 264, 265
    reciprocal of, 69
  histograms, 447
  integers, 253
  least common multiple, 41
  mixed numbers
    dividing, 74
    multiplying, 61
  order of operations, 21, 115
  ordered pairs, 281
  percents, 222
    finding, 229
  powers, 15
  ratios, 194
    comparing and graphing, 214
    equivalent, 201
  solving equations, 312
  solving inequalities, 337
  statistical questions, 394
  statistics
    mean, 400
    measures of center, 407
Numerical expression(s)
  defined, 18
  writing, 20

O

Open-Ended, *Throughout. For example, see:*
  area of a polygon, 163, 171

Associative Property of Addition, 130, 137
Associative Property of Multiplication, 130
box-and-whisker plots, 465
coordinate plane, 281
decimals
  adding, 83
  multiplying, 91
Distributive Property, 137
Division Property of Equality, 312
equations
  in one variable, 298
  in two variables, 321
factoring expressions, 141
fractions, 67
  multiplying, 60
greatest common factor, 35
inequalities, solving, 342, 343
integers, 252
mixed numbers, 59
Multiplication Property of One, 130
order of operations, 21
percents and fractions, 222
polygons in the coordinate plane, 179
reciprocals, 67
relating variables, 320
shapes of distribution, 457
statistics, 401
  measures of variation, 417
Subtraction Property of Inequality, 336
trapezoids, 171
triangles, 163
Operations
  choosing among, 2–3
  order of, 16–21
    defined, 18
    error analysis, 20
    real-life application, 19
    writing, 20
  whole numbers, 2–9
Opposites, defined, 250
Order of operations, 16–21
  defined, 18
  error analysis, 20
  exponents, 18
  real-life application, 19
  writing, 20
Ordered pairs, 276–277

Ordering
  decimals, 260–265
  fractions, 260–265
    real-life application, 263
  integers, 254–259
Origin, defined, 276
Outlier(s)
  checking for, 415
  defined, 399

P

Parallelogram(s)
  area of, 152–157
    error analysis, 156
    formula, 154
    real-life application, 155
Patterns, *Throughout. For example, see:*
  exponents, 15
  powers, 15
  recognition of, 63
  writing expressions, 123
Percent(s), 218–223
  defined, 220
  error analysis, 222
  finding, 224–231
    error analysis, 229, 230
    of a number, 224–226
    real-life application, 228
    the whole, 227
  fractions and, 218–223
  real-life application, 221
  research, 223
  writing, 222
    error analysis, 222
    as fractions, 218–223, 219
    fractions as, 218–223
Perfect square
  defined, 13
  real-life application, 13
Perimeter
  of a rectangle
    finding, 177
    writing, 178
Polygon(s)
  area of, 152–157
    error analysis, 156, 162, 170
    real-life application, 155, 161, 169, 173, 177
    research, 163
    writing, 156, 163

  in the coordinate plane, 174–179
    error analysis, 178
    finding distances, 176
    real-life application, 177
    writing, 178
  defined, 152
Polyhedron(s), *See also* Solids
  defined, 356
  edges of, 356
  faces of, 356
  project, 359
  research, 559
  vertices of, 356
Positive number, defined, 250
Powers, 10–15
  defined, 12
  error analysis, 14
  exponents and, 10–15
  writing, 11
Precision, *Throughout. For example, see:*
  absolute value, 269
  dividing decimals, 99
  equations in two variables, 319
  frequency tables, 441
  polygons in the coordinate plane, 179
  ratios, 195
    comparing, 211
    graphing, 211
  statistics, 401
  surface area of a pyramid, 372
Prime factorization, 24–29
  defined, 26
  error analysis, 28
  factor pairs, 26
  factor tree, 26
  modeling, 29, 31
Prism(s)
  defined, 356
  drawing, 357
  nets for, 360–363
    defined, 362
    a rectangular, 362
    a triangular, 363
  surface area of, 360–365
    a rectangular, 362
    a triangular, 363
  volume of rectangular, 374–379
Problem Solving, *Throughout. For example, see:*
  decimals
    dividing, 99
    ordering, 265

  dividing fractions, 69
  equations in two variables, 321
  greatest common factor, 35
  histograms, 446
  ratios, 203
  statistics
    mean absolute deviation, 423
    measures of center, 408, 423
  volumes of prisms, 379
Process diagram, 366
Properties
  Addition Property of Equality, 303
  Addition Property of Inequality, 334
  Addition Property of Zero, 129
  Associative Property of Addition, 126–131
    error analysis, 130
  Associative Property of Multiplication, 126–131
  Commutative Property of Addition, 126–131
    real-life application, 129
  Commutative Property of Multiplication, 126–131
  Distributive Property, 132–141
    error analysis, 137, 138
    real-life application, 135
    writing, 137
  Division Property of Equality, 311
  Division Property of Inequality, 340
  Multiplication Property of Equality, 310
  Multiplication Property of Inequality, 340
  Multiplication Property of One, 129
  Multiplication Property of Zero, 129
  Multiplicative Inverse Property, 310
  Subtraction Property of Equality, 303
  Subtraction Property of Inequality, 334
Pyramid(s)
  defined, 356
  drawing, 357
  net of, 368–371
    a square, 370
    a triangular, 371

square, 370
surface area of, 368–373
triangular, 371

Q

Quadrant(s), defined, 276
Quartile(s), 414–417
defined, 414

R

Range, 412–417
defined, 414
interquartile, 414
Rate(s), 204–209
defined, 206
double number line diagrams, 205
equivalent
defined, 206
finding, 207
unit
cost, 207
defined, 206
finding, 206
writing, 208
writing, 206
Ratio(s), 190–195
comparing, 210–215
defined, 192
diagramming, 191, 193
equivalent
defined, 198
error analysis, 202
graphing, 210–215
modeling, 191
rates and, 204–209
Ratio table(s), 196–203
defined, 198
error analysis, 202
graphing, 210–211
writing, 214
Reading
check for reasonableness, 317
coordinate plane, 277
inequalities, 328, 335
opposites, 251
symbols of, 302, 327
translating words to equations, 297
quartiles, 414

Real-Life Applications
absolute value, 271
algebraic expressions, 114
area
of composite figures, 173
of parallelograms, 155
of trapezoids, 169
of triangles, 161
Commutative Property of Addition, 129
coordinate plane, 278
decimals
adding, 81
dividing, 96
multiplying, 88
subtracting, 81
Distributive Property, 135
equations
in one variable, 297
solving, 304, 311
in two variables, 318
writing, 297, 304, 317
fractions
multiplying, 57, 58
ordering, 263
greatest common factor, 33
inequalities
graphing, 328
solving, 335, 341
integers, 251
comparing, 257
least common multiple, 39
mean absolute deviation, 421
mixed numbers, dividing, 73
order of operations, 19
percents, 221, 228
finding the whole, 228
perfect square, 13
polygons in the coordinate plane, 177
whole numbers, division, 6
writing expressions, 121, 129
Reasoning, *Throughout. For example, see:*
absolute value, 272
algebraic expressions, 117
area
parallelograms, 153
triangles, 163
coordinate plane, 274, 280
ordered pairs, 281
reflecting points in, 283

data displays
box-and-whisker plots, 465
histograms, 447
shapes of distribution, 455, 457
stem-and-leaf plots, 434, 439, 447, 455
decimals, multiplying, 90
Distributive Property, 138, 139
division, 9
expressions, 122, 123
factoring, 141
fractions
dividing, 67, 68, 69
multiplying, 59
as percents, 223
greatest common factor, 35
integers, 249, 257
comparing, 259
ordering, 258
least common multiple, 41
mixed numbers, dividing, 75
order of operations, 20
outliers, 445
parallelograms, dimensions of, 157
percents, 219, 231
powers, 15
ratios, 195, 203
rectangles, 171
solids, 359
solving
equations, 131, 306
inequalities, 342
statistics, 390, 391, 395
mean, 397, 400, 401, 407, 457
mean absolute deviation, 419, 422, 423, 457
measures of center, 407
mode, 408
outliers, 417
statistical questions, 392
trapezoids, 170, 171
volumes of prisms, 378, 379
Reciprocals, defined, 64
Rectangle(s), perimeter of, 177
Reflecting points in the coordinate plane, 282–283
Repeated Reasoning, *Throughout. For example, see:*
decimals, 91
Distributive Property, 132
integers, 253

measures of variation, 412
perfect squares, 15
ratio tables, 197
solving equations, 301
writing expressions, 118

S

Shape of a distribution, 450–457
  box-and-whisker plots, 462
  choosing appropriate measures, 456–457
  skewed left, 452
  skewed right, 452
  symmetric, 452
Solid(s)
  defined, 356
  drawing, 354–355, 357
  edges of, 356
  faces of, 356
  polyhedron(s), 354–359
    defined, 356
    edges of, 356
    faces of, 356
    project, 359
    research, 559
    vertices of, 356
  prism(s)
    defined, 356
    drawing, 357
    nets for, 360–363
      defined, 362
      a rectangular, 362
      a triangular, 363
    surface area of, 360–365
      a rectangular, 362
      a triangular, 363
    volume of rectangular, 374–379
  project, 359, 379
  pyramid(s)
    defined, 356
    drawing, 357
    net of, 368–371
      a square, 370
      a triangular, 371
    square, 370
    surface area of, 368–373
    triangular, 371
  research, 359
  surface area of
    defined, 362
    prisms, 360–365
    pyramids, 368–373
  vertices of, 356
  volume of, 374–375
Solution, defined, 302
Solution of an equation in two variables, defined, 316
Solution of an inequality, defined, 327
Solution set, defined, 327
Square(s), perfect,
  defined, 13
  real-life application, 13
Statistical question(s), 392–395
  defined, 392
Statistics, *See also* Data displays
  defined, 392
  mean, 396–401
    defined, 398
    modeling, 397
    research, 408
  mean absolute deviation, 418–423
    defined, 420
    error analysis, 422
    real-life application, 421
  measures of center, 396–409
    choosing, 456–457
    defined, 404
    error analysis, 407
    research, 408
  measures of variation, 412–417
    defined, 414
    error analysis, 416
    writing, 417
  median, 402–409
    defined, 404
    error analysis, 407
    research, 408
  mode, 402–409
    defined, 404
    research, 408
  modeling, 390
  outliers
    defined, 399
    writing, 407
  quartiles, 414–417
    defined, 414
    first, 414
    third, 414
  range, 412–417
    defined, 414
    interquartile, 414
  statistical questions, 392–395
    defined, 392
Stem, defined, 436
Stem-and-leaf plot(s), 434–439
  defined, 436
  error analysis, 438
  leaf, 436
  stem, 436
  writing, 438
Structure, *Throughout. For example, see:*
  adding decimals, 83
  box-and-whisker plots, 465
  coordinate plane, 274
    reflecting points in, 283
  exponents, 11
  factoring expressions, 141
  fractions
    dividing by, 63
    multiplying, 55
  greatest common factor, 31
  integers, 249
  interquartile range, 417
  least common multiple, 37
  polygons in the coordinate plane, 175, 179
  properties of multiplication, 131
  ratio tables, 215
  solving equations, 313
  surface area of a pyramid, 373
Study Tip
  algebraic expressions, 113
  base of a triangular pyramid, 371
  box-and-whisker plots, 462
  check for reasonableness, 4
  composite figures, 172
  coordinate plane, 278
  data displays
    box-and-whisker plots, 461, 462
    shapes of distribution, 452, 462
  decimals
    adding, 80
    dividing, 95, 96
    subtracting, 80
  Distributive Property, 135
  dot plots, 392
  equivalent fractions and percents, 220
  expressions
    algebraic, 112
    equivalent, 128
    factoring, 140
    writing, 121

finding the percent of a number, 226
fractions, 66
　adding and subtracting, 43
　multiplying, 56
　simplest form, 42
graphing
　equations in two variables, 317
　ordered pairs from a ratio table, 213
greatest common factor, 32
inverse operations, 334
least common denominators, 66
measures of center, 404
Multiplicative Inverse Property, 64
order of operations, 18
ordering integers, 257
polygons in the coordinate plane, 176, 177
prime factorization, 26, 27
ratio tables, 199, 200
reciprocals, 64
solving
　equations, 303, 304
　inequalities, 334
unit cost, 207
unit rates, 206
variable by itself, 112
variables, 121
volume of a cube, 376
Subtraction
　of decimals, 78–83
　　error analysis, 82
　　real-life application, 81
　　writing, 82
　equations
　　error analysis, 305
　　real-life application, 304
　　solving by, 300–307
　of fractions, 42–43
　inequalities, solving by, 332–337
　of mixed numbers, 43
　Property of
　　Equality, 303
　　Inequality, 334
　whole numbers, 4
Subtraction Property of Equality, 303
　real-life application, 304
　writing, 305
Subtraction Property of Inequality, 334
　real-life application, 335

Summary triangle, 266
Surface Area
　of prisms, 360–365
　of pyramids, 368–373
　of a solid, defined, 362

T

Tables
　frequency, 441
　for graphing equations, 317
Tape diagram(s), 191, 193
Term(s)
　coefficient, 112
　constant, 112
　defined, 112
　variable, 112
Third quartile, defined, 414
Three-dimensional figures, *See* Solids
Trapezoid(s)
　area of, 166–171
　　error analysis, 170
　　formula, 168
　　real-life application, 169
　　writing, 167
Triangle(s)
　area of, 158–163
　　error analysis, 162
　　formula, 160
　　real-life application, 161
　　research, 163
　　writing, 163

U

Unit analysis, defined, 234
Unit cost, 207
Unit rate(s), 206–209
　defined, 206
　finding, 206
　unit cost, 207
　writing, 208
U.S. customary system
　converting units, 232–237
　　to metric units, 232–237
　defined, 234

V

Variable(s)
　coefficient of, 112
　defined, 112

　dependent, 316
　equations in one, 294–299
　　error analysis, 298
　　real-life application, 297
　equations in two, 314–321
　　defined, 316
　　dependent variable, 316
　　error analysis, 319
　　graphing, 317
　　independent variable, 316
　　real-life application, 318
　　solution of, 316
　independent, 316
Venn diagram
　defined, 30
　for identifying
　　factors, 30–31, 37
　　multiples, 36
Vertex
　of a solid, defined, 356
Volume
　defined, 374
　of rectangular prisms, 374–379
　　formula, 29, 376

W

Which One Doesn't Belong?, *Throughout. For example, see:*
　algebraic expressions, 115
　area, 170
　coordinate plane, 279
　Distributive Property, 137
　equations in two variables, 319
　equivalent ratios, 201
　exponents, 14
　factor pairs, 28
　fractions, 67
　measures of center, 407
　percents, 222
　properties, 130
　ratios, 194
　statistics
　　mean absolute deviation, 422
　　measures of center, 407
Whole number(s)
　adding, 4
　dividing, 4–6
　　a decimal by, 94
　　error analysis, 8
　　by a fraction, 65
　　real-life application, 6

multiplying, 4
    by decimals, 86
    error analysis, 8
  operations, 2–9
    choosing among, 2–3
    modeling, 9
  perfect square of, 13
    defined, 13
    real-life application, 13
  subtracting, 4
Word magnet, 448
Writing, *Throughout. For example, see:*
  area
    of a trapezoid, 167
    of a triangle, 163
  converting units, 236
  decimals
    adding, 82
    dividing, 93
    subtracting, 82

Distributive Property, 137
Division Property of Equality, 312
equations, 297
  in two variables, 314
expressions
  algebraic, 112, 120–121
  error analysis, 122
  modeling, 123
  real-life application, 121
fractions, multiplying, 59
greatest common factor, 34
inequalities, 332, 338
  graphing, 329
integers, 249
  comparing, 258
least common multiples, 40
measures of center, 407
measures of variation, 417

numerical expressions, 20
order of operations, 20
percents, 222
perimeter of a rectangle, 178
polygons
  area of, 156
  in the coordinate plane, 178
rates, 206
  unit, 208
ratio tables, 214
ratios, 192, 218
solving
  equations, 305
  inequalities, 336
stem-and-leaf plots, 438

Photo Credits

Front matter
iv Big Ideas Learning, LLC; **viii** *top* ©iStockphoto.com/ALEAIMAGE, ©iStockphoto.com/Ann Marie Kurtz; *bottom* ©iStockphoto.com/Jane norton; **ix** *top* ©iStockphoto.com/ALEAIMAGE, ©iStockphoto.com/Ann Marie Kurtz; *bottom* wavebrekmedia ltd/Shutterstock.com; **x** *top* stephan kerkhofs/Shutterstock.com, Cigdem Sean Cooper/Shutterstock.com, ©iStockphoto.com/Andreas Gradin; *bottom* Odua Images/Shutterstock.com; **xi** *top* ©iStockphoto.com/sumnersgraphicsinc, ©iStockphoto.com/Ann Marie Kurtz; *bottom* James Flint/Shutterstock.com; **xii** *top* ©iStockphoto.com/Lisa Thornberg, ©iStockphoto.com/Ann Marie Kurtz; *bottom* william casey/Shutterstock.com; **xiii** *top* ©iStockphoto.com/Jonathan Larsen; *bottom* Edyta Pawlowska/Shutterstock.com; **xiv** *top* Varina and Jay Patel/Shutterstock.com, ©iStockphoto.com/Ann Marie Kurtz; *bottom* PETER CLOSE/Shutterstock.com; **xv** *top* ©iStockphoto/Michael Flippo, ©iStockphoto.com/Ann Marie Kurtz; *bottom* ©iStockphoto.com/ranplett; **xvi** *top* Chiyacat/Shutterstock.com, Zoom Team/Shutterstock.com; *bottom* ©iStockphoto.com/Noraznen Azit; **xvii** *top* ©iStockphoto.com/Alistair Cotton; *bottom* ©iStockphoto.com/Thomas Perkins; **xviii** Ljupco Smokovski/Shutterstock.com

Chapter 1
1 ©iStockphoto.com/ALEAIMAGE, ©iStockphoto.com/Ann Marie Kurtz; **2** *top right* S.Dashkevych/Shutterstock.com; *bottom right* auremar/Shutterstock.com; **5** yxm2008/Shutterstock.com; **6** ©iStockphoto.com/Santino Ambrogio; **8** Jiang Dao Hua/Shutterstock.com; **9** *top right* Africa Studio/Shutterstock.com; *center left* Podriv Ustoev/Shutterstock.com; **10** Danomyte/Shutterstock.com; **11** Nicole Gordine/Shutterstock.com; **13** MONOPOLY® & ©2012 Hasbro Inc. Used with permission.; **15** ©iStockphoto.com/WestLight; **20** ©iStockphoto.com/clu; **21** ©iStockphoto.com/mladn61; **23** *center right* Wendy Nero/Shutterstock.com; *bottom right* ©iStockphoto.com/craftvision; **24** James Hoenstine/Shutterstock.com; **26** farbeffekte/Shutterstock.com; **29** Michael Mitchell/Shutterstock.com; **33** jmatzick/Shutterstock.com; **34** Glenda M. Powers/Shutterstock.com; **35** Africa Studio/Shutterstock.com; **39** Alinochka/Shutterstock.com; **40** Orla/Shutterstock.com; **41** yuyangc/Shutterstock.com; **44** *center left* Mike Flippo/Shutterstock.com; *bottom right* Nitr/Shutterstock.com; **48** beboy/Shutterstock.com

Chapter 2
52 ©iStockphoto.com/ALEAIMAGE, ©iStockphoto.com/Ann Marie Kurtz; **59** *center left* fivespots/Shutterstock.com; *center right* bluehand/Shutterstock.com; **68** tele52/Shutterstock.com, lithian/Shutterstock.com, Galina Barskaya/Shutterstock.com; **69** *center left* ©iStockphoto.com/Michael Plumb; *center right* g215/Shutterstock.com; **73** ©iStockphoto.com/Karin Lau; **81** Vacclav/Shutterstock.com; **82** Peredniankina/Shutterstock.com; **83** SSSCCC/Shutterstock.com; **89** Luba V Nel/Shutterstock.com; **90** KENCKOphotography/Shutterstock.com; **97** ©iStockphoto.com/bananahuman; **99** ©iStockphoto.com/Janis Litavnieks; **104** ©iStockphoto.com/graham heywood

Chapter 3
108 stephan kerkhofs/Shutterstock.com, Cigdem Sean Cooper/Shutterstock.com, ©iStockphoto.com/Andreas Gradin; **121** Kateryna Larina/Shutterstock.com; **125** Vivid Pixels/Shutterstock.com; **138** Pal Teravagimov/Shutterstock.com; **141** Inhabitant/Shutterstock.com; **142** Fotofermer/Shutterstock.com; **143** Helder Almeida/Shutterstock.com; **146** *center right* Aptyp_koK/Shutterstock.com; *bottom left* Andy Cash/Shutterstock.com

Chapter 4
150 ©iStockphoto.com/sumnersgraphicsinc, ©iStockphoto.com/Ann Marie Kurtz; **162** Terrance Emerson/Shutterstock.com; **163** ©iStockphoto.com/bamse009; **171** Kharidehal Abhirama Ashwin/Shutterstock.com; **177** Christian Musat/Shutterstock.com; **179** Brandon Seidel/Shutterstock.com; **184** *top right* ©iStockphoto.com/AlexMax; *bottom right* U.S. Geological Survey

Chapter 5
188 ©iStockphoto.com/Lisa Thornberg, ©iStockphoto.com/Ann Marie Kurtz; **190** Elnur/Shutterstock.com; **192** Vladimir Wrangel/Shutterstock.com; **193** Lepas/Shutterstock.com; **196** *top* foxie/Shutterstock.com; *center right* Constantinos/Shutterstock.com; **197** Danny Smythe/Shutterstock.com; **199** South 12th Photography/Shutterstock.com; **200** Petr Malyshev/Shutterstock.com; **202** ©iStockphoto.com/otakumania; **203** Anneka/Shutterstock.com; **205** vita khorzhevska/Shutterstock.com; **206** © HO/Reuters/Corbis; **207** Alex Staroseltsev/Shutterstock.com; **208** *top right* ©iStockphoto.com/Gord Horne; *bottom right* leonid_tit/Shutterstock.com; **209** *top left* ©iStockphoto.com/ljpat; *bottom right* ©iStockphoto.com/Birgitte Magnus; **210** Mike Flippo/Shutterstock.com; **212** *top right* ultimathule/Shutterstock.com; *top left* Christopher Kolaczan/Shutterstock.com; **213** ©iStockphoto.com/Bart Wolczyk; **214** Catalin Petolea/Shutterstock.com, Kaspri/Shutterstock.com; **215** R. Gino Santa Maria/Shutterstock.com; **225** Spasiblo/Shutterstock.com; **230** Thank You/Shutterstock.com; **236** Derek Wong / CC-BY-SA-3.0; **237** ©iStockphoto.com/Paul Tessier; **238** *bottom left* Kitch Bain/Shutterstock.com; *bottom* ©iStockphoto.com/Ermin Gutenberger; **242** Joe Gough/Shutterstock.com

Chapter 6
246 ©iStockphoto.com/Jonathan Larsen; **251** ©iStockphoto.com/Egor Mopanko; **253** haveseen/Shutterstock.com; **254** *top right* NASA/Kim Shiflett; *bottom* NASA; **255** NASA; **259** ©iStockphoto.com/Andrew Penner; **260** NASA; **261** *left* Andrey Armyagov/Shutterstock.com; *center* ©iStockphoto.com/Island Effects; *right* U.S. Navy photo by Photographer's Mate 2nd Class Prince Hughes III; **264** ©iStockphoto.com/jclegg, ©iStockphoto.com/spxChrome, ©iStockphoto.com/Laura Eisenberg; **265** A'lya/Shutterstock.com; **267** *top right* NicolasMcComber/Shutterstock.com; *bottom* ©iStockphoto.com/Zuki; **269** *left* ©iStockphoto.com/james steidl; *right* U.S. Navy photo by Photographers Mate 2nd Class Michael Sandberg; **271** Shane W Thompson/Shutterstock.com; **278** Christophe Testi/Shutterstock.com

Chapter 7
292 Varina and Jay Patel/Shutterstock.com, ©iStockphoto.com/Ann Marie Kurtz; **299** *top right* ©iStockphoto.com/Joop Snijder, ©iStockphoto.com/Michael MacFadden, ©iStockphoto.com/Steve Goodwin; *center left* ©iStockphoto.com/Kenneth C. Zirkel; **304** jocic/Shutterstock.com, Marko Poplasen/Shutterstock.com; **306** *top left* ©iStockphoto.com/Jeremy Wee; *top right* ©iStockphoto.com/Jan Will; **307** ©iStockphoto.com/Keith Reicher; **309** ©iStockphoto.com/Leo Blanchette; **312** ©iStockphoto.com/Christopher Futcher; **313** ©iStockphoto.com/Eric Isselée; **314** mangostock/Shutterstock.com; **315** Ilya Andriyanov/Shutterstock.com; **316** violetkaipa/Shutterstock.com; **318** kokandr/Shutterstock.com; **319** discpicture/Shutterstock.com; **320** ©iStockphoto.com/Mutlu Kurtbas; **324** *first* ©iStockphoto.com/Studio-Annika; *second* ©iStockphoto.com/nicholas belton; *third* ©iStockphoto.com/Robert Dant; **328** NASA/Johns Hopkins University Applied Physics Laboratory; **330** ©iStockphoto.com/George Peters; **331** ©iStockphoto.com/Anthony Ladd; **332** *top right* CLS Design/Shutterstock.com; *bottom right* Liquid Productions, LLC /Shutterstock.com; **337** *top left* ©iStockphoto.com/o-che; *center left* ©iStockphoto.com/sunygraphics; **338** *top right* Dennis Owusu-Ansah/Shutterstock.com; *bottom right* Julia Zakharova /Shutterstock.com; **341** Nathan Till/Shutterstock.com; **342** ©iStockphoto.com/rami ben ami; **343** Karin Hildebrand Lau/Shutterstock.com; **344** ©iStockphoto.com/Algimantas Balezentis; **348** ©iStockphoto.com/Jani Bryson

Chapter 8
352 ©iStockphoto.com/Michael Flippo, ©iStockphoto.com/Ann Marie Kurtz; **358** *Exercise 17* ©iStockphoto.com/Rich Koele; *Exercise 21* design56/Shutterstock.com; **359** *top right* ©iStockphoto.com/Hedda Gjerpen; *center* ©iStockphoto.com/rzdeb; **363** Niki Crucillo/Shutterstock.com; **370** PeterG/Shutterstock.com; **372** Itana/Shutterstock.com; **373** Tupungato/Shutterstock.com; **377** *top left* ©iStockphoto.com/William Britten; *center left* Denis Barbulat/Shutterstock.com; **379** *center left* ©iStockphoto.com/Jill Chen; *center right* ©iStockphoto.com/LongHa2006

Chapter 9
388 Chiyacat/Shutterstock.com, Zoom Team/Shutterstock.com; **390** LeventeGyori/Shutterstock.com; **391** *center left and right* Nattika/Shutterstock.com; **392** Eric Isselée/Shutterstock.com; **393** *top right* Iznogood/Shutterstock.com; *bottom left* AISPIX by Image Source/Shutterstock.com; **395** Laralova/Shutterstock.com; **396** Rob Byron/Shutterstock.com; **397** Denis Vrublevski/Shutterstock.com; **399** ©iStockphoto.com/Eric Isselée; **402** *top right* Hein Nouwens/Shutterstock.com; *bottom* ivelly/Shutterstock.com; **403** ©iStockphoto.com/Andrew Rich; **409** *top right* ©iStockphoto.com/suemack; *top left* ©iStockphoto.com/muratkoc; **415** Charlie Hutton/Shutterstock.com; **416** Talvi/Shutterstock.com; **417** ©iStockphoto.com/Jason Lugo; **420** Danny Smythe/Shutterstock.com; **421** *top left* Ganko/Shutterstock.com; *top right* Mark Herreid/Shutterstock.com; **422** alarich/Shutterstock.com; **423** tab62/Shutterstock.com; **424** Racheal Grazias/Shutterstock.com; **426** Jan Martin Will/Shutterstock.com; **427** Kitch Bain/Shutterstock.com

Chapter 10
432 ©iStockphoto.com/Alistair Cotton; **434** Elzbieta Szpak/Shutterstock.com; **435** ©CORBIS; **437** ©iStockphoto.com/Pekka Nikonen; **438** ©iStockphoto.com/Mehmet Salih Guler; **442** stockshoppe/Shutterstock.com; **443** *top right* ©iStockphoto.com/susaro; *bottom left* Arman Zhenikeyev/Shutterstock.com; **444** Tomasz Trojanowski/Shutterstock.com; **447** ©iStockphoto.com/Eric Isselée; **449** ©iStockphoto.com/vincent chien chow chine; **450** Sergey Mironov/Shutterstock.com; **451** fresher/Shutterstock.com; **453** mmaxer/Shutterstock.com; **457** Lightspring/Shutterstock.com; **459** *first* windu/Shutterstock.com; *second* motorolka/Shutterstock.com; *third* Preto Perola/Shutterstock.com; *fourth* nikkytok/Shutterstock.com; *center right* Apollofoto/Shutterstock.com; **460** ©iStockphoto.com/rusm; **461** Sebastian Knight/Shutterstock.com; **462** Garbuzov/Shutterstock.com; **464** *center left* Ffooter/Shutterstock.com; *bottom right* zhuda/Shutterstock.com; **465** Rob Marmion/Shutterstock.com; **469** Nikola Bilic/Shutterstock.com; **470** Monkey Business Images/Shutterstock.com

Appendix A
A0 *background* ©iStockphoto.com/Björn Kindler; *top left and bottom* ©iStockphoto.com/Ralf Hettler; **A1** *top right* Emmer, Michele, ed., The Visual Mind: Art and Mathematics, Plate 2, © 1993 Massachusetts Institute of Technology, by permission of The MIT Press.; *bottom left* ©iStockphoto.com/Andrew Cribb; *bottom right* ©iStockphoto.com/Liz Leyden; **A4** *top right* ©iStockphoto.com/Ralf Hettler; *center right* ©iStockphoto.com/Ragnarocks; *bottom left* ©iStockphoto.com/Linda Steward; *bottom right* ©iStockphoto.com/rackermann; **A5** *top and bottom right* ©iStockphoto.com/rackermann; *bottom left* ©iStockphoto.com/Ralf Hettler; **A6** *top right* Stannered, DTR; *bottom left* Emmer, Michele, ed., The Visual Mind: Art and Mathematics, Plate 2, © 1993 Massachusetts Institute of Technology, by permission of The MIT Press.; **A7** *top right* Elena Borodynkina/Shutterstock.com; *center right* ©iStockphoto.com/Andrew Cribb; *bottom left* ©iStockphoto.com/smokyme; *bottom right* Sculpture by Vladimir Bulatov; **A8** *top right* ©iStockphoto.com/Dan Van Oss; *bottom left* ©iStockphoto.com/Tomasz Tulik; **A9** *top right* ©iStockphoto.com/timoph; *bottom left* ©iStockphoto.com/Liz Leyden; *bottom right* ©iStockphoto.com/Pauline S Mills

Cartoon illustrations Tyler Stout

Learning Progression

Kindergarten

Counting and Cardinality	– Count to 100 by Ones and Tens; Compare Numbers
Operations and Algebraic Thinking	– Understand and Model Addition and Subtraction
Number and Operations in Base Ten	– Work with Numbers 11–19 to Gain Foundations for Place Value
Measurement and Data	– Describe and Compare Measurable Attributes; Classify Objects into Categories
Geometry	– Identify and Describe Shapes

Grade 1

Operations and Algebraic Thinking	– Represent and Solve Addition and Subtraction Problems
Number and Operations in Base Ten	– Understand Place Value for Two-Digit Numbers; Use Place Value and Properties to Add and Subtract
Measurement and Data	– Measure Lengths Indirectly; Write and Tell Time; Represent and Interpret Data
Geometry	– Draw Shapes; Partition Circles and Rectangles into Two and Four Equal Shares

Grade 2

Operations and Algebraic Thinking	– Solve One- and Two-Step Problems Involving Addition and Subtraction; Build a Foundation for Multiplication
Number and Operations in Base Ten	– Understand Place Value for Three-Digit Numbers; Use Place Value and Properties to Add and Subtract
Measurement and Data	– Measure and Estimate Lengths in Standard Units; Work with Time and Money
Geometry	– Draw and Identify Shapes; Partition Circles and Rectangles into Two, Three, and Four Equal Shares

Grade 3

Operations and Algebraic Thinking	– Represent and Solve Problems Involving Multiplication and Division; Solve Two-Step Problems Involving Four Operations
Number and Operations in Base Ten	– Round Whole Numbers; Add, Subtract, and Multiply Multi-Digit Whole Numbers
Number and Operations—Fractions	– Understand Fractions as Numbers
Measurement and Data	– Solve Time, Liquid Volume, and Mass Problems; Understand Perimeter and Area
Geometry	– Reason with Shapes and Their Attributes

Grade 4

Operations and Algebraic Thinking	– Use the Four Operations with Whole Numbers to Solve Problems; Understand Factors and Multiples
Number and Operations in Base Ten	– Generalize Place Value Understanding; Perform Multi-Digit Arithmetic
Number and Operations—Fractions	– Build Fractions from Unit Fractions; Understand Decimal Notation for Fractions
Measurement and Data	– Convert Measurements; Understand and Measure Angles
Geometry	– Draw and Identify Lines and Angles; Classify Shapes

Grade 5

Operations and Algebraic Thinking	– Write and Interpret Numerical Expressions
Number and Operations in Base Ten	– Perform Operations with Multi-Digit Numbers and Decimals to Hundredths
Number and Operations—Fractions	– Add, Subtract, Multiply, and Divide Fractions
Measurement and Data	– Convert Measurements within a Measurement System; Understand Volume
Geometry	– Graph Points in the First Quadrant of the Coordinate Plane; Classify Two-Dimensional Figures

Mathematics Reference Sheet

Conversions

U.S. Customary
1 foot = 12 inches
1 yard = 3 feet
1 mile = 5280 feet
1 acre = 43,560 square feet
1 cup = 8 fluid ounces
1 pint = 2 cups
1 quart = 2 pints
1 gallon = 4 quarts
1 gallon = 231 cubic inches
1 pound = 16 ounces
1 ton = 2000 pounds
1 cubic foot ≈ 7.5 gallons

U.S. Customary to Metric
1 inch = 2.54 centimeters
1 foot ≈ 0.3 meter
1 mile ≈ 1.61 kilometers
1 quart ≈ 0.95 liter
1 gallon ≈ 3.79 liters
1 cup ≈ 237 milliliters
1 pound ≈ 0.45 kilogram
1 ounce ≈ 28.3 grams
1 gallon ≈ 3785 cubic centimeters

Time
1 minute = 60 seconds
1 hour = 60 minutes
1 hour = 3600 seconds
1 year = 52 weeks

Temperature
$$C = \frac{5}{9}(F - 32)$$
$$F = \frac{9}{5}C + 32$$

Metric
1 centimeter = 10 millimeters
1 meter = 100 centimeters
1 kilometer = 1000 meters
1 liter = 1000 milliliters
1 kiloliter = 1000 liters
1 milliliter = 1 cubic centimeter
1 liter = 1000 cubic centimeters
1 cubic millimeter = 0.001 milliliter
1 gram = 1000 milligrams
1 kilogram = 1000 grams

Metric to U.S. Customary
1 centimeter ≈ 0.39 inch
1 meter ≈ 3.28 feet
1 kilometer ≈ 0.62 mile
1 liter ≈ 1.06 quarts
1 liter ≈ 0.26 gallon
1 kilogram ≈ 2.2 pounds
1 gram ≈ 0.035 ounce
1 cubic meter ≈ 264 gallons

Number Properties

Commutative Properties of Addition and Multiplication
$$a + b = b + a$$
$$a \cdot b = b \cdot a$$

Associative Properties of Addition and Multiplication
$$(a + b) + c = a + (b + c)$$
$$(a \cdot b) \cdot c = a \cdot (b \cdot c)$$

Addition Property of Zero
$$a + 0 = a$$

Multiplication Properties of Zero and One
$$a \cdot 0 = 0$$
$$a \cdot 1 = a$$

Distributive Property:
$$a(b + c) = ab + ac$$
$$a(b - c) = ab - ac$$

Properties of Equality

Addition Property of Equality
If $a = b$, then $a + c = b + c$.

Subtraction Property of Equality
If $a = b$, then $a - c = b - c$.

Multiplication Property of Equality
If $a = b$, then $a \cdot c = b \cdot c$.

Multiplicative Inverse Property
$$n \cdot \frac{1}{n} = \frac{1}{n} \cdot n = 1, n \neq 0$$

Division Property of Equality
If $a = b$, then $a \div c = b \div c, c \neq 0$.

Properties of Inequality

Addition Property of Inequality
If $a > b$, then $a + c > b + c$.

Subtraction Property of Inequality
If $a > b$, then $a - c > b - c$.

Multiplication Property of Inequality
If $a > b$ and c is positive, then $a \cdot c > b \cdot c$.

Division Property of Inequality
If $a > b$ and c is positive, then $a \div c > b \div c$.

Perimeter and Area

Square	Rectangle	Parallelogram	Triangle	Trapezoid
$P = 4s$ $A = s^2$	$P = 2\ell + 2w$ $A = \ell w$	$A = bh$	$A = \dfrac{1}{2}bh$	$A = \dfrac{1}{2}h(b_1 + b_2)$

Surface Area

Prism

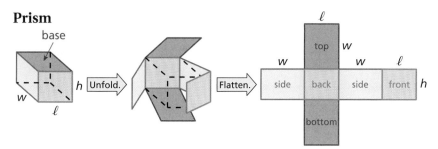

S = areas of bases + areas of lateral faces

Pyramid

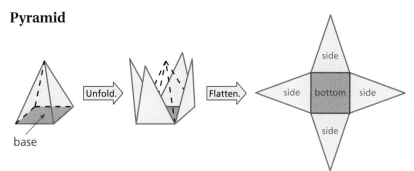

S = area of base + areas of lateral faces

Volume of a Rectangular Prism

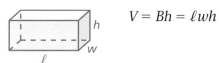

$V = Bh = \ell wh$

The Coordinate Plane

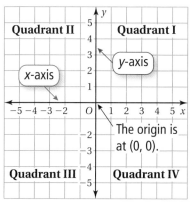

The origin is at (0, 0).